GILROY   MAR 28 '88

S0-AWH-270

THE READING PROGRAM
SANTA CLARA COUNTY LIBRARY
42 MARTIN STREET
GILROY, CA 95020
(408) 848-5366

UNDERSTANDING ELEMENTARY ALGEBRA

THE READING PROGRAM
SANTA CLARA COUNTY LIBRARY
42 MARTIN STREET
GILROY, CA 95020
(408) 848-5366

UNDERSTANDING ELEMENTARY ALGEBRA
Second Edition

Robert G. Moon

Fullerton College

WITHDRAWN

_00_817GI

SANTA CLARA COUNTY LIBRARY
SAN JOSE, CALIFORNIA

Merrill Publishing Company
A Bell & Howell Information Company
Columbus   Toronto   London   Melbourne

To
Marilyn

Cover Art: Michael Linley Illustration
Published by Merrill Publishing Company
A Bell & Howell Information Company
Columbus, Ohio 43216
This book was set in Optima
Administrative Editor: John Yarley
Production Coordinator: Ben Shriver
Cover Designer: Cathy Watterson
Copyright © 1987, 1978, by Merrill Publishing Company. All rights reserved. No part of this book may be reproduced in any form, electronic or mechanical, including photocopy, recording, or any information storage and retrieval system, without permission in writing from the publisher. "Merrill Publishing Company" and "Merrill" are registered trademarks of Merrill Publishing Company.
Library of Congress Catalog Card Number: 86–62992
International Standard Book Number: 0–675–20418–6
Printed in the United States of America
1 2 3 4 5 6 7 8 9—91 90 89 88 87

SANTA CLARA COUNTY LIBRARY
SAN JOSE CALIFORNIA

PREFACE

This book has been developed for use by college students taking a first course in algebra. No prerequisite is assumed except a reasonable proficiency in arithmetic skills.

The main purpose of this book is to develop understanding and computational skill with the techniques usually found in a first-level algebra course. To reach this goal, the author has written a text that students can read and understand. The treatment of each topic is direct, straightforward, and mathematically sound. However, it is not so rigorous that students will be confused. Each concept is carefully developed and followed with illustrative examples.

Each section of the text contains many examples with detailed step-by-step solutions, illustrations, and other aids. Color is used to highlight important steps in text and illustrations. Important concepts and rules are enclosed in boxes for easy identification. Important terms are keyed in the margins and appear in color when first introduced.

Every section contains a set of exercises suitable for homework. Each set of exercises contains numerous problems that range from the simple to the more complex. Answers to the odd-numbered problems are included at the back of the book. Full solutions to all problems are found in the *Solutions Manual* available from the publisher. Due to its readability, this text can easily be adapted to fit a variety of instructional programs, such as self-study centers and learning laboratories.

Each chapter concludes with a summary and a set of self-checking exercises. Students should be encouraged to study these before a chapter test. They also constitute an excellent review for the final examination.

TEXT ORGANIZATION

Real Numbers   Chapter 1 develops the real numbers and the basic terminology of algebra. Algebraic expressions are discussed together with the importance of order of operations and grouping symbols. The four fundamental operations of addition, subtraction, multiplication, and division are discussed as they apply to signed real numbers.

Algebraic Expressions and Equations   Chapter 2 discusses whole number exponents, operations with simple algebraic expressions, and solution of linear equations. Students also receive experience using linear equations to solve applied problems.

Fractional Expressions   Chapter 3 develops expansion, reduction, and the fundamental operations of fractional expressions. Prime numbers and composite numbers are discussed in order to develop a general method of forming the lowest common denominator. The chapter culminates with a discussion of ratio and proportion.

Polynomials   Chapter 4 deals with operations involving polynomial expressions. Special products are developed. Factoring is developed progressively over three sections. Selected quadratic equations are introduced and solved by factoring. Applied problems involving linear and quadratic equations are also solved.

Rational Expressions   Chapter 5 discusses operations with rational expressions. Complex fractions are simplified, and students learn to solve equations and inequalities that contain rational expressions. Fractional equations and inequalities are also used to solve applied problems.

Exponents and Radicals   Chapter 6 develops negative exponents, rational exponents, roots, and radicals. Scientific notation is discussed as an application of positive and negative exponents. Operations are performed on radical expressions. Students are taught to solve simple equations that contain radicals. Quadratic equations are solved by completing the square and by developing the quadratic formula.

Graphing   Chapter 7 introduces students to equations containing two variables and develops the rectangular coordinate system. Equations are graphed with particular emphasis on linear equations, linear inequalities, and quadratic equations. Relations, functions, and functional notation are also discussed.

Systems of Linear Equations and Inequalities   Chapter 8 deals with systems of linear equations and inequalities in two variables. Systems of linear equations are solved by graphing, addition (elimination), and substitution. Systems of linear inequalities are solved by graphing. Applied problems are solved by using systems of linear equations.

NEW TO THIS EDITION

Thanks to the suggestions of many users and reviewers, the Second Edition contains the following improvements:

Real numbers are developed in the first chapter rather than the first three chapters.

Set theory has been minimized and used only where it enhances the explanation.

The development of linear equations has been expanded and strengthened.

The sections on factoring of polynomials have been expanded and strengthened.

A section covering ratio and proportion has been added.

Most exercise sets have been expanded to provide many more problems for in-class practice or homework.

SUPPLEMENTARY MATERIAL

The following supplementary material is available from Merrill Publishing:

Instructor's Manual and Test Bank.
Contains numerous test questions for each chapter. Also has two mid-term examinations and two final examinations.

Teacher's Solutions Manual.
Contains detailed solutions to all sets of exercises.

Answer Key.
Answers to all exercise questions.

Student's Study Manual.
Contains a synopsis and examples covering each section of the text. Detailed solutions of selected exercises are also included.

(IBM-PC)   *Computerized Test Bank.*
(APPLE IIe)   *Computerized Test Bank.*
Contains the questions found in the *Instructor's Manual and Test Bank,* cross-referenced to the section numbers of the text. Available for both Apple and IBM-PC computers. Questions can be either multiple-choice or free-response.

ACKNOWLEDGEMENTS

The authors wish to thank the following reviewers for their helpful comments: John C. Anderson; Dawson Carr, Sandhills Community College; John Fajie; Grace P. Foster, Beaufort County Community College; Marceline C. Gratiaa, Winona State University; Jerry Holland, Everett Community College; Donald Skow, Pan American University; Virginia Strawderman, Georgia State University; Wesley W. Tom; Patricia J. Zeman, Columbus Technical Institute; and Emar Zemgalis, Highline Community College.

Robert G. Moon

CONTENTS

THE SET OF REAL NUMBERS
Its Operations and Properties

1

WHOLE NUMBERS, OPERATIONS, AND EXPONENTS

1.1

set

elements   members

Algebra, being a study of properties and patterns of numbers, involves many terms and symbols. The first term we discuss is the word **set.** A set is a group or collection of objects. For our purposes, these objects are usually numbers and are called **elements** or **members** of the set.

Example 1

Given the set of counting numbers between 5 and 9, what are its elements?

Solution: Counting numbers are those numbers commonly used for counting, namely 1, 2, 3, 4, 5, 6, 7, 8, 9, 10, 11, and so forth. Thus, the set of counting numbers between 5 and 9 consists of the elements 6, 7, and 8.

Sets can be represented by upper case letters and are defined either by a word statement or by listing the elements within braces.

Example 2

Define this set by writing a word statement describing these elements: $A = \{1,2,3,4\}$.

Solution: A is the set containing the first four counting numbers.

Example 3

Define this set by listing its elements: S is the set consisting of the last three letters of the English alphabet.

Solution: Place the elements within braces. Thus, $S = \{x,y,z\}$.

empty set

A set which contains no elements is called the **empty set** and is symbolized by either empty braces, { } or this Scandanavian letter, $\varnothing$. For example, if the set E consists of all counting numbers between 7 and 8, then $E = \{ \ \}$ or $E = \varnothing$.

natural numbers

Two important sets of numbers commonly used in algebra are the set of natural numbers and the set of whole numbers. The set of **natural numbers,** denoted by N, is composed of all the counting numbers. Thus,

$$N = \{1,2,3,4,5,6,7,8,9,10,11,12, \ . \ . \ .\}$$

whole numbers

where the three dots indicate that the established pattern continues indefinitely. The set of **whole numbers,** denoted by W, is composed of all the natural numbers together with zero. Hence,

$$W = \{0,1,2,3,4,5,6,7,8,9,10,11,12, \ . \ . \ .\}$$

Three important symbols enable us to compare two numbers: the equal sign $=$, the less than sign $<$, and the greater than sign $>$. The equal sign states that two symbols represent the same number. For example, $3 + 2 = 5$. The signs for "less than" and "greater than" are called signs of inequality, since they indicate that two symbols represent different numbers. To be used correctly, they must point to the smaller number.

Thus the statement $5 < 8$

is read, "five is less than eight," while the statement

$$12 > 4$$

is read, "twelve is greater than four."

Numbers make use of the four fundamental operations. They are addition, subtraction, multiplication, and division.

Addition is indicated by a plus sign. The terms to be added are called **addends,** while the answer is called the **sum.**

addends   sum

In the addition problem $3 + 5 = 8$, identify the addends and the sum.

Example 4

Solution: The addends are 3 and 5. The sum is 8.

Subtraction is denoted by a minus sign. The first term is the **minuend** and the second term is the **subtrahend.** The answer is called the **difference.**

minuend
subtrahend   difference

In the subtraction problem $6 - 4 = 2$, identify the minuend, the subtrahend, and the difference.

Example 5

Solution: The minuend is 6, the subtrahend is 4, and the difference is 2.

Subtraction is closely related to addition, because the difference, when added to the subtrahend, produces the minuend. So, $6 - 4 = 2$, because $2 + 4 = 6$.

Multiplication is indicated by either a dot or parentheses. For example, $2 \cdot 3$, $(2)(3)$, and $2(3)$ each represent 2 times 3. The numbers being multiplied are called **factors,** while the result is called the **product.**

factors   product

In the multiplication problem $3 \cdot 5 = 15$, identify the factors and the product.

Example 6

Solution: The factors are 3 and 5. The product is 15.

When one of the factors is a natural number, multiplication can be considered as repetitive addition. Thus, $3 \cdot 5$ means that 5 is used as an addend three times, giving

$$3 \cdot 5 = 5 + 5 + 5 = 15$$

It then follows that any number times zero is zero. For example, $2 \cdot 0 = 0$, since $2 \cdot 0 = 0 + 0 = 0$.

Multiplication by Zero
The product of any number and zero is zero.

4

THE SET OF REAL NUMBERS

Often with multiplication, it is necessary to repeat factors. This is done conveniently by using exponents. Thus, *aunque*, *Therefore*

$$2 \cdot 2 \cdot 2 \cdot 2 = 2^4$$

base
exponent
power

where 2 is said to be the **base** and 4 is called the **exponent** or **power**. The expression 2^4 is called an exponential and is read, "two to the fourth power." A natural number exponent indicates the number of times the base is used as a factor.

Example 7

Evaluate the following:
(a) $2^4 = 2 \cdot 2 \cdot 2 \cdot 2 = 16$
(b) $3^2 = 3 \cdot 3 = 9$
(c) $7^1 = 7$
(d) $0^3 = 0 \cdot 0 \cdot 0 = 0$

Division is denoted by the familiar symbol ÷. It is also symbolized by a fraction bar. Thus 8 divided by 2 can be written as $8 \div 2$ or $\frac{8}{2}$. The

dividend
divisor
quotient

number which you are dividing into is called the **dividend,** and the number you are dividing by is called the **divisor.** The answer is said to be the **quotient.**

Example 8

In the division problem $8 \div 2 = 4$ or $\frac{8}{2} = 4$, identify the dividend, divisor, and quotient.

Solution: The dividend is 8, the divisor is 2, and the quotient is 4.

Division is related to multiplication since the quotient when multiplied by the divisor gives the dividend. So, $18 \div 6 = 3$, because $3 \cdot 6 = 18$. Using this relationship, we can show that division by zero is impossible. Let's try to divide 8 by zero and see why it will not work.

Example 9

Show that $8 \div 0$ has no answer.

Solution: $8 \div 0 =$ no answer; because no number exists that can be multiplied times the divisor, 0, to produce the dividend, 8. (The product of any number and zero is zero, not 8.)

Example 10

Show that $0 \div 0$ does not have a definite answer.

Solution: $0 \div 0 =$ any number; because any number times the divisor, 0, equals the dividend, 0. Note the following:
(a) $0 \div 0 = 1$, since $1 \cdot 0 = 0$
(b) $0 \div 0 = 12$, since $12 \cdot 0 = 0$
(c) $0 \div 0 = 167$, since $167 \cdot 0 = 0$

Thus we say that $0 \div 0$ does not have a definite answer

because it has infinitely many solutions—the answer is not unique.

The operation of division has three important properties which we will use throughout the text.

Properties of Division

1. Division by zero is not defined. In other words, the divisor must never be zero.
   Example: None of these indicated divisions can be performed.
   $$8 \div 0, \frac{12}{0}, 0 \div 0$$

2. Any number divided by 1 produces the original number.
   Example: $\frac{12}{1} = 12$.

3. A quotient is zero only if the dividend is zero and the divisor is not zero.
   Example: Each of these quotients is zero.
   $$\frac{0}{9} = 0, 0 \div 12 = 0, \frac{0}{1} = 0$$

Define the following sets by listing the elements within braces.

Exercise 1.1

1. The set consisting of the first six letters of the English alphabet.
2. The set consisting of the last four letters of the English alphabet.
3. The set of counting numbers between 15 and 22.
4. The set of counting numbers between 9 and 10.
5. The set of whole numbers less than 8.
6. The set of natural numbers less than 8.
7. The set of whole numbers greater than 5 but less than 12.
8. The set of whole numbers greater than 17 but less than 18.

In the addition problem $8 + 13 = 21$,
9. 8 is called an _____.
10. 13 is called an _____.
11. 21 is called the _____.

In the subtraction problem $25 - 9 = 16$,
12. 25 is called the _____.
13. 9 is called the _____.
14. 16 is called the _____.

In the multiplication problem $8 \cdot 9 = 72$,

15. 8 is called a _____.

16. 9 is called a _____.

17. 72 is called the _____.

In the division problem $30 \div 10 = 3$,

18. 30 is called the _____.

19. 10 is called the _____.

20. 3 is called the _____.

In the division problem $\dfrac{27}{3} = 9$,

21. 27 is called the _____.

22. 3 is called the _____.

23. 9 is called the _____.

Check the following subtraction problems by adding the difference to the subtrahend to produce the minuend.

24. $8 - 2 = 6$ because _____ + _____ = _____.

25. $12 - 7 = 5$ because _____ + _____ = _____.

26. $6 - 0 = 6$ because _____ + _____ = _____.

27. $5 - 5 = 0$ because _____ + _____ = _____.

28. $47 - 21 = 26$ because _____ + _____ = _____.

Write the meaning of each multiplication problem by relating it to addition.

29. $3 \cdot 6$ means _____.

30. $2 \cdot 0$ means _____.

31. $3 \cdot 1$ means _____.

32. $2 \cdot 100$ means _____.

Check the following division problems by multiplying the quotient times the divisor to produce the dividend.

33. $48 \div 6 = 8$ because _____ $\cdot$ _____ = _____.

34. $5 \div 5 = 1$ because _____ $\cdot$ _____ = _____.

35. $\dfrac{12}{1} = 12$ because _____ $\cdot$ _____ = _____.

36. $\dfrac{24}{8} = 3$ because _____ $\cdot$ _____ = _____.

37. $\dfrac{0}{6} = 0$ because _____ $\cdot$ _____ = _____.

Evaluate the following.

38. 0^4

39. 1^3

40. 2^3　　　　　　　　　　　　**41.** 5^2

42. 10^2　　　　　　　　　　　　**43.** 8^1

44. 3^5　　　　　　　　　　　　**45.** 2^7

True or false.

46. $15 \cdot 0 = 0$　　　　　　　　**47.** $0 \cdot 20 = 0$

48. $\dfrac{0}{3} = 0$　　　　　　　　　**49.** $\dfrac{7}{0} = 0$

50. $\dfrac{0}{0} = 1$　　　　　　　　　**51.** $\dfrac{7}{1} = 1$

52. $\dfrac{13}{13} = 1$　　　　　　　　**53.** $10 \div 0 = 0$

54. $0 < 1$　　　　　　　　　　**55.** $23 > 26$

56. Zero is a natural number.　　**57.** Zero can never be used as a dividend.

58. Any number times zero is zero.

NUMERAL EXPRESSIONS, ORDER OF OPERATIONS, AND GROUPING SYMBOLS

1.2

In section 1.1 we studied the four fundamental operations of addition, subtraction, multiplication, and division, as well as exponents. In this section we are concerned with evaluating expressions containing combinations of these operations. Such expressions are called *numeral expressions*.

A **numeral expression** is a phrase containing numerals and at least one symbol of operation. The phrase must represent a number. Therefore, it cannot contain an indicated division by zero.

numeral expression

These are numeral expressions.
(a)　$2 + 4$
(b)　$2 \cdot 3 + 4^2$
(c)　$5 \cdot 2 + 6 \div 3$

Example 1

These are not numeral expressions, since each indicates a division by zero.
(a)　$8 \div 0$
(b)　$\dfrac{15}{0}$
(c)　$2^3 + 4 \div 0 \cdot 3$

Example 2

A numeral expression is said to be *evaluated* when it is represented by exactly one numeral and no symbols of operation.

Example 3

These numeral expressions have been evaluated.
(a) $5 + 2 = 7$
(b) $8 \cdot 0 = 0$
(c) $\dfrac{0}{12} = 0$

In order to formulate a method of evaluating numeral expressions containing two or more operations, consider the expression $5 \cdot 2 + 6$. Observe that two different answers result by simply changing the order in which the operations are performed.

Solution 1:   Add first; multiply second.

$$5 \cdot \underbrace{2 + 6}$$

$$5 \cdot \quad 8$$
$$40$$

Solution 2:   Multiply first; add second.

$$\underbrace{5 \cdot 2} + 6$$

$$10 \; + 6$$
$$16$$

Certainly it is not correct to have two different results for the same expression. Therefore mathematicians have agreed to perform operations in a certain order. This is called the *order of operations agreement* and always results in exactly one answer.

The Order of Operations Agreement

Unless (as discussed later) the expression specifies otherwise, a numeral expression is evaluated by using this three-step procedure:
(1)   Evaluate all exponentials as they occur, working from left to right.
(2)   Perform all multiplications and divisions as they occur, working from left to right.
(3)   Perform all additions and subtractions as they occur, working from left to right.

Example 4

Evaluate $24 \div 2 - 2 \cdot 3 + 1 \cdot 4$

Solution:

Step (1)   The expression contains no exponentials, so proceed to the next step.
Step (2)   Do multiplications and divisions as they occur, working from left to right.

$$\underbrace{24 \div 2} - \underbrace{2 \cdot 3} + \underbrace{1 \cdot 4}$$
$$12 \quad - \quad 6 \quad + \quad 4$$

Step (3) Do additions and subtractions as they occur, working from left to right.

$$\underbrace{12 - 6} + 4$$
$$6 \quad + 4$$
$$10$$

Thus, $24 \div 2 - 2 \cdot 3 + 1 \cdot 4 = 10$. We can also indicate the order in which the operations were performed by placing circled numerals above the corresponding operations:

$$\overset{①}{24} \div \overset{④}{2} - \overset{②}{2} \cdot \overset{⑤}{3} + \overset{③}{1} \cdot 4 = 10$$

Evaluate $3 \cdot 2^3 \div 2 + 7 - 18 \div 3^2$ **Example 5**

Solution:

Step (1) Evaluate all exponentials as they occur.

$$3 \cdot 2^3 \div 2 + 7 - 18 \div 3^2$$
$$3 \cdot 8 \div 2 + 7 - 18 \div 9$$

Step (2) Perform all multiplications and divisions as they occur.

$$\underbrace{3 \cdot 8 \div 2} + 7 - \underbrace{18 \div 9}$$
$$12 \quad + 7 - \quad 2$$

Step (3) Do all additions and subtractions as they occur.

$$\underbrace{12 + 7} - 2$$
$$19 \quad - 2$$
$$17$$

Thus, $3 \cdot 2^3 \div 2 + 7 - 18 \div 3^2 = 17$. The order in which the operations were performed is given by the circled numerals:

$$\overset{③①④}{3 \cdot 2^3 \div 2} + \overset{⑥}{7} - \overset{⑦}{18} \overset{⑤②}{\div 3^2}$$

By the order of operations agreement, $6 \cdot 4 + 3 = 27$. But suppose we wrote the expression, $6 \cdot 4 + 3$, intending that 4 and 3 be added first, and then multiplied by 6. How could we write the expression to obtain a result of 42? Somehow we must show that 4 and 3 are to be added first. The answer is to use grouping symbols. These are represented by parentheses ( ), brackets [ ], and braces { }. Grouping symbols indicate that the enclosed operations are to be done first.

( ) [ ] { }

Example 6   Evaluate $6 \cdot (4 + 3)$

Solution: $6 \cdot \underbrace{(4 + 3)}$
$6 \cdot \quad 7$
$\quad\quad 42$

Thus, $6 \cdot (4 + 3) = 42$. The order in which the operations were performed is given by the circled numerals:

$6 \cdot \overset{②}{(4} + \overset{①}{3)} = 42$

Example 7   Evaluate $(5 - 2) \cdot (3 + 4)$

Solution: These grouping symbols are independent of each other. That is, one set does not fall inside the other. Therefore, it does not matter which is evaluated first.

$\underbrace{(5 - 2)} \cdot \underbrace{(3 + 4)}$
$\quad 3 \quad\; \cdot \quad 7$
$\quad\quad\quad 21$

Thus, $(5 - 2) \cdot (3 + 4) = 21$. These two choices for performing the order of operations are:

$\overset{①}{(5} - \overset{③}{2)} \cdot \overset{②}{(3} + 4) = 21$ or $\overset{②}{(5} - \overset{③}{2)} \cdot \overset{①}{(3} + 4) = 21$

Grouping symbols sometimes occur inside other grouping symbols. It is then easiest to begin by evaluating the innermost pair.

Example 8   Evaluate $2 \cdot [22 - (4 - 1) \cdot 5]$

Solution: Begin evaluating with the innermost pair of grouping symbols. Then, use the order of operations agreement to evaluate the expression within the brackets.

$2 \cdot [22 - \underbrace{(4 - 1)} \cdot 5]$ [do parentheses first]
$2 \cdot [22 - \underbrace{3 \quad\; \cdot 5]}$ [multiply]
$2 \cdot \underbrace{[22 - \quad\quad 15]}$ [evaluate brackets]
$2 \cdot \quad\quad\quad 7$ [multiply]
14

Thus, $\overset{④}{2} \cdot [\overset{③}{22} - \overset{①}{(4} - 1) \overset{②}{\cdot} 5] = 14$

Recall that the fraction bar symbolizes division. It is also considered to be a grouping symbol applying to both the dividend and divisor.

Example 9   Evaluate $\dfrac{3^2 + 6^2}{3 \cdot 2^3 - 9}$

Solution: Since the fraction bar symbolizes division and also indicates grouping symbols, the problem can be written and evaluated as follows:

$$\frac{3^2 + 6^2}{3 \cdot 2^3 - 9} = \frac{9 + 36}{3 \cdot 8 - 9} \qquad \text{[evaluate exponentials]}$$

$$= \frac{45}{24 - 9} \qquad \text{[add above, multiply below]}$$

$$= \frac{45}{15} \qquad \text{[subtract below]}$$

$$= 3 \qquad \text{[divide]}$$

As we progress through the text it will be convenient to classify an expression as being either a basic sum, a basic difference, a basic product, or a basic quotient. This is accomplished by finding the final operation.

Classify $4 \cdot 5 + 7 \cdot 8$ **Example 10**

Solution: $4 \cdot 5 + 7 \cdot 8$ The final operation is addition, so the expression is a basic sum.

Classify $2 \cdot (3 + 5) - 6$ **Example 11**

Solution: $2 \cdot (3 + 5) - 6$ The final operation is subtraction, so the expression is a basic difference.

Classify $3 \cdot (6 - 2)$ **Example 12**

Solution: $3 \cdot (6 - 2)$ The final operation is multiplication. Thus, we have a basic product.

Classify $\dfrac{12 + 3}{7 - 2}$ **Example 13**

Solution: $\dfrac{12 + 3}{7 - 2}$ means $(12 + 3) \div (7 - 2)$ Since the final operation is division, we have a basic quotient.

Evaluate the following numeral expressions using the order of operations agreement.   **Exercise 1.2**

1. $36 \div 3 \cdot 2$ **2.** $12 \div 2 \cdot 4$

3. $7 - 2 - 4$ **4.** $8 + 6 - 1$

5. $4 \cdot 7 + 5^2$

6. $12 - 3 \cdot 2^2$

7. $3 \cdot 4 + 2 \cdot 5 - 2 \cdot 0$

8. $16 \div 2 \cdot 3 + 8 - 10 \div 2$

9. $3 \cdot 4 + 2^4$ ⁄

10. $3 \cdot 4 + 3 \cdot 2^3$

11. $2 \cdot 6 \div 3 \cdot 2$

12. $0 \div 3 + 6 \cdot 2 \div 3 - 2$

13. $20 \cdot 2 + 3 - 2 \cdot 7$ ⁄

14. $10^2 - 4^2 - 3^2$

Evaluate the following numeral expressions using your knowledge of grouping symbols and the order of operations agreement.

15. $(20 - 8) \cdot 2$

16. $3 \cdot (4 + 2)^2$

17. $4 \cdot [7 - (5 - 2)]$

18. $20 \cdot (2 + 3) - 2 \cdot 7$

19. $20 \cdot 2 + (3^2 - 2^3) \cdot 7$

20. $20 \cdot [2 + 3 - 2] \cdot 7$

21. $(10 - 4) - 3$

22. $10 - (4 - 3)^3$

23. $(40 \div 4) \div 2$

24. $40 \div (4 \div 2)^2$

25. $3 \cdot (8 - 1) + 5 \cdot (2 + 6)$

26. $(9 - 3) + (6 - 4)$

27. $3 \cdot \{22 - [(13 - 2) - 4]\}$

28. $2 \cdot [4 \cdot (6 - 1) + 3 \cdot (8 + 2)]$

29. $\dfrac{4^3 + 2^3}{2 \cdot 3^2}$

30. $\dfrac{6^2 - 5^2}{(3 - 2)^4} + \dfrac{(6 + 4)^2}{5}$

Insert grouping symbols so that each of the following equations is true.

31. $5 \cdot 8 + 2 = 50$

32. $6 + 8 \div 2 = 7$

33. $4 + 5 - 6 - 1 = 4$

34. $3 \cdot 2 + 2 \cdot 5 = 60$

35. $3 \cdot 2 + 2 \cdot 5 = 36$

36. $3 \cdot 2 + 2 \cdot 5 = 16$

37. $60 - 3 + 4 \cdot 8 \div 2 = 244$

38. $60 - 3 + 4 \cdot 8 \div 2 = 2$ ⁄

39. $4 \cdot 2^3 + 1^5 \cdot 3 = 44$

40. $3 - 2^3 + 3^2 - 1 = 32$

Classify each of the following expressions as basically a sum, difference, product, or quotient. Show the order of operations by using circled numerals.

41. $7 + 2 \cdot 9$

42. $(7 + 2) \cdot 9$

43. $3 + 9 - 4$

44. $3 \cdot (7 - 4)$

45. $(4 + 3) \cdot (7 - 2)$

46. $6 \div 3 \cdot 2$

47. $5 - (2 + 3)$

48. $[(8 \div 2 + 5) - 7] \div 2$

49. $2 \cdot [3 + 6 - (1 + 2)]$

50. $16 \div 2 - 12 \div 2$

51. $8 \div 2 + 1$

52. $15 \div 3 \cdot (5 - 2) \cdot 3$

53. $5 \cdot \{22 - [(13 - 2) - 4]\}$

54. $3 \cdot (7 + 8) - [2 \cdot (5 + 1) + 4]$

VARIABLES, CONSTANTS, AND ALGEBRAIC EXPRESSIONS

1.3   variable

In algebra it is convenient for letters to represent numbers. When this is done the letter can be called a variable. A **variable,** then, is defined to be a symbol which represents any number from a given replacement set.

The replacement set is sometimes called the *domain* of the variable and contains more than one number. Almost any symbol can be used for a variable, although we generally use letters from the English alphabet, such as a, b, m, n, x, y, and z. Any number from the variable's replacement set or domain may be substituted for the variable.

If x is a variable whose replacement set or domain is {1, 2, 6}, then $x + 10$ represents each of the following:

(a) $1 + 10$ [1 substituted for x]
(b) $2 + 10$ [2 substituted for x]
(c) $6 + 10$ [6 substituted for x]

Example 1

If y is a variable whose replacement set or domain is {10, 15, 20}, then $3 \cdot y + y$ represents each of the following:

(a) $3 \cdot 10 + 10$ [10 substituted for y]
(b) $3 \cdot 15 + 15$ [15 substituted for y]
(c) $3 \cdot 20 + 20$ [20 substituted for y]

Example 2

Notice that if the same variable occurs more than once in an expression, the same number must be substituted each time.

If a and b are variables, each having a domain of whole numbers, then $2 \cdot a + 5 \cdot b$ can represent each of the following:

(a) $2 \cdot 0 + 5 \cdot 1$ $a = 0, b = 1$
(b) $2 \cdot 50 + 5 \cdot 60$ $a = 50, b = 60$
(c) $2 \cdot 15 + 5 \cdot 15$ $a = 15, b = 15$

Example 3

Any whole number may be substituted for a, and any whole number may be substituted for b. In fact, as seen in (c), a and b might even represent the same number.

Remember, a variable has more than one number in its replacement set. This allows the values to "vary." However, a **constant** is any symbol which has exactly one number in its replacement set.

constant

Any numeral, such as 2, 5, or 10, is a constant.

Example 4

π is a constant since it represents approximately 3.14.

Example 5

If c has a replacement set of {6}, then c is a constant.

Example 6

Variables and constants can be combined to form variable expressions. A **variable expression** is any phrase containing at least one variable (symbols of operation may also be included). Variable expressions are more commonly called **algebraic expressions.**

variable expression

algebraic expression

Each of the following is an algebraic (or variable) expression for a replacement set of whole numbers.

(a) $x + 8$

Example 7

(b) $y^2 + 3$
(c) $10 \cdot m - m$
(d) $3 \cdot a + 2 \cdot b - 5 \cdot c$

terms

In an algebraic expression the quantities which are added or subtracted are called **terms.** Thus, the algebraic expression

$$3 \cdot a + 2 \cdot b - 5 \cdot c$$

has three terms; $3 \cdot a$, $2 \cdot b$, and $5 \cdot c$. Algebraic expressions may be evaluated by substituting a number from each variable's replacement set. The operations are then performed using our knowledge of grouping symbols and the order of operations agreement.

Example 8

Evaluate $2 \cdot (x + 5)$ when $x = 0$

Solution: $2 \cdot (x + 5) = 2 \cdot (0 + 5)$
$$= 2 \cdot 5$$
$$= 10$$

Example 9

Evaluate $3 \cdot m^2 + 3 \cdot n^2$ when $m = 5$ and $n = 0$.

Solution: $3 \cdot m^2 + 3 \cdot n^2 = 3 \cdot 5^2 + 3 \cdot 0^2$
$$= 3 \cdot 25 + 3 \cdot 0$$
$$= 75 + 0$$
$$= 75$$

Example 10

Evaluate $2 \cdot a + 3 \cdot (a - 2 \cdot b)$ when $a = 10$ and $b = 3$

Solution: $2 \cdot a + 3 \cdot (a - 2 \cdot b) = 2 \cdot 10 + 3 \cdot (10 - 2 \cdot 3)$
$$= 2 \cdot 10 + 3 \cdot (10 - 6)$$
$$= 2 \cdot 10 + 3 \cdot 4$$
$$= 20 + 12$$
$$= 32$$

An algebraic expression, like a numeral expression, can be classified as a basic sum, difference, product, or quotient. This is done by simply determining the final operation.

Example 11

Classify $3 \cdot x + 4 \cdot y$

Solution: $\overset{①}{3} \cdot x \overset{③}{+} \overset{②}{4} \cdot y$ Basic sum

Example 12

Classify $5 \cdot (a + b) - c$

Solution: $\overset{②}{5} \cdot \overset{①}{(a + b)} \overset{③}{-} c$ 　　　Basic difference

Classify $\overset{2}{x} \cdot \overset{1}{(y + 2)}$ 　　　　　　　　　　　Example 13

Solution: $\overset{②}{x} \cdot \overset{①}{(y + 2)}$ 　　　Basic product

Classify $\dfrac{\overset{1}{(m + 3)} \cdot \overset{3}{n}}{\underset{2}{2}}$ 　　　　　　　　　　Example 14

Solution: $[\overset{①}{(m + 3)} \cdot \overset{②}{n}] \overset{③}{\div} 2$ 　　　Basic quotient

---

Two Important Agreements Concerning Algebraic Expressions

(1) If an operation symbol is obviously missing in an expression, then the operation is assumed to be multiplication. For example, $2x$ means $2 \cdot x$. And $3(x + y)$ means $3 \cdot (x + y)$.

(2) If the same variable occurs more than once in an expression, then the same number must be substituted for them all. See Example 2 and Example 10.

Algebraic expressions can be used as mathematical models to symbolize real-life situations. This is done by translating key words and phrases into an algebraic expression. For example, the phrase "8 less than a certain number" can be represented by the algebraic expression $x - 8$. "Less than" is translated to subtraction and x represents the unknown number. Any letter can be used to represent an unknown number. If we had chosen m, then the mathematical model would have been $m - 8$.

Tables 1.1 through 1.5 give common word phrases and the corresponding algebraic expressions used as mathematical models. We use x to represent the unknown number, but it is correct to use any letter you wish.

TABLE 1.1 Addition

Word Phrase	Mathematical Model
the sum of a number and 6	$x + 6$
a number plus 4	$x + 4$
a number added to 10	$x + 10$
8 more than a number	$x + 8$
a number increased by 5	$x + 5$
the total of two numbers	$x + y$

TABLE 1.2 Subtraction

Word Phrase	Mathematical Model
5 subtracted from a number	$x - 5$
8 less than a number	$x - 8$
10 minus a number	$10 - x$
a number minus 10	$x - 10$
7 taken away from a number	$x - 7$
2 fewer than a number	$x - 2$

TABLE 1.3 Multiplication

Word Phrase	Mathematical Model
8 times a number	$8x$
a number multiplied by 6	$6x$
the product of a number and 4	$4x$
twice an amount	$2x$
a number is tripled	$3x$
a number is quadrupled	$4x$
half of a number	$\frac{1}{2}x$

TABLE 1.4 Division

Word Phrase	Mathematical Model
a number divided by 7	$\frac{x}{7}$ or $x \div 7$
10 divided by a number	$\frac{10}{x}$
half of a number	$\frac{x}{2}$
the ratio of some number to 6	$\frac{x}{6}$

TABLE 1.5 Combinations

Word Phrase	Mathematical Model
5 more than twice a number	$2x + 5$
3 less than four times a number	$4x - 3$
the ratio of twice a number to 10	$\frac{2x}{10}$
a number added to 5 times itself	$5x + x$
the number of cents in x dimes	$10x$
the number of cents in x quarters	$25x$
the sum of two consecutive whole numbers	$x + (x + 1)$

Evaluate each of the following algebraic expressions by letting $x = 5$, $y = 2$, and $z = 0$.

Exercise 1.3

1. $x + 3$

2. $y^2 - 1$

3. $x \cdot z$

4. $3xz$

5. $xy + z^3$

6. $5x^2 - 2y^2$

7. $2(x + y) - 3z$

8. $(x + 3)(y + 4)^2$

9. $3x - y + z$

10. $z(3x - 4y)$

11. $x[(x + y) - (y + z)]$

12. $3[x - (x + z)] + 5y$

Evaluate each of the following algebraic expressions by letting $a = 4$, $b = 2$, and $c = 0$.

13. $a + b$

14. $3a^2 + 2b^3$

15. $5a - 3b + c$

16. $3(a + b)^2$

17. $3a + 3b$

18. $(a + b)(b + c)$

19. $2a^2 + 3b^2 + 4c^2$

20. $3a(2b - a)$

21. $5 \cdot [a + 2(b - c)]$

Classify each of the following expressions as a basic sum, difference, product, or quotient.

22. $m + 8$

23. $x - y$

24. $2xy$

25. $5(z + w)$

26. $6x + 2y - 1$

27. $5(x + y - z)$

28. $(m + 4)(n + 5)$

29. $7a(b + c)$

30. $7ab + 7ac$

Write a mathematical model in the form of an algebraic expression representing these word statements.

31. The sum of a certain number and 10.

32. 6 less than some number.

33. The product of a number and 6.

34. 10 fewer than a certain number.

35. The ratio of some number to 25.

36. The total of three numbers.

37. 9 more than a number.

38. Half of a number.

39. 15 less than some number.

40. The product of a number and 12.

41. 15 less than 7 times a number.

42. 7 more than 3 times a number.

43. 10 less than twice a certain number.

44. 5 more than 6 times a certain number.

45. The number of cents in n nickels.

46. The number of cents in q quarters.
47. The number of cents in $(x - 4)$ half-dollars.
48. The number of cents in $(2x + 4)$ dollars.
49. 3 times the sum of a number and 5 is decreased by twice the number.
50. The sum of 3 consecutive whole numbers.

SIGNED NUMBERS AND INTEGERS

1.4

signed numbers

In the study of algebra the concept of signed numbers plays an important role. **Signed numbers** are composed of positive numbers, zero, and negative numbers. Positive numbers are greater than zero and are denoted by a plus (+) sign, while negative numbers are less than zero and are denoted by a minus (−) sign. The number zero is neither positive nor negative.

Example 1

Here are some illustrations of signed numbers.
(a) $+7$ Read "positive seven."
(b) $+24$ Read "positive twenty-four."
(c) -9 Read "negative nine."
(d) -1 Read "negative one."

For convenience, the plus sign is often omitted from a positive number. Thus, $+7 = 7$ and $+24 = 24$. Signed numbers may be used to represent concepts in many fields. For example, we may represent:

A temperature of 15° above zero as $+15°$ or simply 15°.

A temperature of 15° below zero as $-15°$.

A gain of 10 dollars as $+10$ dollars or simply 10 dollars.

A loss of 10 dollars as -10 dollars.

20 feet above sea level as $+20$ feet or simply 20 feet.

20 feet below sea level as -20 feet.

integers

In this section we are concerned with signed whole numbers. These are called **integers**. Thus, the set of integers denoted by an uppercase I is:

$$I = \{\ldots, -5, -4, -3, -2, -1, 0, +1, +2, +3, +4, +5, \ldots\}$$

The three dots indicate that the numbers continue on indefinitely. The set of integers consists of three main parts: the positive integers, zero, and the negative integers.

$$I = \{\ldots, \underbrace{-5, -4, -3, -2, -1}_{\text{negative integers}}, 0, \underbrace{+1, +2, +3, +4, +5, \ldots}_{\text{positive integers}}\}$$

zero, which is neither positive nor negative

To represent the set of integers visually, we develop the standard number line. This is done in figure 1.1 with a horizontal straight line infinitely long to the left and right. This is symbolized by the arrows on the left and right ends of the line. An arbitrary point is marked and named by the number zero. This is the beginning point and is therefore called the **origin.** Equal distances are then marked off to the left and right of the origin. The points located to the right of the origin are labeled with positive numbers, and the points to the left of the origin are labeled with negative numbers. Thus, on the standard number line, positive is to the right and negative is to the left.

origin

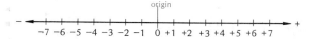

FIGURE 1.1   The Standard Number Line

The standard number line is said to order the integers. As you proceed from left to right along the standard number line, the integers become larger. Thus, the standard number line allows us to compare the sizes of two integers. The larger integer will be located to the right of the smaller integer. Example 2 illustrates the use of the familiar symbols representing "greater than" and "less than." Remember, these symbols must point to the smaller of the two numbers.

These expressions illustrate the "greater than" and "less than" symbols.   **Example 2**
(a) $+6 > +1$
(b) $2 < 4$
(c) $0 > -2$
(d) $-5 < 0$
(e) $-6 > -10$ (Note, -6 lies to the right of -10 on the standard number line. Thus, -6 is greater than -10. Recall that the symbol $>$ must point to the smaller number, -10.)
(f) $-9 < -2$

As shown by the standard number line in figure 1.1, positive and negative integers name points on opposite sides of the origin. Thus, every integer is said to have an opposite. For example, the opposite of $+3$ is -3, the opposite of -5 is $+5$, and the opposite of 0 is 0. In mathematics opposites are called additive inverses. The term **additive inverse** is used because two opposites, when added together, will produce a sum of zero. For example, if you save \$10 (represented by $+10$), and if you then spend \$10 (represented by -10), you have 0 dollars left. Hence, $(+10) + (-10) = 0$. This gives rise to the additive inverse property.

additive inverse

> **The Additive Inverse Property**
>
> For each integer x, there is exactly one opposite, called the additive inverse of x, denoted by $-x$, such that $x + (-x) = 0$ and $(-x) + x = 0$

Example 3

Here are some illustrations of the additive inverse property.
(a)   The additive inverse or opposite of $+7$ is -7, and $(+7) + (-7) = 0$
(b)   The additive inverse of 23 is -23, and $23 + (-23) = 0$
(c)   The additive inverse of -5 is $-(-5)$ or $+5$, and $(-5) + (+5) = 0$
(d)   The additive inverse of 0 is 0, and $0 + 0 = 0$

In algebra it is common to encounter expressions with a series of plus or minus signs, such as $-(+3)$ or $-(-6)$. These expressions can be simplified by changing each of the numbers within the parentheses to its opposite. Thus, $-(+3) = -3$ and $-(-6) = +6$. This procedure can be extended to more complicated problems by remembering the following two properties.

> **Simplifying a Series of Signs**
>
> (1)   An odd number of minus signs will simplify to a single minus sign.
> (2)   An even number of minus signs will simplify to a single plus sign.

Example 4

These equations illustrate simplifying a series of signs.
(a) $-(-8) = +8$ or 8
(b) $-[-(-2)] = -2$
(c) $-\{-[-(-10)]\} = +10$ or 10
(d) $-(+12) = -12$
(e) $-[-(+1)] = +1$ or 1

Notice that a minus sign has a dual role. It can specify a negative number such as -4. But it can also tell us to find the opposite or additive inverse, such as $-(-7)$ which simplifies to $+7$. When a minus sign precedes a variable, such as $-x$, it is telling us to find the opposite. Thus $-x$ means the opposite of whatever number x represents. It does not necessarily specify a negative number. For example, if $x = -5$, then $-x = -(-5)$ or $+5$. For more illustrations study table 1.6.

TABLE 1.6	
If x *equals:*	*Then* −x *equals:*
+3	−3
0	0
−1	+1
−10	+10

Another important idea associated with signed numbers is the concept of absolute value. The **absolute value** of a number gives its distance from the origin on the standard number line. Absolute value is symbolized by two vertical bars. Thus, the absolute value of negative 2 is symbolized by $|-2|$. Since negative 2 is two units from the origin, it follows that $|-2| = 2$. Distance is a physical measurement which is never negative. Therefore, the absolute value of a number is always positive or zero. It is never negative.

absolute value

Here are some illustrations of absolute value.

Example 5

(a) $|+6| = 6$

(b) $|4| = 4$

(c) $|0| = 0$

(d) $|-7| = 7$

Four statements involving a variable are used quite frequently in algebra, and it is important to know their meaning:

(1) $x > 0$ Read "x is greater than zero." This means that x represents only positive numbers.

(2) $x \geq 0$ Read "x is greater than or equal to zero." This means that x represents positive numbers and also zero. (These are sometimes called nonnegative numbers.)

(3) $x < 0$ Read "x is less than zero." This means that x represents only negative numbers.

(4) $x \leq 0$ Read "x is less than or equal to zero." This means that x represents negative numbers and also zero. (These are sometimes called nonpositive numbers.)

We now use these statements to give a formal definition of absolute value.

Definition of Absolute Value
(1)   If $x \geq 0$, then $|x| = x$.
(2)   If $x < 0$, then $|x| = -x$.

The first part of this definition states that when the number inside the absolute value symbol is positive or zero, then the absolute value is equal to the original number. Hence, $|5| = 5$ and $|0| = 0$. The second part of this definition is a little more difficult to understand. It says that if the number inside the absolute value symbol is negative, then the answer is equal to the opposite of the original negative number. Thus, the result is positive. For example, $|-9| = -(-9) = +9$.

Exercise 1.4

Classify the following integers as positive, negative, or neither.

1. $+8$ **2.** -3
3. 0   **4.** 23

Fill in the blanks.

5. If $+50$ means 50 miles north, then -50 means 50 miles _____.

6. If $+15$ means the temperature has increased 15 degrees, then -15 means the temperature has _____15 degrees.

7. If 25 means you have a gain of 25 dollars, then -25 means you have a _____of 25 dollars.

On the standard number line in figure 1.2 several points are labeled with uppercase letters. Give the correct number that belongs at each letter.

FIGURE 1.2

8. Give the correct number for A.
9. Give the correct number for B.
10. Give the correct number for C.
11. Give the correct number for D.
12. Give the correct number for E.

Write the additive inverse or opposite for each of the following.

13. $+2$ **14.** -3
15. 0   **16.** 12

Simplify each of the following expressions.

17. $-(-7)$ **18.** $-(+4)$
19. $-(-1)$ **20.** $-[-(-12)]$
21. $-[-(+1)]$ **22.** -0
23. $-(-a)$ **24.** $-(+b)$

25. $-\{-[-(-13)]\}$ $+13$
26. $(+2) + (-2)$
27. $8 + (-8)$ $- 8$
28. $0 + 0$
29. $(-9) + [-(-9)]$ $- 9$
30. $|-5|$
31. $|+12|$ $- 12$
32. $|0|$
33. $|23|$ 23
34. $|a|$ if $a < 0$

Complete the following table.

	If x equals:	Then $-x$ equals:
35.	$+20$	
36.	13	
37.	0	
38.	-1	
39.	-10	
40.	-30	

True or false.
41. $0 > -3$
42. $0 < -16$
43. $+5 > -6$
44. $-14 > -10$
45. $-200 > 100$
46. $-1 = 1$
47. $+8 = 8$
48. $|-5| = -5$
49. $|+8| = |-8|$
50. If y represents a negative number, then $y < 0$. T
51. If b represents a positive number or possibly zero, then $b \geq 0$. T
52. $-x$ always represents negative numbers. T
53. On the standard number line the point named by the number zero is called the *origin*.

RATIONAL NUMBERS, IRRATIONAL NUMBERS, AND REAL NUMBERS

1.5

In previous sections we developed these three sets of numbers:

(1)   The set of natural numbers. $N = \{1, 2, 3, 4, 5, \ldots\}$
(2)   The set of whole numbers. $W = \{0, 1, 2, 3, 4, 5, \ldots\}$
(3)   The set of integers. $I = \{\ldots, -3, -2, -1, 0, 1, 2, 3, \ldots\}$

You should note that each natural number is also a whole number and each whole number is also an integer. Thus, the set of natural numbers is included within the set of whole numbers, which, in turn, is included within the set of integers.

To continue our study of number sets we next use the integers to develop the set of rational numbers.

rational numbers

The set of **rational numbers** is composed of all numbers which are quotients (or *ratios*) of two integers; thus the name rational.

Example 1

Each of these expressions is a rational number because each is a quotient of two integers.

(a) $1 \div 2$ or $\dfrac{1}{2}$

(b) $1 \div (-2)$ or $\dfrac{1}{-2}$

(c) $2 \div 3$ or $\dfrac{2}{3}$

(d) $(-2) \div 3$ or $\dfrac{-2}{3}$

(e) $6 \div 1$ or 6

(f) $0 \div 1$ or 0

We now state the formal definition of the set of rational numbers.

Definition of Rational Numbers

The set of rational numbers is composed of all numbers that can be written in the form of $a \div b$ or $\dfrac{a}{b}$ where a and b are integers and b is not equal to zero.

The set of rational numbers is represented by an uppercase Q to remind us of "quotient." Some of the numbers belonging to set Q are:

$$Q = \left\{ \ldots, \frac{-12}{5}, \frac{-2}{1}, \frac{-3}{2}, \frac{-1}{1}, \frac{-1}{3}, \frac{0}{1}, \frac{1}{2}, \frac{2}{3}, \frac{1}{1}, \frac{4}{3}, \frac{2}{1}, \ldots \right\}$$

As illustrated below, natural numbers, whole numbers, and integers are also rational numbers because they can be written as quotients of two integers where the divisor is 1.

$$N = \left\{ \frac{1}{1}, \frac{2}{1}, \frac{3}{1}, \frac{4}{1}, \ldots \right\}$$

$$W = \left\{ \frac{0}{1}, \frac{1}{1}, \frac{2}{1}, \frac{3}{1}, \frac{4}{1}, \ldots \right\}$$

$$I = \left\{ \ldots, \frac{-3}{1}, \frac{-2}{1}, \frac{-1}{1}, \frac{0}{1}, \frac{1}{1}, \frac{2}{1}, \frac{3}{1}, \ldots \right\}$$

By definition, every rational number can be symbolized as $\frac{a}{b}$ or $a \div b$, where a and b are integers and $b \neq 0$. Thus, a can actually be divided by b to obtain a decimal numeral.

Example 2

Given the rational number $\frac{3}{4}$, divide 3 by 4 and write the answer as a decimal numeral.

Solution:

$$\begin{array}{r} .75 \\ 4\overline{)3.00} \\ \underline{2\,8} \\ 20 \\ \underline{20} \\ 0 \end{array}$$

$\leftarrow$ {when a remainder is zero, the division process terminates}

Hence, $\frac{3}{4} = .75$

This result is called a *terminating decimal*.

Example 3

Given the rational number $\frac{12}{99}$, divide 12 by 99 and write the answer as a decimal numeral.

Solution:

$$\begin{array}{r} .1212 \ldots \\ 99\overline{)12.0000} \\ \underline{9\,9} \\ 2\,10 \\ \underline{1\,98} \\ 120 \\ \underline{99} \\ 210 \\ \underline{198} \\ 120 \end{array}$$

{when remainders are the same, the quotient repeats indefinitely}

Hence, $\frac{12}{99} = .121212 \ldots$

This result is called a *repeating decimal*.

As illustrated by the next example, a more convenient notation for indicating a repeating decimal is to place a bar directly above the digits which repeat.

Here are some illustrations using a bar to indicate the repetitive pattern.

Example 4

(a) $.121212 \ldots = .\overline{12}$

(b) $.713713713 \ldots = .\overline{713}$

(c)   .2474747 . . . = .2$\overline{47}$
(d)   3.7040404 . . . = 3.7$\overline{04}$

Examples 2 and 3 indicate that a rational number can be divided out and written as either a terminating or a repeating decimal. With more advanced algebra we can actually prove that every terminating or repeating decimal is indeed a rational number. This suggests an alternate definition of a rational number.

Alternate Definition of a Rational Number
A number is said to be rational provided it can be written as either a terminating or a repeating decimal.

Example 5

Each of these numbers is rational.

(a)   .35   (terminating decimal)
(b)   .$\overline{61}$ (repeating decimal)
(c)   5.2$\overline{34}$ (repeating decimal)
(d) -9.0621 (terminating decimal)
(e)   12   (terminating decimal)
(f)   0   (terminating decimal)

Now we are in a position to find some numbers that are not rational. Numbers which are not rational are said to be irrational. Since rational numbers can be written as either terminating decimals or repeating decimals, then **irrational numbers** cannot terminate and cannot repeat. In other words, the digits of an irrational number continue on indefinitely, showing no repetitive pattern.

irrational numbers

Example 6

These are illustrations of irrational numbers.
(a)   .202002000200002 . . .

The number of 0's increases by one each time. The digits do not terminate and do not repeat.
(b)   .123456789101112 . . .

Each digit increases by one each time. The digits do not terminate and do not repeat.
(c) -6.535335333533335 . . .

The number of 3's increases by one each time. The digits do not terminate and do not repeat.
(d) $\pi = 3.14159$ . . .

The digits in the decimal equivalent of π do not terminate and do not repeat. Thus, π is an irrational number.

The preceding examples suggest a formal definition of irrational numbers.

Definition of an Irrational Number
An irrational number is any number whose decimal form is nonterminating and nonrepeating.

The set of irrational numbers is represented by an uppercase H.

The final set of numbers developed for this course is the set of real numbers denoted by R. The set of **real numbers** is the set containing all of the rational numbers and all of the irrational numbers. This includes all numbers having a decimal representation. Thus, we arrive at the definition.

real numbers

Definition of a Real Number
A real number is any number that can be written as a decimal numeral. It can be terminating, repeating, or nonterminating-non-repeating.

Each of these is a real number.

Example 7

(a) $2.3\overline{6}$

(b) $5.\overline{41}$

(c) $3.202002000200002 \ldots$

(d) π

(e) $\dfrac{1}{2}$

(f) $\dfrac{-7}{5}$

(g) 0

(h) -5

(i) 6

Figure 1.3, page 28, illustrates the structure of the real number system.

Real numbers are important because they are used to label every point on the number line. In fact, for every real number, there is a point on the number line, and for every point on the number line, there is a real

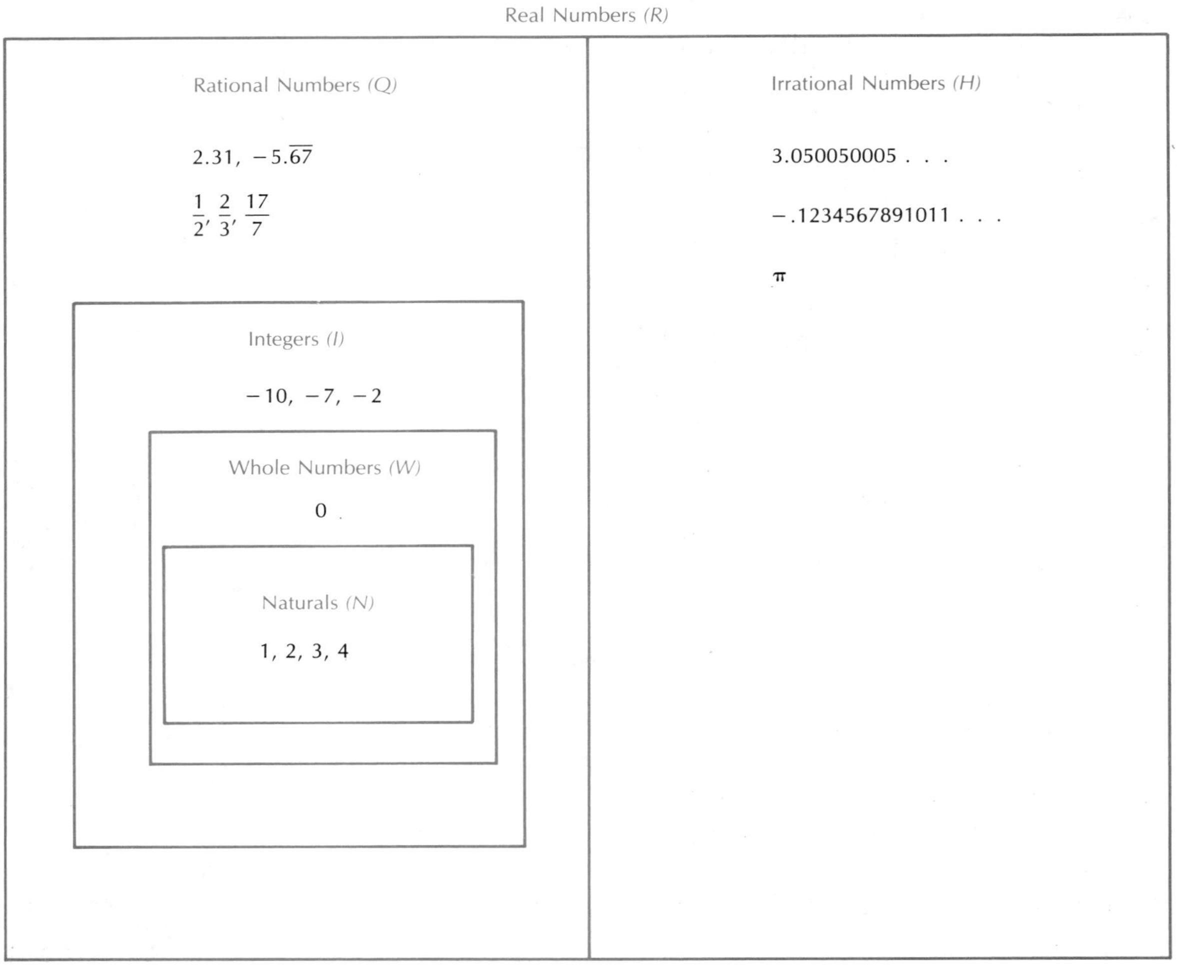
Real Numbers (R)
Rational Numbers (Q)
2.31, −5.$\overline{67}$
$\frac{1}{2}$, $\frac{2}{3}$, $\frac{17}{7}$
Integers (I)
−10, −7, −2
Whole Numbers (W)
0
Naturals (N)
1, 2, 3, 4
Irrational Numbers (H)
3.050050005 . . .
−.1234567891011 . . .
π

FIGURE 1.3

coordinates

number. When real numbers are used to label points, they are called **coordinates.** Figure 1.4 shows some real numbers acting as coordinates.

−
+
−5
−4
−3
−2
−1
0
1
2
3
4
5
−4.5
$-\frac{1}{4}$
$\frac{1}{2}$
1.5
π

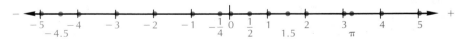

FIGURE 1.4

Exercise 1.5

True or false.

1. Every rational number can be written as a terminating decimal.

2. Every integer is also a rational number.

3. Every rational number is also an integer.

4. Numbers having a decimal representation are known as real numbers.

5. $3 \div 7$ is a rational number.

6. $\frac{4}{5}$ is an irrational number.

7. 0 is a real number.

8. 2.303003000300003 . . . is a rational number.

9. $\frac{2}{5}$ is an integer.

10. -3 is a rational number.

11. π is a rational number.

12. 10 is a real number.

13. The set of natural numbers is included within the set of whole numbers.

14. Every integer is also a real number.

15. Every real number is also irrational.

Use the operation of division to write each of these rational numbers as a decimal. If the decimal repeats, use the bar. (Do not round off.)

16. $\frac{7}{10}$ **17.** $\frac{5}{100}$

18. $\frac{1}{7}$ **19.** $\frac{11}{5}$

20. $\frac{17}{9}$ **21.** $\frac{3}{4}$

Classify each of these numbers as rational or irrational.

22. $\frac{1}{2}$ **23.** $\frac{3}{5}$

24. $\frac{-2}{3}$ **25.** 0

26. -6 **27.** 5.707007000700007 . . .

28. π **29.** $2.6\overline{1}$

30. 0.123456789101112 . . .   **31.** 5.0321

Draw a number line and locate the points named by these coordinates.

32. 0   **33.** -3

34. 2   **35.** 1.5

36. -3.5 **37.** $|-3|$

38. $-|-4|$ **39.** $-\frac{1}{2}$

Define the following terms.

40. Rational number   **41.** Irrational number

42. Real number   **43.** Coordinates

PROPERTIES OF REAL NUMBERS

1.6

In this section we discuss some of the properties of addition and multiplication of real numbers. These properties are useful for evaluating numeral expressions and for simplifying algebraic expressions. In the following statements, a, b, and c all represent real numbers.

---

1. Commutative Properties of Addition and Multiplication

$$a + b = b + a \text{ and } ab = ba$$

The commutative properties state that two real numbers can be added in any order or multiplied in any order.

Example 1

These equations illustrate the commutative property of addition.
(a) $3 + 5 = 5 + 3$
(b) $a + 2b = 2b + a$
(c) $3(m + n) = 3(n + m)$
(d) $2a + 3b + c = 3b + 2a + c$

Example 2

These equations illustrate the commutative property of multiplication.
(a) $6 \cdot 9 = 9 \cdot 6$
(b) $2(ab) = 2(ba)$
(c) $x(y + z) = (y + z)x$

The commutative properties do not hold true for all operations. Division, for example, is not commutative. This is evident if you consider that $6 \div 3 = 2$, and $3 \div 6 = \frac{1}{2}$. Thus, $6 \div 3 \neq 3 \div 6$ ($\neq$ means does not equal). You should also be able to show that subtraction is not commutative.

---

2. Associative Properties of Addition and Multiplication

$$(a + b) + c = a + (b + c) \text{ and } (ab)c = a(bc)$$

The associative properties say that the sum or product of three or more numbers is the same no matter how the numbers are "associated" in groups. For example, according to the associative property of addition,

$(2 + 3) + 4 = 2 + (3 + 4)$. Let's show that this is true by evaluating each side:

$$(2 + 3) + 4 = 2 + (3 + 4)$$
$$5 \quad + 4 = 2 + \quad 7$$
$$9 \quad = \quad 9$$

Each of these equations illustrates the associative property of addition.

Example 3

(a) $(0 + 5) + 2 = 0 + (5 + 2)$
(b) $(2a + 3b) + c = 2a + (3b + c)$
(c) $(1 + 3 + 2) + 5 = (1 + 3) + (2 + 5)$

The commutative and associative properties of addition are especially important when used together. In combination, they allow us to interchange addends and regroup any way we desire. For example, the numeral expression $1 + 3 + 5$ can be rewritten any of the following ways:

$(3 + 1) + 5$	$(1 + 5) + 3$	$5 + (1 + 3)$
$(3 + 5) + 1$	$(1 + 3) + 5$	$5 + (3 + 1)$
$(5 + 1) + 3$	$3 + (1 + 5)$	$1 + (5 + 3)$
$(5 + 3) + 1$	$3 + (5 + 1)$	$1 + (3 + 5)$

Verify that each of these expressions equals 9.

Here are more illustrations combining the commutative and associative properties of addition.

Example 4

(a) $1 + 2 + 5 + 4 = (2 + 1) + (4 + 5)$
(b) $2a + (3b + 4a) = (2a + 4a) + 3b$

These equations illustrate the associative property of multiplication.

Example 5

(a) $(2 \cdot 5) \cdot 6 = 2 \cdot (5 \cdot 6)$
(b) $(a \cdot 2) \cdot 3 = a \cdot (2 \cdot 3)$
(c) $(2 \cdot m \cdot 3) \cdot n = (2 \cdot m) \cdot (3 \cdot n)$

As in addition, the commutative and associative properties of multiplication are particularly helpful when used in combination. They then allow us to interchange factors and regroup any way we choose. Thus, the numeral expression $2 \cdot 4 \cdot 5$ can be rewritten any of the following ways:

$(4 \cdot 2) \cdot 5$	$(2 \cdot 5) \cdot 4$	$5 \cdot (2 \cdot 4)$
$(4 \cdot 5) \cdot 2$	$(2 \cdot 4) \cdot 5$	$5 \cdot (4 \cdot 2)$
$(5 \cdot 2) \cdot 4$	$4 \cdot (2 \cdot 5)$	$2 \cdot (4 \cdot 5)$
$(5 \cdot 4) \cdot 2$	$4 \cdot (5 \cdot 2)$	$2 \cdot (5 \cdot 4)$

Verify that each of these expressions equals 40.

As shown below, neither subtraction nor division is associative:

$$\underbrace{(8 - 5)} - 2 \neq 8 - \underbrace{(5 - 2)} \qquad \underbrace{(36 \div 6)} \div 2 \neq 36 \div \underbrace{(6 \div 2)}$$

$$3 \quad - 2 \neq 8 - \quad 3 \qquad 6 \quad \div 2 \neq 36 \div \quad 3$$

$$1 \quad \neq \quad 5 \qquad\qquad 3 \quad \neq \quad 12$$

---

3. Properties of Identity

There is exactly one real number 0, called the additive identity, such that,

$$0 + a = a, \text{ and } a + 0 = a$$

And, there is exactly one real number 1, called the multiplicative identity, such that,

$$1 \cdot a = a, \text{ and } a \cdot 1 = a$$

The properties of identity state that when adding 0 or when multiplying by 1, the value of the original number is unchanged.

Example 6

These statements illustrate the properties of identity.

(a) $0 + 10 = 10$

(b) $5x + 0 = 5x$

(c) $3(x + 0) = 3x$

(d) $1 \cdot 12 = 12$

(e) $\dfrac{1}{2} \cdot 1 = \dfrac{1}{2}$

(f) $1(a + b) = a + b$

---

4. Inverse Properties

For each real number a, there is exactly one opposite called the additive inverse of a, denoted by $-a$ such that

$$a + (-a) = 0, \text{ and } (-a) + a = 0$$

And for each nonzero real number a, there is exactly one multiplicative inverse (or reciprocal) denoted by $\dfrac{1}{a}$ such that

$$a \cdot \dfrac{1}{a} = 1, \text{ and } \dfrac{1}{a} \cdot a = 1$$

Example 7

These statements illustrate the inverse properties.

(a) The additive inverse (or opposite) of 6 is -6. And, $6 + (-6) = 0$.

(b)   The additive inverse of $-\dfrac{1}{2}$ is $\dfrac{1}{2}$. And, $-\dfrac{1}{2} + \dfrac{1}{2} = 0$.

(c)   The multiplicative inverse (or reciprocal) of 5 is $\dfrac{1}{5}$. And, $5 \cdot \dfrac{1}{5} = 1$.

(d)   For $x \neq 0$, the multiplicative inverse of $3x$ is $\dfrac{1}{3x}$. And,

$(3x) \cdot \dfrac{1}{3x} = 1$.

---

5.   Multiplication by Zero Property

$$0 \cdot a = 0, \text{ and } a \cdot 0 = 0$$

This property states that the product of zero and any real number is zero.

Each of these statements illustrates the multiplication by zero property.

Example 8

(a) $0 \cdot 12 = 0$

(b) $\dfrac{3}{4} \cdot 0 = 0$

(c) $0 \cdot (5x) = 0$

(d) $(2a + b) \cdot 0 = 0$

The final property of this section relates multiplication and addition. It is called the distributive property of multiplication with respect to addition, and it is used to change selected products to sums.

---

6.   The Distributive Property of Multiplication with Respect to Addition

$$a(b + c) = ab + ac, \text{ and } (b + c)a = ba + ca$$

According to the distributive property, a number can be multiplied times a sum by "distributing" it with each addend. Thus, as illustrated by the arrows,

$$3 \cdot (2 + 5) = 3 \cdot 2 + 3 \cdot 5$$

Verify that both sides of the equation give 21. Also note that the distributive property changes a basic product into a basic sum.

basic product   basic sum

$$3 \cdot (2 + 5) = 3 \cdot 2 + 3 \cdot 5$$

Example 9

Use the distributive property to simplify the following.

(a) $5 \cdot (2 + 1) = 5 \cdot 2 + 5 \cdot 1$
$\qquad\qquad = 10 + 5$
$\qquad\qquad = 15$

(b) $(6 + 5) \cdot 2 = 6 \cdot 2 + 5 \cdot 2$
$\qquad\qquad = 12 + 10$
$\qquad\qquad = 22$

(c) $3(a + b) = 3a + 3b$

(d) $(m + 2) \cdot 4 = m \cdot 4 + 2 \cdot 4$
$\qquad\qquad = 4m + 8$

(e) $5(2x + 3y) = 5(2x) + 5(3y)$
$\qquad\qquad = (5 \cdot 2)x + (5 \cdot 3)y$
$\qquad\qquad = 10x + 15y$

The equation describing the distributive property can be reversed to change a basic sum into a basic product. This procedure is called **factoring**. The number common to both addends is removed (or factored out) and multiplied times the remaining group.

factoring

$$\overset{\text{common number}}{\downarrow} \qquad \overset{}{\downarrow} \qquad \overset{\text{removed or factored out}}{\downarrow}$$
$$a \cdot b + a \cdot c = a \cdot (b + c)$$

When the distributive property is used to factor out a common number, the expression is changed from a basic sum to a basic product:

$$\overset{\text{basic sum}}{\downarrow} \qquad\qquad \overset{\text{basic product}}{\downarrow}$$
$$a \cdot b + a \cdot c = a \cdot (b + c)$$

Example 10

The following expressions have been factored using the distributive property.

(a) $a \cdot b + a \cdot c = a \cdot (b + c)$

(b) $3x + 3y = 3(x + y)$

(c) $5m + 15 = 5 \cdot m + 5 \cdot 3$
$\qquad\qquad = 5 (m + 3)$

(d) $4y + 4 = 4 \cdot y + 4 \cdot 1$
$\qquad\qquad = 4(y + 1)$

The properties discussed in this section are summarized in the following chart.

Properties of Real Numbers

For all real numbers a, b, and c:

1. Commutative
    (a) Addition: $a + b = b + a$
    (b) Multiplication: $a \cdot b = b \cdot a$

2. Associative
   (a) Addition: $(a + b) + c = a + (b + c)$
   (b) Multiplication: $(a \cdot b) \cdot c = a \cdot (b \cdot c)$

3. Identities
   (a) The additive identity is 0 and
   $$a + 0 = 0 + a = a$$
   (b) The multiplicative identity is 1 and
   $$a \cdot 1 = 1 \cdot a = a$$

4. Inverses
   (a) The additive inverse of a is $-a$ and
   $$a + (-a) = (-a) + a = 0$$
   (b) For $a \neq 0$, the multiplicative inverse (or reciprocal) is $\frac{1}{a}$ and
   $$a \cdot \frac{1}{a} = \frac{1}{a} \cdot a = 1$$

5. Multiplication by Zero
   $$0 \cdot a = a \cdot 0 = 0$$

6. Distributive Property of Multiplication with Respect to Addition
   $$a(b + c) = ab + ac, \text{ and } (b + c)a = ba + ca$$

Each of the following statements is an example of one of the properties of real numbers. Identify the property.

Exercise 1.6

1. $5 \cdot 3 = 3 \cdot 5$

2. $7 + 10 = 10 + 7$

2. $(3 + 6) + 1 = 3 + (6 + 1)$

4. $(2 \cdot 5) \cdot 3 = 2 \cdot (5 \cdot 3)$

5. $0 \cdot 13 = 0$

6. $\frac{2}{3} \cdot 1 = \frac{2}{3}$

7. $(9 + 7) \cdot 0 = 0$

8. $1 \cdot m = m$

9. $b + 0 = b$

10. $ab + c = c + ab$

11. $3(b + c) = 3b + 3c$

12. $5a + 5b = 5(a + b)$

13. $a(b + 2) = (b + 2)a$

14. $(3 \cdot x) \cdot y = 3 \cdot (x \cdot y)$

15. $(2a + 3b) + c = 2a + (3b + c)$

16. $5(2m + 3n) = 5(3n + 2m)$

17. $a(b + 0) = ab$

18. $\frac{7}{4} \cdot 1 = \frac{7}{4}$

19. $4 \cdot \frac{1}{4} = 1$

20. $a \cdot 1 + 0 = a \cdot 1$

21. $a \cdot 1 + 0 = a + 0$

22. $3(4x) + 3(5y) = 3(4x + 5y)$

23. $7 + (-7) = 0$

24. $(2m) \cdot \frac{1}{2m} = 1$ where $m \neq 0$

25. $-m + m = 0$

26. $(x + y) + 3 = 3 + (x + y)$

Using the indicated property, write a new expression that is equal to the given expression.

27. $3 + x$ (commutative property of addition)

28. $(2x)y$ (associative property of multiplication)

29. $1 \cdot x$ (multiplicative identity)

30. $9(m + n)$ (distributive property)

31. $2(a + b)$ (commutative property of multiplication)

32. $2(a + b)$ (commutative property of addition)

33. $\dfrac{1}{6} \cdot 6$ (multiplicative inverse)

34. $b + (-b)$ (additive inverse)

35. $2x + 2y$ (distributive property)

36. $1 \cdot (3x + y)$ (multiplicative identity)

Use the commutative and/or associative properties to evaluate these expressions two ways. Show your work.

Example

Evaluate $2 + 5 + 1$

Solution (1): $2 + 5 + 1 = 2 + (5 + 1)$
$$= 2 + 6$$
$$= 8$$

Solution (2): $2 + 5 + 1 = (5 + 1) + 2$
$$= 6 + 2$$
$$= 8$$

37. $6 + 0 + 3$ **38.** $1 + 5 + 9$

39. $7 + 12 + 2$ **40.** $6 + 12 + 4$

41. $2 \cdot 3 \cdot 0$ **42.** $6 \cdot 3 \cdot 2$

43. $5 \cdot 1 \cdot 10$ **44.** $6 \cdot 1 \cdot 7$

Evaluate these expressions using the distributive property.

Example

Evaluate $5 \cdot (2 + 3)$

Solution: $5 \cdot (2 + 3) = 5 \cdot 2 + 5 \cdot 3$
$$= 10 + 15$$
$$= 25$$

45. $3 \cdot (4 + 2)$ **46.** $9 \cdot (1 + 3)$

47. $(6 + 10) \cdot 3$ **48.** $(12 + 11) \cdot 5$

49. $8 \cdot (5 + 0)$ **50.** $(10 + 30) \cdot 5$

Use the distributive property to write each expression as a basic sum.

$3 \cdot (x + 4)$ **Example**

Solution: $3 \cdot (x + 4) = 3 \cdot x + 3 \cdot 4$ or $3x + 12$

51. $2(m + n)$ **52.** $3(a + 5)$
53. $5(b + 1)$ **54.** $(x + 4) \cdot 2$
55. $(m + 4) \cdot 7$ **56.** $(a + b) \cdot c$
57. $2x(y + z)$ **58.** $3a(b + c)$

Use the distributive property to factor each expression.

$3m + 3n$ **Example**

Solution: $3m + 3n = 3(m + n)$

59. $5x + 5y$ **60.** $7a + 7b$
61. $ab + ac$ **62.** $3ab + 3ac$
63. $2x + 6$ **64.** $5m + 10$
65. $6x + 6$ **66.** $7x + 7$
67. $5m + 20$ **68.** $6y + 18$
69. $3xy + 9y$ **70.** $4ab + abc$

ADDITION OF REAL NUMBERS

1.7

Addition of signed real numbers can be illustrated by using the standard number line.

Use the number line to find the sum of $(+2) + (+4)$. **Example 1**

Solution: As shown in figure 1.5, we begin at the origin and draw an arrow representing $+2$ by extending it 2 units to the right. Next, we draw an arrow representing $+4$. This arrow begins at the tip of the first arrow and extends 4 units to the right. It terminates at the point having a coordinate of $+6$. Thus, $(+2) + (+4) = 6$.

FIGURE   1.5

Use the number line to find the sum of $(-3) + (-2)$. **Example 2**

Solution: As illustrated in figure 1.6, we again start at the origin and

FIGURE   1.6

draw an arrow representing -3 by extending it 3 units to the left. Then, we draw an arrow representing -2 by beginning at the tip of the first arrow and extending it 2 units to the left. It terminates at the point having a coordinate of -5. Hence, $(-3) + (-2) = -5$.

These two examples illustrate that the sum of two positive numbers is positive, and the sum of two negative numbers is negative. This suggests the following procedure for adding two real numbers having like signs.

> To add two real numbers with like signs, add their absolute values and apply the common sign.

Example 3

These equations illustrate the procedure for adding two numbers with like signs.
(a) $(+6) + (+5) = +11$
(b) $8 + 2 = 10$
(c) $(-5) + (-7) = -12$
(d) $-12 + (-2) = -14$

The sum of a positive number and a negative number can also be illustrated using the standard number line.

Example 4

Use the number line to find the sum of $(+5) + (-3)$.

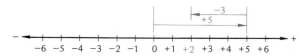

FIGURE   1.7

Solution: As shown in figure 1.7, we begin at the origin and draw an arrow 5 units to the right. Then, from the tip of this arrow we draw a second arrow 3 units to the left, terminating it at the point whose coordinate is $+2$. Thus, $(+5) + (-3) = +2$.

Example 5

Use the number line to find the sum of $-6 + 2$.

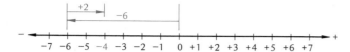

FIGURE 1.8

Solution: As shown in figure 1.8, we draw the first arrow 6 units to the left of the origin. We then draw the second arrow 2 units to the right, terminating it at the point whose coordinate is -4. Hence, $-6 + 2 = -4$.

As illustrated by examples 4 and 5, when two numbers having different signs are added, the sign of the sum is determined by the addend with the greater absolute value. Thus,

$$(+5) + (-3) = +2, \text{ because } |+5| > |-3|$$

and

$$-6 + 2 = -4, \text{ because } |-6| > |+2|$$

We arrive at the following procedure:

To add two real numbers with different signs, follow these steps:
Step (1) Find their absolute values.
Step (2) Subtract the smaller absolute value from the larger.
Step (3) For the sign of the sum, use the sign of the number having the larger absolute value.

Add $(+5) + (-1)$

Example 6

Solution:

Step (1) The absolute values are 5 and 1

Step (2) Subtract 1 from 5 to obtain 4

Step (3) The sum is positive, since $|+5| > |-1|$
Thus, $(+5) + (-1) = +4$

Add $(-6) + (+4)$

Example 7

Solution:

Step (1) The absolute values are 6 and 4

Step (2) Subtract 4 from 6 to obtain 2

Step (3) The sum is negative, since $|-6| > |+4|$
Thus, $(-6) + (+4) = -2$

Example 8

Add $10 + (-15)$

Solution:

> **Step (1)**   The absolute values are 10 and 15
>
> **Step (2)**   Subtract 10 from 15 to obtain 5
>
> **Step (3)**   The sum is negative, since $|-15| > |10|$
> Thus, $10 + (-15) = -5$

The rules for adding signed real numbers are summarized below. Memorize these rules and practice using them to evaluate the examples that follow.

Addition of Real Numbers

(1)   To add two real numbers having like signs, add their absolute values and apply the common sign.

(2)   To add two real numbers having unlike signs, subtract the smaller absolute value from the larger and use the sign of the number with the greater absolute value.

Example 9

Here are several illustrations of adding signed real numbers. Examine each to be sure you understand the rules.
(a) $(+1) + (+7) = +8$
(b) $1 + 7 = 8$
(c) $(-6) + (-9) = -15$
(d) $(+11) + (-3) = +8$
(e) $-7 + 1 = -6$
(f) $-12 + 16 = 4$
(g) $(-7.6) + (-1.2) = -8.8$
(h) $-10.4 + 16.1 = 5.7$

Thus far we have discussed addition of only two integers. It is equally important to be able to add three or more integers. When evaluating expressions of this type, the order of operations agreement instructs us to add from left to right. However, since addition is commutative and associative, we are free to add in any order.

Example 10

Add $(-5) + (+8) + (-6)$

Solution Method 1:   You can use the order of operations agreement.

$$\underline{(-5) + (+8)} + (-6)$$

$$+3 \qquad + (-6)$$

$$-3$$

Solution Method 2: Since addition is commutative and associative, you can also add in any order you wish.

$$(-5) + (+8) + (-6)$$

$$(-11) + (+8)$$

$$-3$$

Evaluate $(-7) + (+4) + (+7) + (-4)$ **Example 11**

Solution: Add in any order you wish.

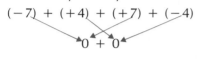

$$(-7) + (+4) + (+7) + (-4)$$

$$0 + 0$$

$$0$$

Evaluate $5 + [(-3) + (-1)] + 2$ **Example 12**

Solution: First evaluate the brackets; then add in any order you wish.

$$5 + [(-3) + (-1)] + 2$$

$$5 + \quad (-4) \quad + 2$$

$$3$$

Use the standard number line to illustrate these sums. **Exercise 1.7**

1. $(+2) + (+3) = +5$ **2.** $(-1) + (-2) = -3$

3. $(-4) + (+1) = -3$ **4.** $(+3) + (-2) = +1$

Find the sums.

5. $(+3) + (+8)$ **6.** $(+1) + (+10)$

7. $2 + 6$ **8.** $3 + 12$

9. $(-4) + (+6)$ **10.** $6 + (-3)$

11. $(-2) + (-1)$ **12.** $(-8) + (-3)$

13. $-10 + 2$ **14.** $5 + (-2)$

15. $-8 + 1$ **16.** $5 + (-5)$

17. $0 + (-2)$ **18.** $-6 + 0$

19. $-23 + 6$ **20.** $(-17) + (-4)$

21. $(-34) + (-18)$ **22.** $12 + (-13)$

23. $(-56) + 23$ **24.** $-18 + 18$

25. $-3.8 + 1.3$ **26.** $15.6 + (-2.9)$

27. $(-2.31) + (-1.04)$ **28.** $-0.2 + 1$

Find these sums using two methods: (a) by using the order of operations agreement, and (b) by adding in any order you wish.

29. $(-6) + (-3) + (+1)$ **30.** $(-5) + (+7) + (-3)$

31. $(-2) + (+3) + (+2) +$ **32.** $17 + (-12) + 5 + 12 +$
     (-3) (-17)

Find the sums.

33. $-1 + [(-3) + (+5)]$ **34.** $(-3 + 4) + (-8 + 2)$

35. $5 + (-6) + [1 + (-3)]$ **36.** $3 + (-4) + 2 + (-8) + 1$

37. $(-3 + 4) + 9 + (-3 + 5)$ **38.** $3 + (-8) + 1 + (-3)$

39. $-16 + [2 + (-6)] + 8$

40. $[-1 + 2] + [3 + (-5)] + [-2 + (-6)]$

41. $(-7 + 0) + (-8 + 1) + (-10 + 1)$

42. $-7 + 2 + 7 + (-2) + 0 + (-1)$

43. $[(-2.6) + (-3.8)] + 10.2$

44. $-6.5 + [5.8 + (-0.7)]$

45. $|-6 + 2| + |-8 + 1|$

46. $-10.1 + |-3 + 1|$

Fill in each blank with a signed real number to make these statements true.

47. _____ $+ 8 = -5$ **48.** _____ $+ (-7) = 6$

49. $-5 +$ _____ $= 11$ **50.** $2 +$ _____ $= -8$

51. _____ $+ (-3) + (-6) = 7$ **52.** $-2 +$ _____ $+ (-5) =$
                                                -14

53. $2 +$ _____ $+ (-7) = 0$ **54.** _____ $+ 8 + (-4) = -1$

Complete each sentence to make a true statement.

55. If x and y represent positive numbers, then $x + y$ represents a _____ number.

56. If x and y represent negative numbers, then $x + y$ represents a _____ number.

57. If $x > 0$, $y < 0$, and $|y| > |x|$, then $x + y$ represents a _____ number.

58. If $x > 0$, $y < 0$, and $|x| > |y|$, then $x + y$ represents a _____ number.

Solve these applied problems.

59. A man owes $175 on his credit card. If he makes a payment of $90, how much does he still owe?

60. The temperature at midnight was $-15°F$. By noon it went up $23°F$. What was the temperature at noon?

61. The temperature at noon was $-8°F$, but by 6:00 P.M. it had decreased $17°F$. What was the temperature at 6:00 P.M.?

62. Jean has $21 in her checking account. What will her balance be if she writes a check for $53?

63. Bob has $192.35 in his checking account. What is his balance if he makes a deposit of $105.83, and then writes checks for $35.04, $18.93, $52.88, and $101.61?

64. The sum of what number and -4 is -11?

65. The sum of what number and 15 is -21?

66. What number must be added to -9 to produce 10?

67. What number must be added to 0 to produce -8?

68. What number must be added to -5 to produce -15?

Calculator problems. With your instructor's approval, evaluate these problems with a calculator.

69. $-3.659 + 8.021 + (-6.919)$

70. $[86.107 + (-91.904)] + (-72.015 + 53.912)$

71. $-200.16 + [-351.29 + (-192.06 + 100.38)]$

72. $[-2.814 + (-3.907)] + [12.914 + (-10.671)] + 3.615$

SUBTRACTION OF REAL NUMBERS

1.8

In this section we develop a procedure for subtracting one signed real number from another. This will be done by using a pattern which converts subtraction problems to addition problems. This pattern will enable us to use the rules of addition to perform subtraction.

You know from previous experience that $5 - 2 = 3$, and $5 + (-2) = 3$. Thus, the basic difference, $5 - 2$, must equal the basic sum, $5 + (-2)$. The following illustrations yield the pattern $a - b = a + (-b)$.

Basic difference		Basic sum
$6 - 4$	$=$	$6 + (-4)$
$10 - 3$	$=$	$10 + (-3)$
$8 - 1$	$=$	$8 + (-1)$
$a - b$	$=$	$a + (-b)$

This pattern provides a definition of subtraction that converts every subtraction problem to an addition problem.

Definition of Subtraction

For all real numbers a and b,

$$a - b = a + (-b)$$

To subtract *b* from *a*, the opposite (or additive inverse) of *b* is added to *a*. This pattern is illustrated below.

subtraction is changed to addition

$$a - b = a + (-b)$$

subtrahend is changed to its opposite

Note that the minuend *a* is not changed. We thus arrive at the following procedure for subtracting signed real numbers.

Subtracting Real Numbers

(1)   Change the subtraction symbol to addition.
(2)   Change the sign of the subtrahend.
(3)   Use the rules of addition.

Example 1

Subtract $(+8) - (-4)$

Solution: subtraction is changed to addition

$$(+8) - (-4) = (+8) + (+4)$$

subtrahend is changed to its opposite

$$= + 12$$

For more practice, study carefully the subtraction problems in the following example.

Example 2

These problems illustrate subtraction of real numbers.

(a) $5 - (-7) = 5 + (+7)$
$$= 12$$

(b) $-3 - (-8) = -3 + (+8)$
$$= 5$$

(c) $-9 - 7 = -9 - (+7)$
$$= -9 + (-7)$$
$$= -16$$

(d) $0 - 5 = 0 - (+5)$
$$= 0 + (-5)$$
$$= -5$$

(e) $1.2 - (-5.7) = 1.2 + 5.7$
$$= 6.9$$

It is often necessary to evaluate expressions containing more than one subtraction. Also, many expressions may contain both subtraction and addition. Such expressions are evaluated by converting each subtraction to addition. Remember, however, only numbers following subtraction symbols have their signs changed.

Evaluate $(-8) - (-2) - (+4)$ **Example 3**

Solution: Convert each subtraction to addition, then perform addition in any order.

$$(-8) - (-2) - (+4) = (-8) + (+2) + (-4)$$
$$= -10$$

Evaluate $(-3) - (-6) + (+5)$ **Example 4**

Solution: Do not change the sign of the addend, $+5$.

$$(-3) - (-6) + (+5) = (-3) + (+6) + (+5)$$
$$= 8$$

Evaluate $-4 - [8 - (-12 + 7)]$ **Example 5**

Solution: First perform operations inside the grouping symbols, working from inside out.

$$-4 - [8 - (-12 + 7)] = -4 - [8 - (-5)]$$
$$= -4 - [8 + 5]$$
$$= -4 - 13$$
$$= -4 + (-13)$$
$$= -17$$

Sometimes expressions are written with the positive signs omitted. For example, $2 - 5 + 6$ means, $(+2) - (+5) + (+6)$. Let's evaluate this expression and we will find a convenient shortcut.

$$2 - 5 + 6 = (+2) - (+5) + (+6)$$
$$= (+2) + (-5) + (+6)$$
$$= +3$$

However, by simply "thinking addition," the subtraction is automatically converted to addition and the original expression can be evaluated much faster.

$$\begin{array}{c} \text{think addition} \\ \downarrow \qquad \downarrow \\ 2 - 5 + 6 = 2 \oplus -5 \oplus +6 \longleftarrow \begin{bmatrix} \text{try to do this step} \\ \text{in your head} \end{bmatrix} \\ = +3 \end{array}$$

Study the following examples until you feel comfortable with the shortcut.

Example 6

Evaluate $8 - 12$

Solution: "Think addition."

$$8 - 12 = 8 \oplus -12$$
$$= -4$$

Example 7

Evaluate $3 - 5 - 1 + 2$

Solution: "Think addition"; then add in any order.

$$3 - 5 - 1 + 2 = 3 \oplus -5 \oplus -1 \oplus +2$$
$$= -1$$

From the preceding examples, it should be evident that subtraction problems are converted to addition problems because addition is commutative and associative, allowing us to add in any order we desire.

Exercise 1.8

Use the pattern $a - b = a + (-b)$ to convert these basic differences to basic sums.

Example: $x - (-3) = x + 3$

1. $x - (-5)$ **2.** $x - (-y)$
3. $m - 4$ **4.** $2n - (-5)$
5. $-a - b$ **6.** $b - (-c)$
7. $2x - (-1)$ **8.** $3x - 4$
9. $2m^2 - (-4)$ **10.** $3x - y$

Evaluate these expressions.

11. $(-2) - (+4)$ **12.** $(+3) - (+2)$
13. $(+5) - (+9)$ **14.** $(-8) - (-15)$
15. $(+10) - (-12)$ **16.** $0 - (-7)$
17. $0 - (+2)$ **18.** $2 - 6$
19. $-12 - 7$ **20.** $25 - (-32)$
21. $7 - 18$ **22.** $14 - 25$
23. $1 - (-6)$ **24.** $18 - 24$
25. $0 - 12$ **26.** $(-1) - (+7) - (-4)$
27. $(-5) - (-2) - (-1)$
28. $(+8) - (+6) - (-12) - (-1)$
29. $6 - 10 - (-5) - (-4)$
30. $0 - 3 - (-2) - 12 - 8 - (-13)$
31. $2 + (-6) + (-8) - 10$
32. $(-2) + (+8) - (-2) - 8$
33. $0 - (-2) + (-3) - (+6) + (+3)$
34. $3 - 7$
35. $2 - (6 - 1)$ **36.** $0 - 2 - 5 + 1$
37. $2 - 3 + 5 - 6$ **38.** $-5 - [6 - (7 - 8)]$
39. $-23 - 108 + 23 - 5 + 108$
40. $-7 + 8 - 2 - 3 + 5$

41. $12 - 10 + [8 + (2 - 6) - 6]$
42. $32 - (7 - 21) + (14 - 62)$
43. $-9 - [(-3 - 1) - (8 - 2)]$
44. $-6 + [(-5 + 1) - (-1 - 7)]$
45. $1.2 - (3.6 - 7.9)$
46. $3.7 - [(1.0 - 2.1) - 0.7]$

Fill in each blank with an integer to make the statement true.

47. _____ $- 7 = 6$ **48.** _____ $- (-2) = -1$
49. _____ $- 3 - 6 = 7$ **50.** $(-2) -$ _____ $+ 4 = 0$
51. _____ $+ 5 - 8 = -6$ **52.** _____ $- 6 + 2 = -4$

Complete each sentence to make a true statement.

53. If a and b represent negative integers and $|b| > |a|$, then $a - b$ represents a _____ integer.
54. A negative integer when subtracted from a positive integer will produce a _____ integer.

Solve these problems.

55. At 6:00 A.M. the temperature was $-7°F$. By noon, the temperature dropped $5°F$. What was the noon temperature?
56. The boiling point of water is $100°C$ and that of nitrogen is $-196°C$. What is the difference in their boiling points?
57. An airplane flying at 3200 feet observes a submarine 600 feet below the surface of the ocean. What is the distance between the airplane and the submarine?
58. One year a company showed a loss of $250,000. However, the next year the company showed a profit of $125,000. How much was the increase?
59. One company showed a profit of $78,000, while another company showed a loss of $12,000. Find the difference.
60. Bob has $75, while his friend John is $95 in debt. Find the difference between these amounts of money.
61. Subtract 5 from the sum of -10 and -8.
62. Subtract -6 from the sun of -4 and 7.
63. From the sum of -8 and 12, subtract -7.
64. From the sum of -1 and -6, subtract the sum of -5 and 2.
65. From the sum of -6, 7, and -3, subtract the sum of -8, -4, and 1.

Calculator problems. With your instructor's approval, evaluate these problems with a calculator.

66. $13.695 - 26.083$ **67.** $-0.0215 - 0.0841$
68. $-3.621 + 8.945 - 10.806$ **69.** $3.45 - (8.62 + 9.01)$

MULTIPLICATION OF REAL NUMBERS

1.9

Rules for multiplying signed real numbers can be formulated by considering these three cases:

(1) Multiplication by zero.
(2) Multiplying two numbers having unlike signs.
(3) Multiplying two numbers having like signs.

The multiplication by zero property states that the product of any number and zero is zero.

Example 1

These equations illustrate multiplication by zero.

(a) $(+5) \cdot 0 = 0$
(b) $0 (-3) = 0$
(c) $0 \cdot \dfrac{2}{3} = 0$
(d) $0 \cdot 0 = 0$

To develop a rule for multiplying two numbers having unlike signs, we recall from section 1.1 that multiplying by a whole number can be considered to be repetitive addition. Thus, $2 \cdot (-6)$ means $(-6) + (-6)$ or -12. This suggests that the product of two real numbers having unlike signs is negative.

Example 2

These equations illustrate multiplication of two numbers having unlike signs.

(a) $4 (-3) = -12$
(b) $(+5) (-2) = -10$
(c) $(-3) (+7) = -21$
(d) $-3 \cdot 2 = -6$
(e) $-7 \cdot 1 = -7$

To multiply two numbers having like signs, we first consider $(+2) \cdot (+3)$. Since

$$(+2) \cdot (+3) = 2 \cdot 3 = 6$$

we observe that the product of two positive numbers is positive. Next we consider the product of two negative numbers such as $(-3) \cdot (-2)$. We will create an expression for which we know the answer. Inside this expression will be the product, $(-3) \cdot (-2)$. Then, indirectly, we will be able to evaluate this product. Consider the following discussion:

$$0 = (-3) \cdot 0 \qquad \text{[multiplication by zero]}$$
$$= (-3) \cdot [(+2) + (-2)] \qquad \text{[additive inverse property]}$$

$$= \underbrace{(-3) \cdot (+2)} \quad + \quad (-3) \cdot (-2) \qquad \text{[distributive property of multiplication]}$$

$$= \qquad -6 \qquad + \underbrace{(-3) \cdot (-2)} \qquad [(-3) \cdot (+2) = -6]$$

$$= \qquad -6 \qquad + \qquad \underbrace{(?)}$$

$$= \qquad -6 \qquad + \qquad (+6) \qquad \text{[additive inverse property. } +6 \text{ is the only number that can be added to } -6 \text{ to produce 0]}$$

Thus, $(-3) \cdot (-2) = +6$

Thus we observe that the product of two negative numbers, like the product of two positive numbers, is positive.

These equations illustrate multiplication of two numbers having like signs.

Example 3

(a) $(+3)(+4) = 12$

(b) $6 \cdot 3 = 18$

(c) $(-2)(-5) = +10$

(d) $-1 \cdot (-7) = 7$

The following rule summarizes the procedure for multiplying signed real numbers.

> **Multiplication of Real Numbers**
>
> The product of two numbers having like signs is positive, and the product of two numbers having unlike signs is negative.

Study these problems to be sure you understand the rule for multiplying real numbers.

Example 4

(a) $(-7) \cdot (-3) = +21$

(b) $-4 \cdot 2 = -8$

(c) $(-1)(-1) = 1$

(d) $-3 \cdot \left(-\dfrac{1}{3}\right) = 1$

(e) $-12 \cdot 0 = 0$

Thus far we have only discussed multiplication of two signed numbers. It is equally important to be able to multiply three or more signed numbers and to combine multiplication with addition and subtraction.

Multiply $(-2) \cdot (-5) \cdot (-6)$

Example 5

Solution Method 1: You can use the order of operations agreement.

$$(-2) \cdot (-5) \cdot (-6)$$

$$(+10) \quad \cdot (-6)$$

$$-60$$

Solution Method 2: Since multiplication is commutative and associative, you can multiply in any order you wish.

$$(-2) \cdot (-5) \cdot (-6)$$

$$(+12) \cdot (-5)$$

$$-60$$

Example 6

Evaluate $(-5) \cdot (+2) \cdot (-2) \cdot (+6)$

Solution: Multiply in any order you wish.

$$(-5) \cdot (+2) \quad \cdot \quad (-2) \cdot (+6)$$

$$(-10) \qquad \cdot \qquad (-12)$$

$$+120$$

Example 7

Evaluate $-3[5 - 2(3 + 1)]$

Solution: Begin within the grouping symbols, working from the inside out.

$$-3[5 - 2(3 + 1)] = -3[5 - 2(4)]$$
$$= -3[5 - 8]$$
$$= -3[5 + (-8)]$$
$$= -3(-3)$$
$$= +9$$

The next example shows that an even number of negative factors produces a positive product while an odd number of negative factors produces a negative product.

Example 8

The following have been evaluated.
(a) $(-3)(-2) = +6$ [two negative factors]
(b) $(-1)(-5)(-2)(-3) = +30$ [four negative factors]
(c) $(-2)(-3)(-5) = -30$ [three negative factors]
(d) $(-1)(-3)(-2)(-5)(-2) = -60$ [five negative factors]

Knowing that an even number of negative factors is positive while an odd number is negative helps us to evaluate powers of signed numbers.

Example 9

Evaluate $(-3)^4$

Solution: Since the exponent is 4, there are an even number of negative factors, making the answer positive.

$$(-3)^4 = (-3)(-3)(-3)(-3) = 81$$

Evaluate $(-2)^5$

Example 10

Solution: Since the exponent is 5, there are an odd number of negative factors, making the answer negative.

$$(-2)^5 = (-2)(-2)(-2)(-2)(-2) = -32$$

Evaluate $(+2)^3$

Example 11

Solution: As shown, a positive number raised to any power will result in a positive answer.

$$(+2)^3 = (+2)(+2)(+2) = +8$$

The foregoing examples suggest this summary of powers:

Powers of Signed Numbers

(1)   A negative base to an even power gives a positive answer.

Example $(-2)^4 = +16$

(2)   A negative base to odd power gives a negative answer.

Example $(-2)^3 = -8$

(3)   A positive base to either an even or an odd power gives a positive answer.

Examples $(+3)^2 = +9$ and $(+2)^3 = +8$

Sometimes there is confusion about a power that does not include grouping symbols, such as -3^2. Mathematicians assume, in the absence of grouping symbols, that an exponent applies only to the first symbol on its immediate left. Thus, -3^2 means the opposite of the square of 3 and is evaluated as follows:

$$\begin{aligned}-3^2 &= -(3)^2 \\ &= -(3 \cdot 3) \\ &= -9\end{aligned}$$

In other words, the exponent 2 applies only to the symbol 3 and not to the minus sign. For the exponent to apply to -3, grouping symbols are needed. Thus,

$$(-3)^2 = (-3)(-3) = +9$$

For more practice with these concepts, study the remaining examples.

Example 12 These powers have been evaluated.

(a) $-5^2 = -(5 \cdot 5) = -25$

(b) $(-5)^2 = (-5)(-5) = +25$

(c) $-1^4 = -(1 \cdot 1 \cdot 1 \cdot 1) = -1$

(d) $(-1)^4 = (-1)(-1)(-1)(-1) = +1$

Example 13 Evaluate $-6^2 + (-1)^2 - 2 \cdot [6 - 7]$

Solution: Use the order of operations agreement.

$$\begin{aligned}
-6^2 + (-1)^2 - 2 \cdot [6 - 7] &= -6^2 + (-1)^2 - 2 \cdot [-1] \\
&= -36 + 1 - 2 \cdot [-1] \\
&= -36 + 1 - (-2) \\
&= -36 + 1 + 2 \\
&= -33
\end{aligned}$$

Exercise 1.9 Find these products.

1. $(+2) \cdot (+5)$

2. $3 \cdot 6$

3. $(-2) \cdot (+4)$

4. $0 \cdot (-8)$

5. $(+7) \cdot 0$

6. $(-4) \cdot (-5)$

7. $4 \cdot 5$

8. $(-7) \cdot (-8)$

9. $(-10) \cdot (+10)$

10. $(-1) \cdot (+1)$

11. $6 \cdot (-8)$

12. $(-9) \cdot (-7)$

13. $(-2.1)(-1.2)$

14. $(0.3)(-0.4)$

15. $2\left(-\dfrac{1}{2}\right)$

16. $(-5)\left(-\dfrac{1}{5}\right)$

Evaluate these expressions.

17. $(-2) \cdot (+3) \cdot (-1)$

18. $(-6) \cdot (-2) \cdot (-3)$

19. $(+1) \cdot (-3) \cdot (+2) \cdot (-1)$

20. $(-3) \cdot (-2) \cdot (+4) \cdot (-5)$

21. $(-1) \cdot 7 \cdot (-2) \cdot 4$

22. $3 \cdot (-2) \cdot (-3) \cdot 5 \cdot 1$

23. $(-1)^2$

24. $(-1)^3$

25. $(-1)^4$

26. $(-6)^2$

27. -6^2

28. $(-3)^4$

29. -3^4

30. $(+5)^2$

31. $(-1)^{234}$

32. $(-1)^{235}$

33. $(-10)^3$

34. $(-2)(-3)^2$

35. $(-2)^2(-3)^2$

36. $-2^2(-3)^2$

37. $(-1)^2 + 1$

38. $-1^2 + 1$

39. $2 - 2 \cdot 3^2$

40. $-1 + 2(-2)^3$

41. $5 \cdot 3^2 - 3 \cdot 5^2$

42. $2(3^2 - 5^2)$

43. $-5^2 - 2^2 - 3^2$

44. $3 \cdot 4^2 - 2 \cdot 4 - 1$

45. $3^2[2 - (4^2 + 1)]$

46. $(2 - 3)^2 - (4 - 1)^2$

47. $3|2 - 6|$

48. $-2|-5| - 3$

49. $|-3 \cdot 5| - 2|-4 \cdot 3|$

Evaluate these expressions, letting $x = -3$, $y = 2$, and $z = 0$.

50. $2x - 3y + z$

51. $-4x + 2y - 5z$

52. $3x(x - y)$

53. $(x + 2)(x - 5)$

54. $-x^2 + y^2$

55. $2y^3 - x^3$

Fill in the blanks to make each statement true.

56. $(-12) \cdot$ _____ $= -12$

57. $(-7) \cdot$ _____ $= -21$

58. _____ $\cdot (+2) = -18$

59. $(-24) \cdot$ _____ $= +24$

60. $(+3) \cdot$ _____ $\cdot (-6) = 36$

61. $(-10) \cdot$ _____ $= 0$

62. A negative base to an _____ power gives a negative answer.

63. A negative base to an _____ power gives a positive answer.

64. If the product of two numbers is positive and one of the numbers is negative, then the other is _____.

65. If the product of two numbers is negative and one of the numbers is negative, then the other is _____.

66. If the product of two numbers is zero, then at least one of the numbers is _____.

Solve these problems.

67. What number must be multiplied times -3 to produce -15?

68. What number must be multiplied times -7 to produce $+28$?

69. Add the product of -6 and 4 to -10.

70. Subtract -8 from the product of -3 and -5.

71. Add the product of -3 and 4 to the product of -5 and -7.

Calculator problems. With your instructor's approval, evaluate these problems with a calculator.

72. $8.96(-5.01)$

73. $(-1.06)(-1.35)(0.2)$

74. $2.6(8.9 - 12.4)$

75. $(-3.01)(2.06) - (7.12)(-3.85)$

DIVISION OF REAL NUMBERS

1.10

In section 1.1 we found that division and multiplication are closely related. The product of the quotient and divisor equals the dividend. Thus,

$$\frac{10}{5} = 2 \text{ because } 2 \cdot 5 = 10$$

This relationship will be used to show that division of signed real numbers follows essentially the same rules as multiplication. We analyze division by considering these three cases:

(1) Division involving zero.
(2) Dividing two numbers having unlike signs.
(3) Dividing two numbers having like signs.

It was shown in section 1.1 that division by zero is impossible. Also, we demonstrated that a quotient is zero only if the dividend is zero and the divisor is not zero.

Example 1

The following expressions are meaningless, since division by zero is impossible.

(a) $7 \div 0$

(b) $\dfrac{-10}{0}$

(c) $0 \div 0$

Example 2

These statements illustrate that zero divided by a non-zero real number is zero.

(a) $0 \div 8 = 0$, since $0 \cdot 8 = 0$

(b) $\dfrac{0}{-3} = 0$, since $0 \cdot (-3) = 0$

(c) $\dfrac{0}{-8.6} = 0$, since $0 \cdot (-8.6) = 0$

To divide two numbers having unlike signs, we consider $(+6) \div (-3)$ and $(-6) \div (+3)$. As shown below, both expressions equal -2.

$$(+6) \div (-3) = -2, \text{ since } (-2)(-3) = +6, \text{ and}$$
$$(-6) \div (+3) = -2, \text{ since } (-2)(+3) = -6$$

Thus, the quotient of two numbers having unlike signs is negative.

Example 3

These equations illustrate division of two numbers having unlike signs.

(a) $\dfrac{10}{-2} = -5$

(b) $\dfrac{-21}{7} = -3$

(c) $-18 \div 3 = -6$
(d) $14 \div (-2) = -7$

To divide two numbers having like signs, we consider

$$(+8) \div (+2) = +4, \text{ since } (+4)(+2) = +8, \text{ and}$$
$$(-8) \div (-2) = +4, \text{ since } (+4)(-2) = -8$$

We observe that the quotient of two numbers having like signs is positive.

These equations illustrate division of two numbers having like signs.

Example 4

(a) $\dfrac{-20}{-5} = 4$

(b) $\dfrac{16}{2} = 8$

(c) $-24 \div (-4) = 6$

The following rule summarizes the procedure for dividing signed real numbers.

Division of Real Numbers

As with multiplication, the quotient of two numbers having like signs is positive, and the quotient of two numbers having unlike signs is negative. If zero is divided by a non-zero number, the quotient is zero. But, division by zero is meaningless and thus impossible.

These equations illustrate the rules for dividing real numbers.

Example 5

(a) $\dfrac{21}{-7} = -3$

(b) $\dfrac{-15}{-3} = 5$

(c) $\dfrac{0}{-4} = 0$

(d) $\dfrac{-12}{0}$ is impossible

(e) $-6 \div (-6) = 1$

(f) $(-2.4) \div 2 = -1.2$

Expressions involving combinations of addition, subtraction, multiplication, division, and powers are evaluated using the order of operations agreement. (Recall that subtraction and division are neither commutative nor associative, so they cannot be evaluated by any other means.)

The Order of Operations Agreement

Step (1) Perform any operations inside grouping symbols.
Step (2) Evaluate all exponentials.
Step (3) Perform all multiplications and divisions as they occur, working from left to right.

> Step (4)   Perform all additions and subtractions as they occur, working from left to right.

Example 6

Evaluate $(-3 - 2)^2 + 3(-4)^2 \div (-2)$

Solution: First evaluate the groups, then the exponentials. Next do the multiplication and the division. Perform addition last.

$$
\begin{aligned}
(-3 - 2)^2 + 3(-4)^2 \div (-2) &= (-5)^2 + 3(-4)^2 \div (-2) \\
&= 25 + 3(16) \div (-2) \\
&= 25 + 48 \div (-2) \\
&= 25 + (-24) \\
&= 1
\end{aligned}
$$

This final example reminds us that the fraction bar acts as a grouping symbol for both the dividend and divisor.

Example 7

Evaluate $\dfrac{-4(-3)^2 + 2(6)}{3(5 - 7)}$

Solution: First evaluate the dividend and divisor, then find the quotient.

$$
\begin{aligned}
\frac{-4(-3)^2 + 2(6)}{3(5 - 7)} &= \frac{-4(9) + 12}{3(-2)} \\
&= \frac{-36 + 12}{-6} \\
&= \frac{-24}{-6} \\
&= 4
\end{aligned}
$$

Exercise 1.10

Evaluate these quotients.

1. $-24 \div (-6)$
2. $0 \div (-2)$
3. $-12 \div 3$
4. $-10 \div (-2)$
5. $-7 \div 0$
6. $30 \div (-5)$
7. $\dfrac{-6}{-2}$
8. $\dfrac{-15}{3}$
9. $\dfrac{0}{-2}$
10. $\dfrac{-5}{0}$
11. $\dfrac{-100}{-10}$
12. $\dfrac{-1}{1}$

13. $\dfrac{-1}{-1}$

14. $\dfrac{48}{-6}$

15. $\dfrac{26}{-13}$

16. $\dfrac{-46}{23}$

17. $\dfrac{-8.4}{4.2}$

18. $\dfrac{0.6}{-0.2}$

19. $\dfrac{-9.6}{-1.2}$

20. $\dfrac{-4}{0.2}$

21. $\dfrac{3}{-0.6}$

22. $\dfrac{-150}{-5}$

23. $\dfrac{280}{-40}$

24. $\dfrac{-175}{-5}$

Evaluate these expressions, using the order of operations agreement.

25. $100 \div (-2) \div (-5) \div 2$

26. $(-18) \div (-3) \div (+2) \div (+1)$

27. $48 \div (-12) \div (-2)$

28. $(-8) \div (-2) \div (-1)$

29. $3 \cdot (-6) \div (-2)$

30. $(-1) \cdot (-1) \div (+1) \cdot (-9) \div (+3)$

31. $(-2)^3 \cdot (-3)^2$

32. $(-1)^4 \div (-1)^5$

33. $-3(-2)^3 - 5(-2)^2 - 4$

34. $(-3 + 1)^2 + (-5 + 5)^3$

35. $4(-6)^2 - 3(-6) - 4$

36. $-3^2 \cdot (-2)$

37. $(-5 + 1)^2 + (-3 + 3)^2$

38. $(-2)(-3)^2 + 3(-3) - 5$

39. $(-3 - 1)^2 + (-2 + 1)^5$

40. $(-6 - 4)^2 + 5(-3) \div 5$

Evaluate these expressions, remembering that the fraction bar acts as a grouping symbol for both the dividend and divisor.

41. $\dfrac{6}{1 - 3}$

42. $\dfrac{2 - 10}{-4}$

43. $\dfrac{-15}{6 - 9}$

44. $\dfrac{13 - 3}{3 - 8}$

45. $\dfrac{17 - 3}{2 - (-5)}$

46. $\dfrac{2(3) + 4}{-1 - 1}$

47. $\dfrac{-100}{-3 - (-8)}$

48. $\dfrac{2(-3) - 4(-2)}{2 - 5}$

49. $\dfrac{-3(10) + 6}{-2(4)}$

50. $\dfrac{3^2 - 2^2}{5(1 - 2)}$

51. $\dfrac{4^2 + 6^2}{3^2 + 2^2}$

52. $\dfrac{-3^2 - 5^2}{-1^2 - 4^2}$

53. $\dfrac{10^2 - 3^3}{8^2 + 3^2}$ **54.** $\dfrac{-2(8 + 2)^2}{2 \cdot 5^2}$

55. $\dfrac{2(-8) + 4^2}{3^2 - 5^2}$

Fill in the blanks to make each statement true.

56. _____ $\div (-5) = -20$ **57.** $(+12) \div$ _____ $= -12$

58. _____ $\div (-7) = 0$ **59.** $(-16) \div$ _____ $= 8$

60. _____ $\div (-1) = 1$ **61.** $(-81) \div$ _____ $= -9$

62. The only way a quotient can be zero is for the _____ to be zero.

Solve these problems.

63. The quotient of what number and 3 is -15?

64. Add -7 to the quotient of -18 and 2.

65. Add 10 to the quotient of -30 and 6.

66. Subtract 6 from the quotient of 18 and -2.

67. Subtract -1 from the quotient of 6 and -6.

68. Raise the quotient of -8 and 4 to the third power.

69. Raise the quotient of 10 and -10 to the fifth power.

70. If a number is divided by -3, the quotient is 12. Find the number.

71. If a number is divided by -7, the quotient is 0. Find the number.

Calculator problems. With your instructor's approval, evaluate these problems with a calculator.

72. $\dfrac{1.32 - 6.04}{8.61}$ **73.** $\dfrac{12.05 - (-8.61)}{-7.23}$

74. $\dfrac{(0.91)^2 - (0.86)^2}{-1.03}$ **75.** $\dfrac{(-0.41)^2 + 3(0.54)}{(0.31)^2 - (0.72)^3}$

Summary

SYMBOLS

$\{a, b\}$	The set containing the elements a and b
$\{\ \}$	Empty set
N	Set of natural numbers: $\{1,2,3,4,5, \ldots\}$
W	Set of whole numbers: $\{0,1,2,3,4,5, \ldots\}$
I	Set of integers: $\{\ldots -3,-2,-1,0,1,2,3, \ldots\}$
Q	Set of rational numbers
H	Set of irrational numbers
R	Set of real numbers
$=$	Equals
$\neq$	Is not equal to
$<$	Is less than

$\leq$	Is less than or equal to
$>$	Is greater than
$\geq$	Is greater than or equal to
$\|x\|$	Absolute value of x

(1) If $x \geq 0$, then $\|x\| = x$ **DEFINITION OF**
(2) If $x < 0$, then $\|x\| = -x$ **ABSOLUTE VALUE**

The set of rational numbers is composed of all numbers that can be written in the form of $a \div b$ or $\frac{a}{b}$ where a and b are integers and $b \neq 0$. **DEFINITION OF RATIONAL NUMBERS**

A number is said to be rational provided it can be written as either a terminating or a repeating decimal. **ALTERNATE DEFINITION OF A RATIONAL NUMBER**

An irrational number is any number whose decimal form is nonterminating and nonrepeating. **DEFINITION OF AN IRRATIONAL NUMBER**

A real number is any number that can be written as a decimal. It can be terminating, repeating, or nonterminating-nonrepeating. **DEFINITION OF A REAL NUMBER**

For all real numbers a, b, and c: **PROPERTIES OF REAL NUMBERS**
(1) Commutative
    (a) Addition: $a + b = b + a$
    (b) Multiplication: $a \cdot b = b \cdot a$
(2) Associative
    (a) Addition: $(a + b) + c = a + (b + c)$
    (b) Multiplication: $(a \cdot b) \cdot c = a \cdot (b \cdot c)$
(3) Identities
    (a) The additive identity is 0 and
$$a + 0 = 0 + a = a$$
    (b) The multiplicative identity is 1 and
$$a \cdot 1 = 1 \cdot a = a$$
(4) Inverses
    (a) The additive inverse of a is $-a$ and
$$a + (-a) = (-a) + a = 0$$
    (b) For $a \neq 0$, the multiplicative inverse (or reciprocal) is $\frac{1}{a}$ and
$$a \cdot \frac{1}{a} = \frac{1}{a} \cdot a = 1$$
(5) Multiplication by zero
$$0 \cdot a = a \cdot 0 = 0$$

(6) Distributive property of multiplication with respect to addition

$$a(b + c) = ab + ac \text{ and } (b + c)a = ba + ca$$

ADDITION OF REAL NUMBERS

(1) If two numbers to be added have like signs, add their absolute values and apply the common sign.
(2) If two numbers to be added have unlike signs, subtract the smaller absolute value from the larger and use the sign of the number with the greater absolute value.

CHANGING SUBTRACTION TO ADDITION

$$a - b = a + (-b)$$

(1) Subtraction is changed to addition.
(2) The subtrahend is changed to its opposite.
(3) The sign of the minuend remains the same.

MULTIPLICATION OF REAL NUMBERS

(1) The product of two numbers having like signs is positive.
(2) The product of two numbers having unlike signs is negative.

RAISING A NUMBER TO A POWER (SIGNS)

(1) A negative base to an even power gives a positive answer: $(-3)^2 = +9$
(2) A negative base to an odd power gives a negative answer: $(-2)^3 = -8$
(3) A positive base to either an even or odd power gives a positive answer: $(+2)^4 = +16$ and $(+2)^3 = +8$

DIVISION OF REAL NUMBERS

(1) The quotient of two numbers having like signs is positive, and the quotient of two numbers having unlike signs is negative.
(2) Division by zero is impossible.
(3) A quotient is zero only if the dividend is zero and the divisor is not zero.

ORDER OF OPERATIONS AGREEMENT

(1) Perform any operations inside grouping symbols.
(2) Evaluate all exponentials.
(3) Perform all multiplications and divisions as they occur, working from left to right.
(4) Perform all additions and subtractions as they occur, working from left to right.

Define these sets by listing the elements within braces.

1. The set consisting of the counting numbers between 5 and 11.
2. The set of counting numbers between 9 and 10.

Fill in the blanks to make true statements.

3. In the addition problem $12 + 10 = 22$, the numbers 12 and 10 are called _____.
4. In the subtraction problem $9 - 3 = 6$, the number 3 is called the _____.
5. In the multiplication problem $6 \cdot 5 = 30$, the number 30 is called the _____.
6. In the division problem $\frac{15}{3} = 5$, the number 3 is called the _____.
7. In a division problem the _____ can never be zero.

Evaluate the following numeral expressions.

8. $24 \div 6 \cdot 2$
9. $40 \div 2^3 - 0 \cdot 5 + 6 \cdot 4 \div 2$
10. $5 \cdot (8 - 2)^2$
11. $(6 - 2) \cdot [10 - (8 - 3)]$

Classify each as a basic sum, difference, product, or quotient.

12. $10 - 2 \cdot 3 + 4$
13. $(9 - 1) \cdot (4 + 5)$
14. $3x + 4y$
15. $6x(y + 2z)$

Insert grouping symbols to make the following equations true.

16. $12 - 4 \cdot 2 + 3 = 40$
17. $12 - 4 \cdot 2 + 3 = 19$

Evaluate the following variable expressions by letting $x = 0$, $y = 5$, and $z = -8$.

18. $3x + y + 2z$
19. $y(z - x)$

Give a variable expression representing these word statements.

20. Four more than twice a certain number.
21. The number of cents in x quarters.

Each of the following equations illustrates one of the properties of real numbers. Name the property.

22. $3 \cdot (a + b) = 3 \cdot (b + a)$
23. $3 \cdot (a + b) = (a + b) \cdot 3$
24. $(a + 2b) + 3c = a + (2b + 3c)$
25. $1 \cdot (2x) = 2x$

26. $2(a + b) = 2a + 2b$ **27.** $3n + 0 = 3n$

28. $3 \cdot \dfrac{1}{3} = 1$ **29.** $2(-x + x) = 2 \cdot 0$

Factor the following expressions using the reverse of the distributive property.

30. $3x + 6$ **31.** $7m + 7$

Give the opposite (additive inverse) for each of the following expressions.

32. $+6$ **33.** 0

34. -4 **35.** x

36. $-y$

Simplify these expressions.

37. $-(-3)$ **38.** $-[-(-4)]$

39. $-[-(+8)]$ **40.** $|+3|$

41. $|-7|$ **42.** $|0|$

Insert $>$, $<$, or $=$ to make these statements true.

43. -3 _____ -1 **44.** 0 _____ -6

45. -7 _____ -5 **46.** 0.634 _____ 0.63

47. -9.6 _____ -10.8 **48.** $|-7.8|$ _____ $|7.8|$

Evaluate each of these expressions.

49. $(+6) + (-6)$ **50.** $(-8) + (+2)$

51. $(-3) + (+5)$ **52.** $-2^3 + 5^2 - 7$

53. $6^2 - 7^2 - 10^2$ **54.** $(-3) - (-2)$

55. $(+1) - (-5) - (+8)$ **56.** $(-2)(+3)$

57. $(-5)^2(-4)^2$ **58.** $(-20) \div (-2)$

59. $\dfrac{-15}{+3}$ **60.** $\dfrac{0}{-2}$

61. $(-3)(+2)(-4)$ **62.** $(-1)(-3)(-6)(-2)$

63. $(-3)^2$ **64.** -3^2

65. $(-1)^{158}$ **66.** $(-1)^{159}$

67. $\dfrac{18 - 3}{-1 - 4}$ **68.** $\dfrac{4^2 - 2^2}{6(1 - 2)}$

69. $\dfrac{4^2 - 5^2}{2^3 - 3^2}$ **70.** $\dfrac{12^2 - 11^2 - 10^2}{6^2 - 5^2}$

71. $\dfrac{-3 \cdot 2^2 + 4 \cdot 3^2}{(5 - 3)^3}$ **72.** $\dfrac{-3^2 - 3^3}{-2^2 - 2^3}$

Classify each of these numbers as being rational or irrational.

73. 0

74. $-6.\overline{34}$

75. $\dfrac{2}{3}$

76. 9.040040004 . . .

Graph each set of numbers on the standard number line.

77. $\{-3, 0, 2, 4.5\}$

78. $\{|-1|, \dfrac{1}{2}, -|-2|, 3\}$

Solve these problems.

79. The boiling point of water is 100°C and that of oxygen is -183°C. What is the difference of their boiling points?

80. Subtract -12 from the product of -3 and -5.

81. Raise the quotient of -6 and 3 to the fourth power.

82. The temperature at midnight was -18°F. By noon it went up 29°F. What was the temperature at noon?

---

1. $\{6, 7, 8, 9, 10\}$

2. the empty set, $\{\ \}$

3. addends

4. subtrahend

5. product

6. divisor

7. divisor

8. $\underbrace{24 \div 6} \cdot 2$
  $\quad\ 4\ \ \ \cdot 2$
  $\qquad\qquad 8$

9. $40 \div 2^3 - 0 \cdot 5 + 6 \cdot 4 \div 2$
  $\qquad\quad\downarrow$
  $\underbrace{40 \div 8} - \underbrace{0 \cdot 5} + \underbrace{6 \cdot 4 \div 2}$
  $\quad\ \ 5\quad -\quad 0\quad +\quad 12\quad = 17$

10. $5 \cdot (8 - 2)^2$
  $5 \cdot\quad 6^2$
  $5 \cdot\quad 36$
  $\quad\ 180$

11. $\underbrace{(6 - 2)} \cdot [10 - \underbrace{(8 - 3)}]$
  $\quad\ 4\quad \cdot [10 -\quad 5]$
  $\quad\ 4\quad \cdot 5$
  $\qquad\ 20$

12. $10 \overset{②}{-} 2 \overset{①}{\cdot} 3 \overset{③}{+} 4$
  basic sum

13. $(9 \overset{①}{-} 1) \overset{③}{\cdot} (4 \overset{②}{+} 5)$
  basic product

14. $3 \overset{①}{\cdot} x \overset{③}{+} 4 \overset{②}{\cdot} y$
  basic sum

15. $6 \overset{③}{\cdot} x \overset{④}{\cdot} (y \overset{②}{+} 2 \overset{①}{\cdot} z)$
  basic product

16. $(12 - 4) \cdot (2 + 3) = 40$

17. $(12 - 4) \cdot 2 + 3 = 19$

18. $3x + y + 2z = 3 \cdot 0 + 5 + 2(-8)$
  $\qquad\qquad\quad = 0 + 5 + (-16)$
  $\qquad\qquad\quad = -11$

19. $y(z - x) = 5(-8 - 0)$
  $\qquad\qquad = 5(-8)$
  $\qquad\qquad = -40$

20. $2x + 4$

21. 25x cents.

22. Commutative property of addition.

23. Commutative property of multiplication.

24. Associative property of addition.

25. Multiplicative identity property.

26. Distributive property of multiplication with respect to addition.

27. Additive identity property.

28. Multiplicative inverse property.

29. Additive inverse property.

30. $3x + 6 = 3(x + 2)$

31. $7m + 7 = 7(m + 1)$

32. -6

33. 0

34. 4

35. $-x$

36. y

37. 3

38. -4

39. 8

40. 3

41. 7

42. 0

43. $-3 < -1$

44. $0 > -6$

45. $-7 < -5$

46. $0.634 > 0.63$

47. $-9.6 > -10.8$

48. $|-7.8| = |7.8|$

49. $(+6) + (-6) = 0$

50. $(-8) + (+2) = -6$

51. $(-3) + (+5) = 2$

52. $-2^3 + 5^2 - 7 = -8 + 25 - 7$
$$= 10$$

53. $6^2 - 7^2 - 10^2 = 36 - 49 - 100$
$$= -113$$

54. $(-3) - (-2) = -3 + 2$
$$= -1$$

55. $(+1) - (-5) - (+8) = 1 + 5 + (-8)$
$$= -2$$

56. $(-2)(+3) = -6$

57. $(-5)^2(-4)^2 = (25)(16)$
$$= 400$$

58. $(-20) \div (-2) = 10$

59. $\dfrac{-15}{+3} = -5$

60. $\dfrac{0}{-2} = 0$

61. $(-3)(+2)(-4) = 24$

62. $(-1)(-3)(-6)(-2) = 36$

63. $(-3)^2 = 9$

64. $-3^2 = -9$

65. $(-1)^{158} = 1$

66. $(-1)^{159} = -1$

67. $\dfrac{18 - 3}{-1 - 4} = \dfrac{15}{-5}$
$$= -3$$

68. $\dfrac{4^2 - 2^2}{6(1 - 2)} = \dfrac{16 - 4}{6(-1)}$
$$= \dfrac{12}{-6}$$
$$= -2$$

69. $\dfrac{4^2 - 5^2}{2^3 - 3^2} = \dfrac{16 - 25}{8 - 9}$
$$= \dfrac{-9}{-1}$$
$$= 9$$

70. $\dfrac{12^2 - 11^2 - 10^2}{6^2 - 5^2} = \dfrac{144 - 121 - 100}{36 - 25}$

$\qquad\qquad = \dfrac{-77}{11}$

$\qquad\qquad = -7$

71. $\dfrac{-3 \cdot 2^2 + 4 \cdot 3^2}{(5 - 3)^3} = \dfrac{-3 \cdot 4 + 4 \cdot 9}{2^3}$

$\qquad\qquad = \dfrac{-12 + 36}{8}$

$\qquad\qquad = \dfrac{24}{8}$

$\qquad\qquad = 3$

72. $\dfrac{-3^2 - 3^3}{-2^2 - 2^3} = \dfrac{-9 - 27}{-4 - 8}$ **73.** rational

$\qquad\quad = \dfrac{-36}{-12}$

$\qquad\quad = 3$

74. rational   **75.** rational

76. irrational

77.

78.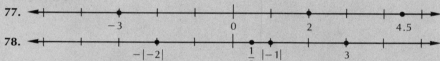

79. $100 - (-183) = 100 + 183 = 283°C$

80. $(-3)(-5) - (-12) = 15 - (-12)$

$\qquad\qquad\qquad\quad = 15 + 12$

$\qquad\qquad\qquad\quad = 27$

82. $-18 + 29 = 11°F$

81. $\left(\dfrac{-6}{3}\right)^4 = (-2)^4$

$\qquad\quad = 16$

ALGEBRAIC EXPRESSIONS AND EQUATIONS

2

PROPERTIES OF WHOLE NUMBER EXPONENTS

2.1

base
exponent
exponential

Throughout the first chapter you evaluated expressions containing exponents. In this section we will study whole number exponents in greater detail and develop several useful properties.

A natural number exponent tells the number of times the base is used as a factor. Thus, the symbol 3^4 means $3 \cdot 3 \cdot 3 \cdot 3$. The **base** is 3 and the **exponent** (or power) is 4. The symbol 3^4 is called an **exponential**. We now write the formal definition for a natural number exponent.

Definition of a Natural Number Exponent

If n is a natural number, then

$$a^n = \overbrace{a \cdot a \cdot a \ldots a}^{n \text{ times}}$$

where a is used as a factor n times.

The exponent zero was not included in this definition. (Remember, zero is not a natural number.) The zero exponent is a special case and will be discussed separately later in this section.

Example 1

Here are several illustrations of exponential notation.
(a) $x \cdot x = x^2$; x is the base and 2 is the exponent or power. The exponential is x^2 and is read "x squared" or "x to the second power."
(b) $(a + b)(a + b)(a + b) = (a + b)^3$; $(a + b)$ is the base and 3 is the exponent or power. The exponential is $(a + b)^3$ and is read "the group (a plus b) cubed" or "the group (a plus b) to the third power."
(c) $a = a^1$; a is the base and 1 is the exponent or power. The exponential is a^1 and is read "a to the first power."

exponential expressions

Expressions where exponentials are combined with the operations of addition, subtraction, multiplication, or division are called **exponential expressions**.

Example 2

Each of the following has been written as an exponential expression.
(a) $2 \cdot x \cdot x \cdot x \cdot x = 2x^4$; read "two times x to the fourth power."
(b) $(2x)(2x)(2x)(2x) = (2x)^4$; read "the group 2x to the fourth power."
(c) $m \cdot m + n \cdot n = m^2 + n^2$; read "m squared plus n squared."
(d) $(m + n)(m + n) = (m + n)^2$; read "the group (m + n) squared."
(e) $3 \cdot x \cdot x \cdot y \cdot y \cdot y = 3x^2y^3$; read "three x squared, y cubed."
(f) $(5 \cdot y \cdot y \cdot y \cdot y) \div (2 \cdot x \cdot x \cdot x \cdot x \cdot x) = \dfrac{5y^4}{2x^5}$; read "five y to the fourth power divided by two x to the fifth power"; or "five y to the fourth power over two x to the fifth power."

We now proceed to develop five properties of exponents which will enable us to simplify many exponential expressions. First we examine multiplication where the exponentials have identical bases. Consider $x^3 \cdot x^5$.

$$
\begin{aligned}
x^3 \cdot x^5 &= (x \cdot x \cdot x)(x \cdot x \cdot x \cdot x \cdot x) \\
&= x \cdot x \cdot x \cdot x \cdot x \cdot x \cdot x \cdot x \\
&= x^8
\end{aligned}
$$

The exponent 8 is the sum of the exponents 3 and 5. Thus, in general, we can state that to multiply exponentials with identical bases, we add the exponents.

The Addition Property of Exponents

For every whole number m and n,

$$a^m \cdot a^n = a^{m+n}$$

These equations illustrate the addition property of exponents.

(a) $5^2 \cdot 5^4 = 5^{2+4} = 5^6$

(b) $x \cdot x^2 = x^1 \cdot x^2 = x^3$

(c) $a^3 \cdot a^4 = a^7$

(d) $(x + y)^2(x + y)^4 = (x + y)^6$

Example 3

Remember, to use the addition property of exponents the bases must be identical. We cannot use this property to simplify $x^2 \cdot y^3$.

Next, we examine division of exponentials having identical bases. Consider $\dfrac{x^7}{x^5}$.

$$
\begin{aligned}
\frac{x^7}{x^5} &= \frac{x \cdot x \cdot x \cdot x \cdot x \cdot x \cdot x}{x \cdot x \cdot x \cdot x \cdot x} \\
&= \frac{x \cdot x \cdot \boxed{x \cdot x \cdot x \cdot x \cdot x}}{\boxed{x \cdot x \cdot x \cdot x \cdot x}} \\
&= x \cdot x \cdot 1 \\
&= x^2
\end{aligned}
$$

The difference of the exponents, $7 - 5$, gives the exponent 2. Thus, to divide exponentials with identical bases, simply subtract their exponents.

> **The Subtraction Property of Exponents**
> For every whole number m and n, where $m \geq n$,
> $$\frac{a^m}{a^n} = a^{m-n}, \ (a \neq 0)$$

The statement $m \geq n$ says that the exponent m must be greater than or equal to the exponent n. We will wait until a later chapter to discuss problems where $m < n$, such as $\dfrac{x^2}{x^5}$.

Example 4

The following equations illustrate the subtraction property of exponents.

(a) $\dfrac{3^7}{3^2} = 3^{7-2} = 3^5$

(b) $\dfrac{x^4}{x} = \dfrac{x^4}{x^1} = x^3$

(c) $\dfrac{a^3}{a^2} = a^1 = a$

Next we evaluate a number raised to the zero power by using the subtraction property of exponents. For example,

$$\frac{2^3}{2^3} = 2^{3-3} = 2^0$$

But, we also know that

$$\frac{2^3}{2^3} = 1$$

because any non-zero number divided by itself produces a quotient of 1. Therefore, we conclude that

$$2^0 = 1$$

The same argument applies to exponentials with a variable base:

$$\text{If } x \neq 0, \text{ then } \frac{x^m}{x^m} = x^{m-m} = x^0$$

$$\text{But, } \frac{x^m}{x^m} = 1$$

$$\text{Thus, } x^0 = 1, \text{ where } x \neq 0$$

> **The Zero Power Property**
> $$\text{If } x \neq 0, \text{ then } x^0 = 1.$$

This property states that any number except zero, when raised to the zero power, produces 1. The zero power means a number has been divided by itself. Thus, 0^0 does not represent a number because division by zero is impossible.

These problems illustrate the zero power property.

Example 5

(a) $3^0 = 1$

(b) $m^0 = 1$, where $m \neq 0$

(c) $(2x)^0 = 1$, where $x \neq 0$

(d) $2x^0 = 2 \cdot 1 = 2$, where $x \neq 0$

Often we need to raise an exponential to a power. Consider, for example, $(x^3)^2$. Note that the exponential x^3 is raised to the second power. This is called a power to a power. To evaluate we use the definition of exponents together with the addition property of exponents:

$$(x^3)^2 = x^3 \cdot x^3$$
$$= x^{3+3}$$
$$= x^6$$

This process can be shortened by observing that the product of the exponents, $3 \cdot 2$, produces the new exponent 6. We summarize by stating the following property.

The Power to a Power Property of Exponents

If m and n represent whole numbers, then

$$(x^m)^n = x^{m \cdot n}$$

This property tells us that a power to a power may be simplified by multiplying the exponents.

These equations illustrate the power to a power property.

Example 6

(a) $(3^4)^5 = 3^{4 \cdot 5} = 3^{20}$

(b) $(r^3)^5 = r^{15}$

(c) $(y^2)^0 = y^0 = 1$, where $y \neq 0$

(d) $[(x + y)^2]^5 = (x + y)^{10}$

The final property of this section, called the distributive property of exponents, allows us to share an exponent with each factor of a basic product or basic quotient. To develop this property, we use the definition of exponents, together with the addition or subtraction property, to simplify these examples, $(x^3 y^4)^2$ and $\left(\dfrac{a^3}{b^4} \right)^2$.

$$(x^3y^4)^2 = (x^3y^4)(x^3y^4)$$

$$\left(\frac{a^3}{b^4}\right)^2 = \frac{a^3}{b^4} \cdot \frac{a^3}{b^4}$$

$$= (x^3x^3)(y^4y^4)$$

$$= \frac{a^3 \cdot a^3}{b^4 \cdot b^4}$$

$$= x^6y^8$$

$$= \frac{a^6}{b^8}$$

Notice that this process can be shortened by multiplying the exponent, 2, times each exponent of the basic product or quotient.

$$(x^3y^4)^2 = x^{3\cdot2}y^{4\cdot2}$$

$$\left(\frac{a^3}{b^4}\right)^2 = \frac{a^{3\cdot2}}{b^{4\cdot2}}$$

$$= x^6y^8$$

$$= \frac{a^6}{b^8}$$

The Distributive Property of Exponents

If a, b, and c represent whole numbers, then $(x^a \cdot y^b)^c = x^{a\cdot c} \cdot y^{b\cdot c}$ and $\left(\dfrac{x^a}{y^b}\right)^c = \dfrac{x^{a\cdot c}}{y^{b\cdot c}}$ where $y \neq 0$.

This property states that an exponent may be shared and then multiplied times each exponent of a basic product or a basic quotient. This property does NOT hold true for basic sums and differences. As in the following example, an exponent must never be distributed over a basic sum or a basic difference.

$$\underbrace{(2 + 3)^2} \neq 2^2 + 3^2$$

$$5^2 \quad \neq 4 + 9$$

$$25 \quad \neq 13$$

Example 7

These equations illustrate the distributive property of exponents.

(a) $(a^4b^2)^4 = a^{4\cdot4}b^{2\cdot4} = a^{16}b^8$

(b) $(xy)^3 = (x^1y^1)^3 = x^3y^3$

(c) $(2x)^3 = 2^3 \cdot x^3 = 8x^3$

(d) $\left(\dfrac{3m}{2n}\right)^2 = \dfrac{3^2 \cdot m^2}{2^2 \cdot n^2} = \dfrac{9m^2}{4n^2}$, where $n \neq 0$

(e) $\left(\dfrac{2a^4}{3bc^2}\right)^3 = \dfrac{8a^{12}}{27b^3c^6}$ where $b, c \neq 0$

The following summarizes the five properties of exponents developed in this section.

Properties of Exponents

For whole number exponents *m, n, p* and for real number bases *a, b:*

(1) Addition Property of Exponents: to multiply exponentials with identical bases, add the exponents.

$$a^m \cdot a^n = a^{m+n}$$

(2) Subtraction Property of Exponents: to divide exponentials with identical bases, subtract the exponents.

$$\frac{a^m}{a^n} = a^{m-n}, \text{ where } a \neq 0 \text{ and } m \geq n$$

(3) Zero Power Property: any number except zero, when raised to the zero power, is 1.

$$a^0 = 1, \text{ where } a \neq 0$$

(4) Power to a Power Property of Exponents: to raise an exponential to a power, multiply the exponents.

$$(a^m)^n = a^{m \cdot n}$$

(5) Distributive Property of Exponents: an exponent may be shared and then multiplied times each exponent of a basic product or a basic quotient.

$$(a^m b^n)^p = a^{m \cdot p} b^{n \cdot p} \text{ and } \left(\frac{a^m}{b^n}\right)^p = \frac{a^{m \cdot p}}{b^{n \cdot p}} \text{ where } b \neq 0$$

Indicate the base and the exponent for the following exponentials.

Exercise 2.1

1. y^6

2. 6^x

3. $5x^2$

4. $(5x)^2$

5. $(x + 4)^5$

6. $(3m)^7$

7. -5^2

8. $(-5)^2$

9. $-x^4$

10. $(-x)^4$

Write a word statement describing how each of these expressions should be read.

11. y^2

12. z^3

13. $(a + 2b)^6$

14. $5x^4$

Change each to exponential notation.

15. $x \cdot x$

16. $y \cdot y \cdot y \cdot y$

17. $3 \cdot 3 \cdot 3 \cdot 3 \cdot 3$

18. $(a + 2)(a + 2)$

19. $a \cdot a \cdot a \cdot b \cdot b$

20. $(xy)(xy)(xy)$

21. $x \cdot x \cdot y \cdot y \cdot y \cdot z \cdot z$

22. $(x + 5y)(x + 5y)$

23. $4 \cdot b \cdot b \cdot b \cdot b \cdot b$

24. $(-1)(-1)(-1)(-1)$

25. $(-x)(-x)(-x)$

26. $a \cdot a \cdot a \cdot (-b)(-b)(-b)$

27. $\dfrac{3 \cdot x \cdot x}{4 \cdot y \cdot y \cdot y}$

28. $\dfrac{(x + y)(x + y)}{(-xy)(-xy)(-xy)}$

Evaluate these exponential expressions.

29. 5^1

30. 6^0

31. $(-3)^0$

32. $(-8)^0$

33. -8^0

34. -10^0

35. $2 \cdot 6^0$

36. $(2 \cdot 6)^0$

37. 2^2

38. $3^2 + (-3)^2$

39. $2^3 + (-2)^3$

40. $(2 + 3)^2$

41. $2^2 + 3^2$

42. $4^2 + 1^3$

43. $0^3 + 10^1$

44. $2^3 \cdot 3^2 + 4^2$

45. $4^0 + 3^0$

46. $2 \cdot 3^3 + 2 \cdot 3^2$

47. $5(6 - 2)^3$

48. $3(5 - 8)^3$

Identify the property of exponents illustrated by each of these equations.

Example $m^3 \cdot m^7 = m^{10}$ [addition property of exponents]

49. $y^0 = 1$, where $y \neq 0$

50. $x^2 \cdot x^4 = x^6$

51. $\dfrac{m^5}{m^2} = m^3$, where $m \neq 0$

52. $(ab)^3 = a^3 b^3$

53. $(a^3)^5 = a^{15}$

54. $(2x^2 y^3 z)^3 = 8x^6 y^9 z^3$

55. $\left(\dfrac{3a^4}{5}\right)^2 = \dfrac{9a^8}{25}$

56. $(2xy)^4 = 16x^4 y^4$

57. $\dfrac{a^7}{a} = a^6$, where $a \neq 0$

58. $3 \cdot (x^5 \cdot x^2) = 3x^7$

Use the properties of exponents to simplify these expressions.

Example $a^5 \cdot a = a^6$

59. $\dfrac{y^9}{y^2}$

60. $x^7 \cdot x^4$

61. 9^0

62. $(m^2)^6$

63. $(2a)^3$

64. $(3x^4 y^5 z^2)^2$

65. $(m + 3n)^7 (m + 3n)^2$

66. $\dfrac{x^4}{x}$

67. $\left(\dfrac{2a^3}{3}\right)^2$

68. $n^5 \cdot n^{10}$

69. $\left(\dfrac{x}{3}\right)^2$ **70.** $(5mn)^2$

Which of the following are incorrect uses of the distributive property of exponents?

71. $(2m^2)^3 = 2m^6$

72. $(3x^4)^2 = 9x^8$

73. $(x + y)^2 = x^2 + y^2$

74. $\left(\dfrac{3a^4}{2}\right)^2 = \dfrac{9a^6}{4}$

75. $(a^2b^3c)^4 = a^8b^{12}c^4$

76. $(x^3 - y^2)^4 = x^{12} - y^8$

Calculator problems.

77. $(2.634)^3$

78. $(-3.21)^2$

79. $3.8^2 - 2.4^3$

80. $\dfrac{8.91^2 - 3.64^3}{2.85^3 - 1.06^4}$

MULTIPLICATION AND DIVISION OF ALGEBRAIC EXPRESSIONS

2.2

You learned in sections 1.9 and 1.10 that the product (and quotient) of a positive number and a negative number is a negative number. For example,

$$3(-2) = -6 \text{ and } \dfrac{10}{-5} = -2$$

A similar property holds true for variables. It is called the opposite law of multiplication and division.

Opposite Law for Multiplication and Division

For all real numbers a and b,

$$a \cdot (-b) = (-a) \cdot b = -(a \cdot b)$$

and for $b \neq 0$,

$$\dfrac{a}{-b} = \dfrac{-a}{b} = -\dfrac{a}{b}$$

These equations illustrate the opposite law.   **Example 1**

(a) $(-3)(x) = -3x$

(b) $2(-b) = -2b$

(c) $(-m)(m) = -m^2$

(d) $\dfrac{-x^3}{x} = -x^2$

Also, you learned that the product (and quotient) of two negative numbers is a positive number. Thus,

$$(-4)(-2) = 8 \text{ and } \frac{-12}{-3} = 4$$

A similar property, called the double opposite law, is true for variables.

Double Opposite Law for Multiplication and Division

For all real numbers a and b,
$$(-a) \cdot (-b) = ab$$
and for $b \neq 0$,
$$\frac{-a}{-b} = \frac{a}{b}$$

Example 2

These equations illustrate the double opposite law.

(a) $(-3)(-x) = 3x$

(b) $(-m)(-m) = m^2$

(c) $\dfrac{-x^5}{-x^2} = x^3$

The properties of exponents, together with the commutative and associative properties, can be used to develop a procedure for multiplying exponential expressions. Consider the product of $(2a^3)(3a^4)$:

$$(2a^3)(3a^4) = (2 \cdot 3)(a^3 \cdot a^4) \quad \text{[commutative and associative properties]}$$
$$= 6a^7 \quad \text{[addition property of exponents]}$$

numerical coefficients

This example suggests a shortcut. Instead of rearranging the factors, simply multiply the numbers (called the **numerical coefficients**) and add the exponents appearing on identical bases.

Example 3

Multiply $(-5a^2)(2a^3)$

Solution:
multiply numerical coefficients
$$(-5a^2)(2a^3) = -10a^5$$
add exponents

Example 4

Multiply $(6mn^2)(3m^3n)$

Solution: Remember, $m = m^1$ and $n = n^1$
$$(6mn^2)(3m^3n) = (6m^1n^2)(3m^3n^1)$$
$$= 18m^4n^3$$

Multiply $(-3x^2y)(-xy^2)(2x^3y)$ **Example 5**

Solution:

multiply numerical coefficients

$$(-3x^2y)(-xy^2)(2x^3y) = 6x^6y^4$$

add exponents

Similarly, to divide exponential expressions, we divide the numerical coefficients and subtract exponents appearing on identical bases.

Divide $(-10x^5) \div (2x^3)$, where $x \neq 0$.   **Example 6**

Solution:

divide numerical coefficents

$$(-10x^5) \div (2x^3) = -5x^2$$

subtract exponents

Divide $\dfrac{10a^5}{5a^2}$, where $a \neq 0$ **Example 7**

Solution: Divide 10 by 5 and subtract the exponents, since the bases are identical.

$$\frac{10a^5}{5a^2} = 2a^{5-2} = 2a^3$$

Divide $\dfrac{-24a^3b^2}{-3ab^2}$, where $a, b \neq 0$.   **Example 8**

Solution: $\dfrac{-24a^3b^2}{-3ab^2} = 8a^2b^0 = 8a^2$ [remember, $b^0 = 1$]

The distributive property of multiplication with respect to addition is used to multiply an exponential times a basic sum. Consider the product of $3x^2 (2x^3 + 4)$.

$$3x^2(2x^3 + 4) = (3x^2)(2x^3) + (3x^2) \cdot 4 \qquad \begin{bmatrix} \text{distributive} \\ \text{property} \end{bmatrix}$$
$$= 6x^5 + 12x^2$$

Remember, the multiplier must be distributed to both addends.

Multiply $5mn(3m^2 + 2n^2)$ **Example 9**

Solution: $5mn(3m^2 + 2n^2) = 15m^3n + 10mn^3$

As illustrated by the following argument, multiplication is also distributive with respect to subtraction:

$$a \cdot [b - c] = a[b + (-c)] \qquad \text{[definition of subtraction]}$$
$$= a \cdot b + a \cdot (-c) \qquad \text{[dist. prop. of mult. over add.]}$$
$$= a \cdot b + [-(a \cdot c)] \qquad \text{[opposite law of multiplication]}$$
$$= a \cdot b - a \cdot c \qquad \text{[definition of subtraction]}$$

The Distributive Property of Multiplication with Respect to Subtraction

For all real numbers a, b, and c,
$$a(b - c) = ab - ac \text{ and } (b - c)a = ba - ca$$

Example 10

Here are some illustrations of the distributive property of multiplication
with respect to subtraction.

(a) $2(x - y) = 2x - 2y$
(b) $3m(m - n) = 3m^2 - 3mn$
(c) $-5x^2(2x - 3y) = -10x^3 + 15x^2y$

The distributive properties of addition and subtraction can be extended
to any number of terms. This allows us to multiply a number times a
group containing combinations of addition and subtraction.

Example 11

Multiply $2(a + b - c)$

Solution: Share the multiplier, 2, with each term inside the group.
$$2(a + b - c) = 2a + 2b - 2c$$

Example 12

Multiply $3a(2a^2 - 4a - 1)$

Solution: $3a(2a^2 - 4a - 1) = 6a^3 - 12a^2 - 3a$

Example 13

Multiply $-5mn^2(2m^2n - mn^2 + 3)$

Solution: $-5mn^2(2m^2n - mn^2 + 3) = -10m^3n^3 + 5m^2n^4 - 15mn^2$

Exercise 2.2

Multiply the following expressions according to the opposite law or the
double opposite law. Indicate the property used.

1. $(-4)(-x)$ 2. $(-a) \cdot c$
3. $(-y)(-y)$ 4. $(-m)(m)$
5. $(-2)(a)$ 6. $(x)(-y)$

7. $(-3)(-x)$ **8.** $(-n)(-n)$

9. $(-5)(x)$ **10.** $(-5)(-b)$

Multiply or divide as indicated. Assume that no divisor represents zero.

11. $4(2x)$ **12.** $-5(-3y)$

13. $(-3)(-5a)$ **14.** $(-5x)(-7x)$

15. $(-3a^2)(2a^3)$ **16.** $(-4x^2y)(-3xy^2)$

17. $-3(2a)$ **18.** $5(3m)$

19. $(3x)(4x)$ **20.** $(2n)(5n)$

21. $(2a^2)(5a^3)$ **22.** $(-n^2)(2n^4)$

23. $(3x^2y)(2xy^3)$ **24.** $(-a^2bc^3)(2ab^3)$

25. $(-xy^2z)(5x^3yz^4)$ **26.** $(2xy^2)(5xy^3)$

27. $(-a^2bc)(2a^4b^2)$ **28.** $(3x)(2x^3)(4x^2)$

29. $(5a^2)(2a)(3a^3)$ **30.** $(2m^3n)(mn^2)(3m^2n^3)$

31. $(ab^2c)(3a^2bc)(4abc^3)$ **32.** $(-3xy^2)(-2yz^3)(-4z)$

33. $(-4x^2)(5x^2y)(2xy)$ **34.** $(2m^2n)(-6mn)(3mn^3)$

35. $(-4a^2b^3)(-2abc)(-3b^3c^2)$ **36.** $(-mn^2)(6m^2)(-m)$

37. $(-7m^2n)(2mn^3)$ **38.** $(10m) \div (-2)$

39. $(-4a^2) \div 2a$ **40.** $(9a^2) \div (-3a)$

41. $\dfrac{8m}{2}$ **42.** $\dfrac{9x}{-3}$

43. $\dfrac{-12m}{-4m}$ **44.** $\dfrac{-10a}{-2a}$

45. $\dfrac{6r}{2r}$ **46.** $\dfrac{15m^3}{3m^2}$

47. $\dfrac{6x^3}{-2x}$ **48.** $\dfrac{12a^3b^4}{-2ab^2}$

49. $\dfrac{15a^3b^2c}{-3ab}$ **50.** $\dfrac{9m^5n^2}{3m^2n^2}$

51. $\dfrac{-9x^2}{3}$ **52.** $\dfrac{-9x^2}{3x}$

53. $\dfrac{10xy^3}{-10xy^3}$ **54.** $\dfrac{-20a^3b^2}{-10ab^2}$

55. $\dfrac{15x^5y^7z^3}{-3x^2y^4z}$ **56.** $\dfrac{-18m^3n^2}{6mn^2}$

Multiply these expressions using the distributive properties of multiplication or their extensions.

57. $2(x + 6)$ **58.** $3(m - n)$

59. $5(x - 3)$ **60.** $-3a(7a - 5)$

61. $-2m(3m^2 - 1)$ **62.** $2(a - 3b)$

63. $5x(x^2 - y)$ **64.** $3(a + b - c - d)$

65. $2a(3a - b + c)$ **66.** $3(2x + 3y - 4z)$

67. $-2(3a - 4b + 2c - 5d)$ **68.** $5ab(2a^2 - 3ab + c)$

69. $4xy^2(2x^2y - 3xy^2 + 1)$ **70.** $-3x(2x^2 - x + 2)$

71. $2x(3x + 4y - 3)$ **72.** $3a(5a^2 + 2ab - b^2)$

73. $2mn(m^3 + n^3)$ **74.** $3m^2(2m^3 + 5)$

75. $2ab(3a^2 + 4b^2 - 1)$ **76.** $6x^2y^3(2x^2 + 3y^2)$

77. $5ab^2(2a^3b + 3b)$ **78.** $mn^2(m^5 - n^4)$

79. $4a^2bc^3(2a^2b + 3b^3c)$

80. $-3xy^2(-4x^3 + 3xy^2 - x^2y^3 + y^4)$

ADDITION AND SUBTRACTION OF ALGEBRAIC EXPRESSIONS

2.3

Using basic factoring studied in section 1.6, we will develop a method for adding and subtracting certain types of algebraic (or variable) expressions.

Example 1

Add $2x + 3x$

Solution: Factor out an x

$$2x + 3x = x \cdot (2 + 3)$$
$$= x \cdot 5$$
$$= 5x$$

Therefore, $2x + 3x = 5x$

Example 2

Combine $7a^2 + 9a^2 - 15a^2$

Solution: Factor out an a^2

$$7a^2 + 9a^2 - 15a^2 = a^2 \cdot (7 + 9 - 15)$$
$$= a^2 \cdot 1$$
$$= 1 \cdot a^2 \quad \text{or} \quad a^2$$

Therefore, $7a^2 + 9a^2 - 15a^2 = a^2$

The expressions in examples 1 and 2 could be combined because the same variable and exponent were present in each term. This allowed us to factor and then combine the remaining numerical coefficients. This procedure suggests the following shortcut.

Algebraic expressions can be added or subtracted by combining their respective numerical coefficients, provided the variables and exponents are identical. Terms having identical variables and exponents are called **like terms**. Thus, the following rule holds for adding and subtracting like terms.

like terms

> Like terms can be added or subtracted by combining their respective numerical coefficients.

Example 3

Add $2x + 3x$

Solution: Combine numerical coefficients of like terms.

$$2x + 3x = 5x$$

Combine $7a^2 + 9a^2 - 15a^2$

Example 4

Solution: Combine numerical coefficients of like terms.

$$7a^2 + 9a^2 - 15a^2 = 1a^2 \text{ or } a^2$$

Combine $r + 2rt + 5r - 6rt$

Example 5

Solution: Separate out the like terms and combine their respective numerical coefficients. Remember, r means $1r$

$$r + 2rt + 5r - 6rt = 6r - 4rt$$

Combine $7a^2b + 4ab^2 - 3a^2b - 5ab^2$

Example 6

Solution: Separate out like terms and combine their respective numerical coefficients.

$$7a^2b + 4ab^2 - 3a^2b - 5ab^2 = 4a^2b - 1ab^2$$
$$= 4a^2b - ab^2$$

Remember, only like terms can be combined. Variable expressions which do not contain like terms cannot be simplified.

These expressions cannot be further simplified.

Example 7

(a) $5a + 3b$ cannot be simplified because the variables are different.
(b) $6x^2y - 4xy^2$ cannot be simplified because the exponents differ, even though the variables are the same.
(c) $3m + 1$ cannot be combined because the second term does not contain the variable m.

Variable expressions can be added vertically if like terms are arranged in the same columns.

Add $(3x^2 - 5x + 2) + (7x^2 - 3x - 8)$

Example 8

Solution:　Arrange like terms in the same columns; then add each column.

$$3x^2 - 5x + 2$$
$$\underline{7x^2 - 3x - 8}$$
$$10x^2 - 8x - 6$$

Example 9　　Add $(6y^2 + 7y - 3) + (3y^2 - 5y) + (-y - 10)$

Solution:　Arrange like terms in the same columns; then add each column.

$$6y^2 + 7y - 3$$
$$3y^2 - 5y$$
$$\underline{ - y - 10}$$
$$9y^2 + 1y - 13 \qquad \text{or} \qquad 9y^2 + y - 13$$

Subtraction of variable expressions can also be performed vertically by recalling how subtraction is changed to addition:

$$(-8) - (-2) = (-8) + (+2)$$

(1)　The minuend is not changed.
(2)　Subtraction is changed to addition.
(3)　The subtrahend has its sign changed.

Therefore, variable expressions may be subtracted vertically by changing the sign of each term in the subtrahend.

Example 10　　Combine, $(5x^2 - 6x + 7) - (2x^2 + 3x - 9)$

Solution:　Arrange vertically, change each sign in the subtrahend, then add each column.

$$5x^2 - 6x + 7 \qquad\qquad\qquad 5x^2 - 6x + 7$$
$$\underline{2x^2 + 3x - 9}\;\text{——change signs——}\;\underline{-2x^2 - 3x + 9}$$
$$3x^2 - 9x + 16$$

Example 11　　Subtract $(2m^3 - 4m)$ from $(4m^2 + m - 5)$.

Solution:　Arrange vertically, change each sign in the subtrahend, then add each column.

$$4m^2 + m - 5 \qquad\qquad\qquad 4m^2 + m - 5$$
$$\underline{2m^2 - 4m}\;\text{——change signs——}\;\underline{-2m^2 + 4m}$$
$$2m^2 + 5m - 5$$

Why is it incorrect to combine the following expressions by addition or subtraction of the numerical coefficients?

Exercise 2.3

1. $3x + 7y$ **2.** $4m^2 - 3m$

3. $5x^2 - 4x$ **4.** $7x + 2$

5. $2a^3b + 3ab^2$ **6.** $3rs + 2r$

Simplify by combining like terms.

7. $6a + 4a$ **8.** $5x + 2x$

9. $3y - 2y$ **10.** $6m - 7m$

11. $12n + 2n$ **12.** $3a + a$

13. $5y - y$ **14.** $5x + x$

15. $3a - 5a$ **16.** $2r + 5 - 4r - 6$

17. $3x + 1 - 2x + 2$ **18.** $12m + 3m - 5$

19. $7b - 3b + 2b$ **20.** $3x + 2x - x$

21. $11xy + 4xy$ **22.** $2mn - 5mn + mn$

23. $4r^2t - 3r^2t$ **24.** $2m^2n - 3m^2n - 4m^2n$

25. $3ab^2 - 4ab^2 + ab^2$ **26.** $3m + 4 - 7m - 6$

27. $5pq - 7pq + 3p$ **28.** $2a - 7 + 3b - 7$

29. $8x^2 + 7x^2 - 2x^2$ **30.** $4m^2 + 3m^2 - 8m^2$

31. $2x^2 + x^2 - x$ **32.** $-mn + mn - 4mn$

33. $12ab - 10ab + 5$ **34.** $6x^3y^2 - 7x^3y^2 + x^3y^2$

35. $p^3 + p - 3p + 7$ **36.** $7mn - m + 2m - mn$

37. $2x - 5y + 4x - 2y + 3y$ **38.** $6ab - 7a + 2b + 8ab$

39. $2m^3n^2 + 3m^2n^3 - 6m^3n^2 + 4m^2n^3$

40. $3xy^3 + 2x^3y + 4x^3y - xy^3$

41. $r^2 + 3 - 2r^2 + 7 - r$

42. $3rs - r^2 + 2rs + s^2 - 2r^2$

43. $5a - b + 2c - 6a + b - 3c$

44. $-mn + m^2 - 3 + mn - 4m^2 + 5$

45. $2a^2 - 3a + 4 + 3a^2 - 5a + 1$

46. $2x^2 - 3x - 5 + 4x^2 - 7x + 12$

Combine the following expressions using vertical addition or subtraction.

47. $(5x^2 + 2x - 7) + (6x^2 - 7x + 10)$

48. $(5r^2 + 2r - 1) + (2r^2 - 5r + 2)$

49. $(6a^2 - 2a + 7) + (-2a^2 - 6)$

50. $(2y^2 + 1) + (3y - 8) + (y^2 - y)$

51. $(7x^2 - 2x + 3) - (2x^2 - 6x + 5)$

52. $(3a^2 - 4a + 1) - (5a^2 + 2a)$

53. $(4x^2 - 5x + 1) - (7x^2 - 2x - 8)$

54. $(3x^2 + x) - (7x + 1)$
55. $(6x^2 + 7x - 4) + (2x^2 - 3x + 2) + (5x^2 - 7x + 1)$
56. $(2m^2 + m - 5) + (3m^2 + 6) + (4m^2 - 2m) + (m^2 - m + 2)$
57. $(2n^2 - 7n) + (n + 2) + (3n^2 - 4)$
58. $(5b^2 - 7b + 2) + (b^2 - b + 3) + (4b^2 - 6) + (b - 3)$
59. $(x^2 - x - 1) - (x^2 + x + 1) - (2x^2 + x - 5)$
60. $(3a^2 - 6a + 1) - (2a^2 + 8a - 5) - (4a^2 + a - 2)$
61. $(5p^2 - p + 1) - (p^2 + p + 3) + (2p^2 - 3p - 4)$
62. $(2n^2 - 3n) + (n^2 + 4) - (3n - 7) + (2n + 4) - (5n^2 + n)$

Solve the following.
63. Subtract $(3x^2 - 5x + 2)$ from $(9x^2 - 3x - 7)$.
64. Subtract $(-2b^2 - 3b + 4)$ from $(b^2 - 7b + 6)$.
65. Subtract $(2y^2 + 5y + 1)$ from the sum of $(6y^2 - 7y + 3)$ and $(2y^2 + 4y - 5)$.
66. Subtract $(h^2 - h - 4)$ from the sum of $(3h^2 - 2h + 8)$, $(4h^2 - 7h - 5)$, and $(-3h^2 - 5h + 1)$.
67. From the sum of $(k^3 + 2k^2)$ and $(3k^3 + 4k^2 + k)$, subtract the sum of $(4k^3 - k^2 - k)$ and $(2k^3 + 2k^2)$.
68. Subtract the sum of $(5x^2 + 2x - 7)$ and $(x^2 - 3x + 2)$ from the sum of $(7x^2 - x + 1)$ and $(-5x^2 + 6x + 3)$.

REMOVING GROUPING SYMBOLS

2.4

Often when simplifying variable expressions, you will need to remove grouping symbols. These grouping symbols will be preceded by either a plus sign or a minus sign. Therefore, we need to examine such expressions and see what effect a plus or a minus sign has with respect to each term inside the group. First, consider a group preceded by a plus sign. The expression $+(a - b)$ can be thought of as $+1 \cdot (a - b)$. This, in turn, equals $a - b$ because of the multiplicative identity property. Thus

$$+(a - b) = a - b$$

In short, if a plus sign is in front of a group, then the grouping symbol can simply be discarded without affecting any term within the group.

> Removing a grouping symbol preceded by a plus sign does not affect any term within the group.

Example 1   Simplify $+(3x - 5)$

Solution: Make no change within the group.

$$+(3x - 5) = 3x - 5$$

Simplify $3x + (4y - 7)$ **Example 2**

Solution: Make no changes within the group.

$$3x + (4y - 7) = 3x + 4y - 7$$

Simplify $(2x^2 + x) + (3y^2 - y)$ **Example 3**

Solution: If no sign precedes a group, we treat it as a plus sign.

$$(2x^2 + x) + (3y^2 - y) = 2x^2 + x + 3y^2 - y$$

Next, consider a group preceded by a minus sign, such as $-(-a + b)$. This expression can be thought of as $-1 \cdot (-a + b)$ and can be simplified using the distributive property as follows:

$$\begin{aligned} -1 \cdot (-a + b) &= (-1)(-a) + (-1)(b) \\ &= +a + (-b) \\ &= +a - b \end{aligned}$$

Thus,

$$-(-a + b) = +a - b$$

Notice, the sign of each term has been changed. This leads to the following generalization:

To remove a grouping symbol preceded by a minus sign, change the sign of each term within the group.

Simplify $-(-x - 4)$ **Example 4**

Solution: Change the sign of each term.

$$-(-x - 4) = +x + 4 \qquad \text{or} \qquad x + 4$$

These expressions have been simplified.   **Example 5**

(a) $-(y + 5) = y - 5$
(b) $3x - (2y + z) = 3x - 2y - z$

Some expressions have groups within groups. When this occurs, it is easiest to remove the innermost grouping symbols first.

Simplify $3a - [2b - (c + 5)]$ **Example 6**

Solution: Remove the innermost grouping symbols first.

$$3a - [2b - (c + 5)] = 3a - [2b - c - 5]$$
$$= 3a - 2b + c + 5$$

Example 7

Simplify $2x - \{3y + [2z - (-w + 2)]\}$

Solution: Remove the innermost grouping symbols first and progress to the outside.

$$2x - \{3y + [2z - (-w + 2)]\} = 2x - \{3y + [2z + w - 2]\}$$
$$= 2x - \{3y + 2z + w - 2\}$$
$$= 2x - 3y - 2z - w + 2$$

After removing grouping symbols, some expressions can be further simplified by combining like terms.

Example 8

Simplify $(5x + 2) + (3x - 8)$

Solution: Remove grouping symbols; then combine like terms.

$$(5x + 2) + (3x - 8) = 5x + 2 + 3x - 8$$
$$= 8x - 6$$

Example 9

Simplify $(3x - 7) - (4x + 3)$

Solution: Remove grouping symbols, then combine like terms. Remember, the minus sign will change the sign of each term within the group.

$$(3x - 7) - (4x + 3) = 3x - 7 - 4x - 3$$
$$= -1x - 10$$
$$= -x - 10$$

Example 10

Simplify $3x - \{8x - (4 + 2x)\}$

Solution: Remove grouping symbols, beginning on the inside; then combine like terms.

$$3x - \{8x - (4 + 2x)\} = 3x - \{8x - 4 - 2x\}$$
$$= 3x - 8x + 4 + 2x$$
$$= -3x + 4$$

Exercise 2.4

Remove grouping symbols from each of the following expressions.

1. $+(3x - y)$ 2. $-(3x - y)$
3. $+(2a + 7b)$ 4. $-(7m + 1)$
5. $-(-2y + 3)$ 6. $+(7a - 3b + c)$

7. $-(6a - 2b + 3c)$

8. $7a + (2b + 3)$

9. $7a - (2b + 3)$

10. $5m - (3n + 4)$

11. $(6a - 2b) + (7c - d)$

12. $(2a + 3b) - (c + 2d)$

13. $(x + y) + (z - w)$

14. $(2x - 3y) - (4z + 7w)$

15. $6x - [5y - (3z + 7)]$

16. $8x - [4y - (2z + 3)]$

17. $3x - \{4b - [3c + (d - 3) - 3e]\}$

18. $(5a - b) + [6c - (4d - 5e) + 7f]$

Remove grouping symbols and simplify by combining like terms.

19. $2x + (4x + 3)$

20. $2x^3 - (4x^3 + 3)$

21. $(5a^2 + b) - a^2$

22. $(6m - 2n) + 3n$

23. $3x - (5x - 6)$

24. $2x - (7x - 2)$

25. $(3a^2 + 4) + (5a^2 - 6)$

26. $(5m^3 - 1) + (2m^3 + 4)$

27. $(a - 5) - (a + 6)$

28. $(7x - 6) - (8x + 2)$

29. $(6a + b) - (7a + b)$

30. $(m - 2n) - (3m + n)$

31. $(6a - 2) - (2a + 1)$

32. $-[3a - (2b - a)]$

33. $-[5x + (3x - 1)]$

34. $7x - [8x + (x - 4)]$

35. $8a - [3a - (5 - 2a)]$

36. $7m - [3m + (m - 2)]$

37. $4y - [2y - (6 - 5y)]$

38. $4a - [2a - (3a - 4)]$

39. $(4y - 1) - [2y - (8 + 7y)]$

40. $3x - \{2x + [2z - (3z + 2x)]\}$

41. $(p^2 - 2) - (p^2 - 3) + (p^2 + 4)$

42. $5h - [4h - (2h + 1) - 3]$

43. $(x - 2y) - (3x - 2y)$

44. $-[2a - (3a + 4) - 1] + 2$

45. $2m - \{5m + [3n - (2m + 4n)] - 3\} + 2$

46. $[2a - (3 + 4b)] - [5a + (2 - 3b)]$

47. $-(p + (2q - 3) - (p + 3q)] - (p + q)$

48. $[3r^2 + (2s^2 - 1)] - [r^2 - (3s^2 + 7)] - [4r^2 - (s^2 + 3)]$

49. $(2t - 3) - [3t + (t - 1) - 6t] + [7t - (t - 2)]$

50. $3x^2 - [x + (2x^2 - 3) + x^2] - [5x^2 - (x + 2)]$

Simplify by applying the distributive properties of multiplication and combining like terms.

Simplify $2(x - y) - (3x - y)$ **Example**

Solution: $2(x - y) - (3x - y) = 2x - 2y - 3x + y$

$$= -x - y$$

51. $4(a - b) - (2a - b)$

52. $5(2m - n) - (3m + 2n)$

53. $2(3p + 2q) - (4p - 3q)$

54. $-(t + 2s) + 2(t - 5s)$
55. $-(3m - 2n) + 4(m - 3n) + 2(3m + n)$
56. $-(x^2 + 3x + 4) + 2(3x^2 - x - 5)$
57. $-3[a^2 - (a^2 + a - 5)]$
58. $-2(x^2 + 7) - (x^2 - 1) + (5x^2 + 4)$
59. $3(m - 3n) + 2(3m - n)$
60. $3(a + 2b) - 4(a - 3b)$
61. $5(x - 2y + 3) - 4(x + 3y - 1) - (3x + y - 1)$
62. $2[x - 3(y + x) + 3x]$
63. $-3[x - (2y + 1) - (3x + y) + 2(x + 3y)]$
64. $2(m^2 + m - 1) - 3(2m^2 - m + 2) + (m^2 - m + 3)$
65. $3\{x^2 - [x + 2(x^2 + 3x - 1)] + (2x - 1)\}$

THE MULTIPLICATION-DIVISION
PROPERTY OF EQUATIONS

2.5

equation   An **equation** is a symbolic sentence stating an equality relationship between two expressions. Every equation has three parts: a left member, an equal sign, and a right member.

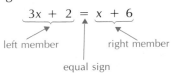

solution   A **solution** (or root) of an equation is any value from the variable's domain that makes the equation true. In this text we assume, unless specified otherwise, that the domain is the set of real numbers. To determine whether a given number is a solution, we substitute it for the variable and see if the resulting equation is true.

Example 1   In each of the following, the given number has been substituted in the equation to determine if it is a solution.
(a) -5 is a solution of $3x = -15$ because $3(-5) = -15$ is true.
(b) 0 is a solution of $6y + 2 = 2$ because $6(0) + 2 = 2$ is true. Both members equal 2.
(c) -1 is not a solution of $4m = 12$ since $4(-1) = 12$ is false, $-4 \neq 12$.
(d)   Both 2 and -2 are solutions of $x^2 = 4$ because $2^2 = 4$ and $(-2)^2 = 4$ are each true.

equivalent   Equations are said to be **equivalent** when they have exactly the same solutions.

Example 2   These equations are equivalent because each has exactly the same solution, 2.

$$2x + 1 = 5$$
$$2x = 4$$
$$x - 2 = 0$$

These equations are not equivalent because their solutions differ.   **Example 3**

$2y = 6$ [solution is 3]

$5y = 10$ [solution is 2]

$y^2 = 9$ [solutions are 3 and -3]

An important property that will help us simplify certain equations is the multiplication-division property of equivalent equations.

The Multiplication-Division Property of Equivalent Equations

If the same number or expression, other than zero, is multiplied times (or divided into) both sides of an equation, then the resulting equation is equivalent to the original.

As shown by the next two examples, this property can be used to change certain equations into a simpler equivalent form.

Change the equation $\frac{1}{2}x = 5$ into the simpler equivalent equation   **Example 4**
$x = 10$.

Solution: Use the multiplication–division property to multiply both members by 2 $\left(\text{the reciprocal of } \frac{1}{2}\right)$.

$$\frac{1}{2}x = 5$$

$$2\left(\frac{1}{2}x\right) = 2(5) \qquad \text{[multiply by 2]}$$

$$1x = 10$$

$$x = 10$$

Change the equation $-4m = 12$ into the simpler equivalent equation   **Example 5**
$m = -3$.

Solution: Use the multiplication-division property to divide both members by -4.

$$-4m = 12$$

$$\frac{-4m}{-4} = \frac{12}{-4} \qquad \text{[divide by } -4]$$

$$1m = -3$$
$$m = -3$$

solve an equation

To **solve an equation** means to find all of its solutions (or roots). In this section we use the multiplication-division property to solve equations of the form

$$ax = b$$

where a, b are real constants and x is the variable. This is done by following this procedure:

(1) If a (the numerical coefficient of the variable) is a fraction, multiply both members of the equation by the reciprocal of a.
(2) If a is not a fraction, divide both members of the equation by a.
(3) Be sure to check each solution by substituting it for the variable in the original equation.

Example 6

Solve $\frac{1}{5}x = 4$

Solution: Since the numerical coefficient of x is a fraction, multiply both members by its reciprocal, 5.

$$\frac{1}{5}x = 4$$

$$5\left(\frac{1}{5}x\right) = 5(4) \qquad \text{[multiply by 5]}$$

$$1x = 20$$

$$x = 20$$

The solution is 20. Check by replacing it for x in the original equation as follows:

$$\frac{1}{5}x = 4$$

$$\frac{1}{5}(20) = 4$$

$$4 = 4 \qquad \text{[true]}$$

Solve $\frac{y}{8} = -2$

Example 7

Solution: $\frac{y}{8}$ means $\frac{1}{8}y$. Multiply both members by 8.

$$\frac{y}{8} = -2$$

$$\frac{1}{8}y = -2$$

$$8\left(\frac{1}{8}y\right) = 8(-2)$$

$$y = -16$$

The solution is -16. You may check this answer by substituting into the original equation.

Solve $3x = -12$. **Example 8**

Solution: Since the coefficient of x is not a fraction, divide both members by 3.

$$3x = -12$$

$$\frac{3x}{3} = \frac{-12}{3} \qquad \text{[divide by 3]}$$

$$1x = -4$$

$$x = -4$$

The solution is -4. Check by substituting into the original equation.

Solve $-n = 7$. **Example 9**

Solution: $-n$ means $-1 \cdot n$. Divide both members by -1.

$$-n = 7$$

$$-1n = 7$$

$$\frac{-1n}{-1} = \frac{7}{-1} \qquad \text{[divide by } -1\text{]}$$

$$n = -7$$

Or, you may choose to multiply both members of the equation by -1 as follows:

$$-n = 7$$

$$(-1)(-n) = (-1)(7) \qquad \text{[multiply by } -1\text{]}$$

$$n = -7$$

Either method shows the solution to be -7. Be sure to check this answer.

As illustrated by the final example, it is often necessary to combine like terms before applying the multiplication-division property.

Solve $4m - 6m = 1$. **Example 10**

Solution: Combine like terms, then divide both members by the coefficient of *m*.

$$4m - 6m = 1$$
$$-2m = 1$$
$$\frac{-2m}{-2} = \frac{1}{-2}$$
$$m = -\frac{1}{2}$$

The solution is $-\frac{1}{2}$. Check by replacing it for *m* in the given equation.

$$4m - 6m = 1$$
$$4\left(-\frac{1}{2}\right) - 6\left(-\frac{1}{2}\right) = 1$$
$$-2 + 3 = 1$$
$$1 = 1 \qquad \text{[true]}$$

Exercise 2.5   Solve each equation and check your answer.

1. $\frac{1}{3}x = 2$

2. $\frac{1}{6}y = 1$

3. $\frac{1}{4}n = 2$

4. $5m = 10$

5. $6n = 18$

6. $8a = 24$

7. $-\frac{1}{2}x = 5$

8. $\frac{m}{3} = -4$

9. $\frac{k}{-5} = 6$

10. $-3x = 24$

11. $-8y = 56$

12. $7t = -49$

13. $-y = 4$

14. $-x = -3$

15. $3n = 0$

16. $\frac{-p}{7} = 1$

17. $\frac{-r}{2} = 0$

18. $\frac{k}{4} = 1$

19. $2x = 1$

20. $3y = 1$

21. $5k = -1$

22. $4m = 2$

23. $3a = -2$

24. $-6b = -3$

25. $2.1x = 4.2$

26. $0.2h = 4$

27. $0.3k = 2.1$

28. $-1.5x = 4.5$

29. $3x + 4x = 14$ **30.** $5x + x = 12$
31. $7m - 2m = -10$ **32.** $2r - 3r = 5$
33. $17r - 5r = 0$ **34.** $15a - 19a = 16$
35. $12y - 18y = -6$ **36.** $8p - 10p = 18$
37. $4z + 5z = -72$ **38.** $11p - p = 20$
39. $-3x - 2x = 10$ **40.** $-7y - 8y = +5$
41. $-n - n = 6$ **42.** $3a - 4a = 0$
43. $-y + 7y + 2y = 16$ **44.** $3x - 5x + 7x = -25$
45. $m - 5m + 2m = 4$ **46.** $2a + 3a - 6a = -13$
47. $-3p + 7p - 6p = 1$ **48.** $13x - 17x + x = -12$
49. $-1.2y + 3.4y = -4.4$ **50.** $1.3n - n + 2.4n = 5.4$
51. $0.12x - 0.15x + 0.43x = 0.2$
52. $m - 0.7m - 0.2m = 1.6$

Calculator problems.

53. $2.065x = 9.013$ **54.** $23.9m = -82.6$

55. $\dfrac{y}{2.31} = 9.64$ **56.** $\dfrac{a}{0.603} = -2.15$

57. $-1.634a = 9.827$ **58.** $-5.613p = -2.054$

Write an equation for each problem and solve.
59. If a number is divided by 3, the result is 15. Find the number.
60. If a number is divided by -4, the result is -5. Find the number.
61. If a number is multiplied by 3, the result is -18. Find the number.
62. If twice a number is added to three times the same number, the result is -35. Find the number.

THE ADDITION-SUBTRACTION PROPERTY OF EQUATIONS

2.6

In this section we continue our study of equations by learning to solve the form

$$ax + b = c$$

where a, b are real constants and x is the variable. To do this, we use the following additon-subtraction property.

The Addition-Subtraction Property of Equivalent Equations

If the same number or expression is added to (or subtracted from) both sides of an equation, then the resulting equation is equivalent to the original.

As illustrated by the next two examples, the addition-subtraction property is used to simplify equations containing certain sums or differences.

Example 1

Change the equation $x - 3 = 2$ into the simpler equivalent equation, $x = 5$.

Solution: Use the addition-subtraction property to add 3 to both sides.

$$x - 3 = 2$$
$$x - 3 + 3 = 2 + 3 \qquad \text{[add 3]}$$
$$x + 0 = 5$$
$$x = 5$$

Example 2

Change the equation $2m + 5 = 9$ into the simpler equivalent form $2m = 4$.

Solution: Either add -5 to both sides or subtract 5 from both sides.

$$2m + 5 = 9$$
$$2m + 5 + (-5) = 9 + (-5) \qquad \text{[add} -5\text{]}$$
$$2m + 0 = 4$$
$$2m = 4$$

To solve equations of the form $ax + b = c$, we go through a series of steps using the addition-subtraction property first and then the multiplication-division property to obtain an equivalent form of

$$x = \text{a real number}$$

Example 3

Solve $x + 4 = 1$.

Solution: Add -4 to both members of the equation.

$$x + 4 = 1$$
$$x + 4 + (-4) = 1 + (-4) \qquad \text{[add} -4\text{]}$$
$$x + 0 = -3$$
$$x = -3$$

The solution is -3. Check this answer by substituting into the given equation.

Example 4

Solve $3m - 2 = 10$.

Solution: Add 2 to both sides, then divide by the coefficient of m.

$$3m - 2 = 10$$
$$3m - 2 + 2 = 10 + 2 \qquad \text{[add 2]}$$

$$3m = 12$$

$$\frac{3m}{3} = \frac{12}{3} \qquad \text{[divide by 3]}$$

$$m = 4$$

The solution is 4. Be sure to check it in the original equation.

Solve $\dfrac{n}{2} + 5 = 1$.

Example 5

Solution: Add -5 to both sides.

$$\frac{n}{2} + 5 = 1$$

$$\frac{n}{2} + 5 + (-5) = 1 + (-5)$$

$$\frac{n}{2} = -4$$

Multiply both sides by 2.

$$2 \cdot \frac{n}{2} = 2 \cdot (-4)$$

$$n = -8$$

The solution is -8. Check the answer.

As illustrated by the following two examples, it is often necessary to perform indicated multiplications and combine like terms before applying the properties of equations.

Solve $2(3y - 1) = 1$

Example 6

Solution: Multiply out the left member using the distributive property.

$$2(3y - 1) = 1$$

$$6y - 2 = 1$$

Add 2 to both members of the equation.

$$6y - 2 + 2 = 1 + 2$$

$$6y = 3$$

Divide both sides by 6.

$$\frac{6y}{6} = \frac{3}{6}$$

$$y = \frac{1}{2}$$

The solution is $\dfrac{1}{2}$. Check the answer.

Example 7   Solve $3(2m + 5) - 2(m - 3) = 1$

> *Solution:* Multiply out the left member using the distributive property.
>
> $$3(2m + 5) - 2(m - 3) = 1$$
> $$6m + 15 - 2m + 6 = 1$$
>
> Combine like terms.
>
> $$4m + 21 = 1$$
>
> Add -21 to both sides.
>
> $$4m + 21 + (-21) = 1 + (-21)$$
> $$4m = -20$$
>
> Divide both sides by 4.
>
> $$\frac{4m}{4} = \frac{-20}{4}$$
> $$m = -5$$
>
> The solution is -5. Be sure to check the answer.

Exercise 2.6   Solve these equations using the addition-subtraction property of equivalent equations. Check each solution.

1. $x - 2 = 5$
2. $y - 4 = 9$
3. $m + 2 = 6$
4. $p - 10 = 1$
5. $n - 6 = 0$
6. $x + 1 = 0$
7. $k + 12 = 10$
8. $h + 15 = 5$
9. $x + 17 = -3$
10. $n - 7 = -1$
11. $p + 1 = -3$
12. $m + 7 = 7$
13. $k - 1 = 1$
14. $3 + x = 3$
15. $5 + n = 2$
16. $6 + m = -3$
17. $-1 + y = 2$
18. $-17 + x = 14$
19. $y - 2.3 = 5.8$
20. $x + 6.1 = 5.8$
21. $p - 12.7 = 9.3$
22. $k + 0.62 = 1$
23. $n - 1 = 0.024$
24. $6.23 + x = 8.96$

Solve these equations using the addition-subtraction property together with the multiplication-division property. Check each solution.

25. $2y + 1 = 5$
26. $3x - 4 = 8$
27. $5m + 1 = 11$
28. $3y + 2 = -1$
29. $7x + 13 = -1$
30. $2m - 1 = 1$
31. $3n + 4 = 4$
32. $-y + 6 = 1$
33. $4 - x = 7$
34. $7 - 2p = 1$
35. $9t - 7 = 11$
36. $1 - y = 1$
37. $3p - 2 = 4$
38. $-4n + 5 = -7$

39. $-6h + 5 = -1$

40. $7x - 6 = -20$

41. $-3 - 4k = 5$

42. $2m - 1 = 0$

43. $4n - 1 = 1$

44. $10y + 7 = 8$

45. $13k + 17 = -35$

46. $3x + 7 = 1$

47. $9h + 7 = -11$

48. $8 - 2p = 0$

49. $7 - 5x = -23$

50. $8 - y = 12$

51. $13 - r = 0$

52. $13p + 40 = -12$

53. $2x + 1.2 = 3.6$

54. $3y - 2.9 = 4.3$

55. $2.2m - 8.6 = 4.6$

56. $3.1y - 1.21 = 2.2$

57. $1.5x + 2.1 = 5.1$

Use the distributive properties of multiplication and the concept of combining like terms to help solve these equations. Be sure to check each solution.

58. $2(x + 1) = 8$

59. $3(y - 2) = -18$

60. $4(m - 1) = 0$

61. $5(n + 1) = 5$

62. $-(3m + 4) = 5$

63. $3(5 - x) = -6$

64. $3x - 2x = 4$

65. $4p - 5p = 3$

66. $y - 10y + 2 = -79$

67. $n + 3n - 5 = 11$

68. $2k + 1 - 5k + 2 = 12$

69. $3(p - 1) - p = 1$

70. $5(x + 1) - 4x = 7$

71. $3(y - 1) - 2(y + 3) = 6$

72. $2(5m - 1) - (3m - 4) = -33$

73. $(p + 9) - 2(p + 5) = 1$

Write an equation for each problem and solve.

74. Three times the sum of a number and 5 is -6. Find the number.

75. Twice the sum of a number and 7 is 20. Find the number.

76. If twice a number is subtracted from 6, the result is zero. Find the number.

77. If five times a number is added to three times the same number, the result is 56. Find the number.

78. If four times a number is subtracted from three times the same number, the result is 10. Find the number.

79. The sum of ten times a number and three times the same number is reduced by 5. The result is 47. Find the number.

SOLVING LINEAR EQUATIONS

2.7

In this section we use the properties of equivalent equations to form a technique for solving linear equations of one variable. A **linear (or first degree) equation in one variable** has an equivalent form fitting the pattern,

linear equation

$$ax + b = c$$

first degree equation

where a, b, and c represent real constants and $a \neq 0$. Linear equations are also said to be **first degree** because the variable has an exponent of 1.

Example 1

Each of these equations is linear (or first degree).

(a) $2x + 3 = 7$

(b) $5y - 1 = 9$

(c) $5(m + 2) = 3m + 2$

Example 2

None of these equations is linear (or first degree).

(a) $x^2 = 4$

(b) $y^2 + 5y + 6 = 0$

(c) $3m^3 = 24$

To solve linear equations, we combine the methods studied in preceding sections and follow this procedure.

Solving Linear Equations

Step (1)   On each side of the equation, perform indicated multiplications and, if possible, combine like terms.

Step (2)   Use the addition-subtraction property, if necessary, to obtain an equivalent form where the variable is on one side and the number is on the other side.

Step (3)   Use the multiplication-division property, if necessary, to obtain an equivalent equation of the form x = a number.

Step (4)   Check the solution by replacing it for the variable in the original equation.

Example 3

Solve $5(x - 2) - 2x = 2$

Solution:

Step (1)   Multiply and combine like terms.

$$5(x - 2) - 2x = 2$$
$$5x - 10 - 2x = 2$$
$$3x - 10 = 2$$

Step (2)   Use the addition-subtraction property to add 10 to both sides.

$$3x - 10 + 10 = 2 + 10$$
$$3x = 12$$

Step (3) Use the multiplication-division property to divide both sides by 3.

$$\frac{3x}{3} = \frac{12}{3}$$

$$x = 4$$

Step (4) Check the solution. Replace x with 4 in the original equation.

$$5(x - 2) - 2x = 2$$
$$5(4 - 2) - 2(4) = 2$$
$$5 \cdot 2 - 2 \cdot 4 = 2$$
$$10 - 8 = 2$$
$$2 = 2 \quad \text{[true]}$$

Thus, the solution is 4.

Solve $4m - 3 + 2m = 3m - 9$.   **Example 4**

Solution:

Step (1) Combine like terms.

$$4m - 3 + 2m = 3m - 9$$
$$6m - 3 = 3m - 9$$

Step (2) Use the addition-subtraction property to obtain the variable on one side and the number on the other. We add 3 to both sides, then subtract $3m$ from both sides.

$$6m - 3 + 3 = 3m - 9 + 3 \quad \text{[add 3]}$$
$$6m = 3m - 6$$
$$6m - 3m = 3m - 6 - 3m \quad \text{[subtract } 3m\text{]}$$
$$3m = -6$$

Step (3) Use the multiplication-division property to divide both sides by 3.

$$\frac{3m}{3} = \frac{-6}{3}$$

$$m = -2$$

Step (4) Substitute -2 for m in the original equation.

$$4m - 3 + 2m = 3m - 9$$
$$4(-2) - 3 + 2(-2) = 3(-2) - 9$$
$$-8 - 3 - 4 = -6 - 9$$
$$-15 = -15 \quad \text{[true]}$$

Thus, the solution is -2.

Example 5

Solve $5y - (6 - 3y) + 1 = 2y + 13$

Solution:

Step (1) Remove grouping symbols and combine like terms.

$$5y - (6 - 3y) + 1 = 2y + 13$$
$$5y - 6 + 3y + 1 = 2y + 13$$
$$8y - 5 = 2y + 13$$

Step (2) Add 5 to both sides and subtract 2y from both sides.

$$8y - 5 + 5 = 2y + 13 + 5 \quad \text{[add 5]}$$
$$8y = 2y + 18$$
$$8y - 2y = 2y + 18 - 2y \quad \text{[subtract 2y]}$$
$$6y = 18$$

Step (3) Divide both sides by 6.

$$\frac{6y}{6} = \frac{18}{6}$$
$$y = 3$$

Step (4) Check by replacing y with 3 in the original equation.

$$5y - (6 - 3y) + 1 = 2y + 13$$
$$5 \cdot 3 - (6 - 3 \cdot 3) + 1 = 2 \cdot 3 + 13$$
$$5 \cdot 3 - (6 - 9) + 1 = 2 \cdot 3 + 13$$
$$15 - (-3) + 1 = 6 + 13$$
$$15 + 3 + 1 = 6 + 13$$
$$19 = 19 \quad \text{[true]}$$

The solution is 3.

With experience, your work can be condensed as illustrated by the following examples.

Example 6

Solve $3(n - 5) - 4 = 2(3n + 1)$

Solution:

$$3(n - 5) - 4 = 2(3n + 1)$$
$$3n - 15 - 4 = 6n + 2 \quad \text{[multiply]}$$
$$3n - 19 = 6n + 2 \quad \text{[combine like terms]}$$
$$3n - 19 + 19 = 6n + 2 + 19 \quad \text{[add 19]}$$
$$3n = 6n + 21$$
$$3n - 6n = 6n + 21 - 6n \quad \text{[subtract 6n]}$$
$$-3n = 21$$
$$\frac{-3n}{-3} = \frac{21}{-3} \quad \text{[divide by -3]}$$
$$n = -7$$

The solution is -7. Be sure to check it in the original equation.

Solve $p - (10p - 3) = 4(5 - 2p)$

Example 7

Solution:

$$p - (10p - 3) = 4(5 - 2p)$$
$$p - 10p + 3 = 20 - 8p \qquad \text{[simplify]}$$
$$-9p + 3 = 20 - 8p \qquad \text{[combine like terms]}$$
$$-9p + 3 - 3 = 20 - 8p - 3 \qquad \text{[subtract 3]}$$
$$-9p = 17 - 8p$$
$$-9p + 8p = 17 - 8p + 8p \qquad \text{[add 8p]}$$
$$-p = 17$$
$$\frac{-p}{-1} = \frac{17}{-1} \qquad \text{[divide by } -1\text{]}$$
$$p = -17$$

The solution is -17. Check it in the original equation.

Solve these equations and check each solution.

Exercise 2.7

1. $-5x = 50$ **2.** $-3y = -9$

3. $12n = 48$ **4.** $2m = 2$

5. $11t = 55$ **6.** $3r = 0$

7. $-y = 7$ **8.** $-x = -5$

9. $-k = 0$ **10.** $-2p = 0$

11. $\frac{1}{2}x = -3$ **12.** $\frac{1}{10}y = 3$

13. $\frac{k}{4} = 1$ **14.** $\frac{-n}{6} = 5$

15. $x - 2 = 6$ **16.** $y + 4 = 7$

17. $m - 3 = 1$ **18.** $n + 4 = 4$

19. $k - 1 = 0$ **20.** $2x + 1 = 0$

21. $3 - p = 0$ **22.** $4 - k = 7$

23. $2y + 1 = 5$ **24.** $3y - 4 = 8$

25. $-5m + 1 = -9$ **26.** $2n + 8 = 0$

27. $4m - 3 = 6m + 5$ **28.** $4x + 3 = 2x + 11$

29. $2y + 3 + y = y + 3$ **30.** $5m + 2 - 3m = m + 3$

31. $3(a + 2) = 21$ **32.** $5(x - 4) = 5$

33. $4y + 3 - y = y + 13$ **34.** $10 + 2x + 5 = x + 20$

35. $2(3b + 1) = 5(b - 1)$ **36.** $8(x + 1) = 3(x + 2) + 12$

37. $7a - 2(a - 2) = 3(a + 2)$ **38.** $2(5y - 3) = 4(2y + 3) + 4$

39. $13 + 4(2r - 3) = 2(r - 3) + 7$
40. $2(p + 3) + 3(4p - 1) = 4p + 13$
41. $5 + 2(a - 1) = 7 + a$
42. $10(r + 1) = 2(3r + 7)$
43. $-4(t - 8) + 3(2t + 1) = 7$
44. $x - (2x + 1) = 8 - 3(x + 1)$
45. $-4 - 3(2y + 1) = -1$
46. $6(3k + 2) = 5(4k + 3) + 1$
47. $5(2t - 1) - 7 = 4(2t + 1)$
48. $2s - 9 + 2(3s + 2) = 4 + 3(s - 3)$
49. $3z - 6 = 2z + 2(3z + 2) + 5$
50. $6(2h - 8) - 3(5h - 6) = -30$
51. $2[3(x - 2) - 2x] = 7x - 2$
52. $2(4a - 3) - 7 = 17a - 7(a + 2)$
53. $3(5y - 1) + 28 = 5(-3y - 6) + 2y - 1$
54. $2(k - 1) - (3k + 2) = 5k - 2(7k + 1) + 2$
55. $2[7 - (x + 1)] - (3 - x) = 3(7x - 4) - 1$

Calculator problems.
56. $3.45x - 7.21 = 1.06x + 5.86$
57. $16.3 - 8.9x = 2.7 + 1.3x$
58. $1.021p - 3.068 - 0.259p = 6.704 - 3.658p$
59. $2.1(m - 0.1) = 3.2(0.3 - m)$
60. $1.3(2t + 1) = 2.7(3t - 1)$

USING EQUATIONS TO SOLVE WORD PROBLEMS

2.8

Equations ease the difficulty of solving word problems. This is not to say that solving word problems is easy. Most students find that success with word problems comes after much practice. Therefore, don't be discouraged if you do not completely understand each of the following problems. You will obtain more practice as you progress through the text.

In section 1.3 we found that variable expressions can be used as mathematical models to describe key word phrases. As a review, tables 1.1 through 1.5 are repeated here. We use x to represent the unknown number, but it is correct to use any letter you wish.

TABLE 1.1   Addition

Word Phrase	Mathematical Model
the sum of a number and 6	$x + 6$
a number plus 4	$x + 4$
a number added to 10	$x + 10$
8 more than a number	$x + 8$
a number increased by 5	$x + 5$
the total of two numbers	$x + y$

TABLE 1.2   Subtraction

Word Phrase	Mathematical Model
5 subtracted from a number	$x - 5$
8 less than a number	$x - 8$
10 minus a number	$10 - x$
a number minus 10	$x - 10$
7 taken away from a number	$x - 7$
2 fewer than a number	$x - 2$

TABLE 1.3   Multiplication

Word Phrase	Mathematical Model
8 times a number	$8x$
a number multiplied by 6	$6x$
the product of a number and 4	$4x$
twice an amount	$2x$
a number is tripled	$3x$
a number is quadrupled	$4x$
half of a number	$\frac{1}{2}x$

TABLE 1.4   Division

Word Phrase	Mathematical Model
a number divided by 7	$\frac{x}{7}$ or $x \div 7$
10 divided by a number	$\frac{10}{x}$
half of a number	$\frac{x}{2}$
the ratio of some number to 6	$\frac{x}{6}$

TABLE 1.5   Combinations

Word Phrase	Mathematical Model
5 more than twice a number	$2x + 5$
3 less than four times a number	$4x - 3$
the ratio of twice a number to 10	$\dfrac{2x}{10}$
a number added to 5 times itself	$5x + x$
the number of cents in x dimes	$10x$
the number of cents in x quarters	$25x$
the sum of two consecutive integers	$x + (x + 1)$

To solve word problems we will translate word phrases into variable expressions. Then the variable expressions will be related by an equation. It is helpful to follow these steps.

Step (1)   Let some letter represent one of the unknown numbers.
Step (2)   Translate the word phrases into variable expressions and form an equation. (Consult Tables 1.1−1.5.)
Step (3)   Solve the equation.
Step (4)   Interpret the solution. You must watch this step closely. There are times when the solution to the equation is not the answer to the written problem. Something else may have to be done.

Example 1

Answers to word problems should be checked using the wording of the given problem.

If 4 times the sum of a certain number and 5 is decreased by 32, the result is twice the number. Find the number.

Solution:

Step (1)   Let x represent the unknown number.
Step (2)   Translate the word phrases into variable expressions and form an equation.

4 times the sum of a certain number and 5	decreased by	32	the result is	twice the number
↓	↓	↓	↓	↓
$4(x + 5)$	$-$	32	$=$	$2x$

Step (3)   Solve the equation.

$$4(x + 5) - 32 = 2x$$
$$4x + 20 - 32 = 2x$$
$$4x - 12 = 2x$$
$$4x - 12 + 12 = 2x + 12 \qquad \text{[add 12]}$$

$$4x = 2x + 12$$
$$4x - 2x = 2x + 12 - 2x \quad \text{[subtract 2x]}$$
$$2x = 12$$
$$\frac{2x}{2} = \frac{12}{2} \quad \text{[divide by 2]}$$
$$x = 6$$

Step (4) Interpret the solution. The unknown number is 6. To check, we go back to the original problem and substitute 6 for the phrase "certain number." This gives the following statement: "If 4 times the sum of 6 and 5 is decreased by 32, the result is twice 6." Thus,

$$4 \cdot (6 + 5) - 32 = 2 \cdot 6$$
$$4 \cdot 11 - 32 = 2 \cdot 6$$
$$44 - 32 = 12$$
$$12 = 12 \quad \text{[true]}$$

The sum of two numbers is 39. The larger number is 3 less than twice the smaller number. Find both numbers.

Example 2

Solution:

Step (1) Let x represent the smaller number.

Step (2) Translate the word phrases into variable expressions and form an equation.

smaller number	the sum of two numbers	larger number is 3 less than twice the smaller	is	39
↓	↓	↓	↓	↓
x	+	2x − 3	=	39

Step (3) Solve the equation.

$$x + 2x - 3 = 39$$
$$3x - 3 = 39$$
$$3x - 3 + 3 = 39 + 3 \quad \text{[add 3]}$$
$$3x = 42$$
$$\frac{3x}{3} = \frac{42}{3} \quad \text{[divide by 3]}$$
$$x = 14$$

Step (4) Interpret the solution. x represents the smaller number, which is 14. 2x − 3 represents the larger number, which is 2 · 14 − 3, or 25. To check, we substitute 14 for the smaller number and 25 for the larger. This produces the following statements: "The sum of 14 and 25 is 39." "25 is 3 less than twice 14." Thus,

$$14 + 25 = 39 \quad \text{[true]}$$

and $25 = 2 \cdot 14 - 3$

$25 = 28 - 3$

$25 = 25$ [true]

Example 3

A wire 22 inches long is cut into two parts. One piece is 6 inches longer than the other. Find the length of each piece.

Solution:

Step (1)   Let x represent the length of the shorter piece. Then $x + 6$ represents the length of the longer piece.

Step (2)   Translate the word phrases into variable expressions and form an equation. Sometimes it is helpful to draw a diagram.

22 inches

x $x + 6$

$x + (x + 6) = 22$

Step (3)   Solve the equation.

$$x + (x + 6) = 22$$
$$x + x + 6 = 22$$
$$2x + 6 = 22$$
$$2x + 6 - 6 = 22 - 6 \qquad \text{[subtract 6]}$$
$$2x = 16$$
$$\frac{2x}{2} = \frac{16}{2} \qquad \text{[divide by 2]}$$
$$x = 8$$

Step (4)   Interpret the solution. x represents the smaller piece, which is 8 inches. $x + 6$ represents the larger piece, which is $8 + 6$, or 14 inches. Be sure to check these answers in the original problem.

Example 4

A collection of nickels and dimes has a value of $7.00. How many nickels and dimes are in the collection if there are 40 more dimes than nickels?

Solution:

Step (1)   Let x represent the number of nickels. Since there are 40 more dimes than nickels, then $x + 40$ would represent the number of dimes.

Step (2)   Form an equation. The value of nickels plus the value of dimes must result in 7 dollars, or 700 cents. The value of x nickels is $5x$ cents. The value of $x + 40$ dimes, then, is $10(x + 40)$ cents.

Value of nickels in cents	+	Value of dimes in cents	=	Value of collection in cents

$$5x \quad + \quad 10(x + 40) \quad = \quad 700$$

Step (3) Solve the equation.

$$5x + 10(x + 40) = 700$$
$$5x + 10x + 400 = 700$$
$$15x + 400 = 700$$
$$15x + 400 - 400 = 700 - 400 \qquad \text{[subtract 400]}$$
$$15x = 300$$
$$\frac{15x}{15} = \frac{300}{15} \qquad \text{[divide by 15]}$$
$$x = 20$$

Step (4) Interpret the solution. x represents the number of nickels, which is 20. $x + 40$ represents the number of dimes, which is $20 + 40$, or 60. Check these answers in the original problem.

The perimeter (P) of a rectangle is 120 meters and the width (w) is 15 meters. Find the length (l) of the rectangle.

Example 5

Solution:

Step (1) The perimeter is the distance around the rectangle and is given by the formula

$$P = 2l + 2w$$

Step (2) We form an equation by drawing a diagram and replacing 120 for P and 15 for w.

$w = 15$ [rectangle diagram] $P = 120$

l

Thus, the formula $P = 2l + 2w$ becomes the equation

$$120 = 2l + 2(15)$$

Step (3) Solve the equation.

$$120 = 2l + 30$$
$$120 - 30 = 2l + 30 - 30 \qquad \text{[subtract 30]}$$
$$90 = 2l$$
$$\frac{90}{2} = \frac{2l}{2} \qquad \text{[divide by 2]}$$
$$45 = l$$

Step (4)   The length (*l*) of the rectangle is 45 meters. Be sure to check this answer.

Exercise 2.8

Write an expression describing each phrase. Use x as the variable.

1. The sum of a number and 3
2. -3 added to a number
3. A number plus -2
4. A number added to 16
5. 3 more than a number
6. A number increased by 7
7. A number decreased by 10
8. The total of some number and 2
9. 4 less than a number
10. 10 taken away from a number
11. 12 fewer than a number
12. Subtract a number from 15
13. 5 times a number
14. The product of a number and 7
15. Twice a number
16. Triple a number
17. A number divided by 10
18. -6 divided by a number
19. Half of a number
20. The ratio of some number to 12
21. The quotient of a number and 8
22. 6 more than twice a number
23. 5 less than six times a number
24. A number added to twice the same number
25. 10 less than four times a number
26. The sum of three consecutive integers
27. The number of cents in x nickels
28. The number of cents in x half dollars

Write an equation for each word problem and then solve. Be sure to check your answers in the original problem.

29. If twice a certain number is decreased by 3, the result is 9. Find the number.
30. If 10 is added to 3 times a number the result is 31. Find the number.
31. If 3 times the sum of a certain number and 5 is decreased by twice the number, the result is 20. Find the number.
32. If 5 times the sum of a certain number and 3 is decreased by 21, the result is twice the number. Find the number.

33. The sum of two numbers is 48. The larger number is 3 times the smaller. Find both numbers.

34. The sum of two numbers is 38. The larger number is 18 more than the smaller. Find both numbers.

35. The sum of two numbers is 46. The smaller number is 14 less than the larger. Find both numbers.

36. The sum of two numbers is 15. The larger number is 7 more than 3 times the smaller. Find both numbers.

37. The sum of two numbers is 66. The larger number is 6 more than 3 times the smaller number. Find both numbers.

38. The sum of two numbers is 45. The larger number is 5 more than 4 times the smaller number. Find both numbers.

39. A wire 20 centimeters long is cut into two pieces so that one piece is 6 centimeters longer than the other. How long is each piece?

40. A rope 50 feet long is cut into two pieces. One piece is 8 feet longer than the other. How long is each piece?

41. An electric current of 23 amperes is branched off into three circuits. The second branch carries 3 times the current of the first branch. The third branch carries 2 amperes less than the first branch. Find the amount of current carried in each branch.

42. The sum of two consecutive intergers is 43. Find both integers.

43. The sum of three consecutive integers is 31 more than the first integer. Find all three integers.

44. A wire 24 inches long is cut into two pieces. The larger piece is 4 inches longer than the smaller. Find the length of both pieces.

45. A wire 48 centimeters long is cut into three pieces. The second piece is twice the length of the first, and the third piece is 3 times the length of the first. Find the length of each piece.

46. Three resistors connected in series have a total resistance of 50 ohms. The second resistor has a resistance of 5 ohms more than the first, while the third resistor has a resistance of 12 ohms more than the first. Find the resistance of each resistor.

47. The flow capacity of a pipeline is 800 liters per minute. The pipeline separates into three branches. The second branch has 3 times the capacity of the first, while the third branch has a capacity of 100 liters per minute more than the second. Find the capacity of each branch.

48. A coin collection consisting of nickels and dimes is worth $4.00. There are 10 more dimes than nickels. Find the number of nickels and dimes.

49. A coin collection consisting of nickels, quarters, and half dollars is worth $10. There are 6 times as many nickels as quarters and 1 less half dollar than quarters. Find the number of nickels, quarters, and half dollars.

50. A collection of dimes and quarters has a value of $10.00. How many dimes and quarters are in the collection if there are 12 fewer dimes than quarters?
51. A collection of nickels and quarters is worth $5.00. How many nickels and quarters are in the collection if there are 10 more nickels than quarters?
52. In a basketball game one team made twice as many points as the other. The total number of points made by both teams was 87. How many points did each team score?
53. At a football game, the price of admission was $4.00 for regular seats and $6.00 for box seats. The total receipts were $75,600 for a total of 16,300 paid admissions. Find the number of regular seats and box seats that were sold.
54. A house and lot cost $66,000. The cost of the house is 10 times the cost of the lot. Find the cost of the house.
55. The perimeter (P) of a rectangle is 120 meters and the width (w) is 10 meters. Find the length (l) of the rectangle using the formula $P = 2l + 2w$.
56. The area (A) of a rectangle is 40 square feet, while the width (w) is 5 feet. Find the length (l) using the formula $A = l \cdot w$.
57. The area (A) of a triangle is 130 square meters. The base (b) of the triangle is 26 meters. Find the height (h) using the formula $A = \frac{1}{2}bh$.
58. The perimeter (P) of a square is 100 inches. Find the length of a side (s) using the formula $P = 4s$.
59. The perimeter of a triangle is 36 centimeters. If one side is 12 centimeters and another is 18 centimeters, find the length of the third side.
60. The perimeter of a rectangle is 84 inches. The length is 18 inches more than the width. Find the width.
61. A triangle has a perimeter of 29 feet. The second side is twice the first, while the third side is 5 feet longer than the first. Find the length of each side.
62. The circumference (C) of a circle is 31.4 square centimeters. Find its radius (r) using the formula $C = 2\pi r$. (Use 3.14 as an approximation for π.)

Summary

DEFINITION OF EXPONENTS

If n is a natural number, then $a^n = \overbrace{a \cdot a \cdot a \ldots a}^{n \text{ times}}$.
The exponent n states that the base a is used as a factor n times.

PROPERTIES OF EXPONENTS

If m, n, and p are whole numbers, then each of the following is true:
(1) Addition property of exponents: $a^m \cdot a^n = a^{m+n}$

(2)   Subtraction property of exponents: $\dfrac{a^m}{a^n} = a^{m-n}$, where $m \geq n$ and

$a \neq 0$

(3)   Zero power property: $a^0 = 1$, where $a \neq 0$
(4)   Power to a power property of exponents: $(a^m)^n = a^{m \cdot n}$
(5)   Distributive property of exponents: $(a^m b^n)^p = a^{mp} \cdot b^{np}$ and

$\left(\dfrac{a^m}{b^n}\right)^p = \dfrac{a^{mp}}{b^{np}}$ for $b \neq 0$

For all real numbers a and b,

$$a(-b) = (-a)b = -(ab)$$

and for $b \neq 0$

$$\frac{a}{-b} = \frac{-a}{b} = -\frac{a}{b}$$

THE OPPOSITE LAW OF
MULTIPLICATION

For all real numbers a and b,

$$(-a)(-b) = ab$$

and for $b \neq 0$

$$\frac{-a}{-b} = \frac{a}{b}$$

THE DOUBLE OPPOSITE
LAW OF
MULTIPLICATION

For all real numbers a, b, and c,

$$a(b - c) = ab - ac \text{ and } (b - c)a = ba - ca$$

THE DISTRIBUTIVE
PROPERTY OF
MULTIPLICATION WITH
RESPECT TO
SUBTRACTION

Like terms can be added or subtracted by combining their respective numerical coefficients.

COMBINING ALGEBRAIC
EXPRESSIONS

(1)   Removing grouping symbols preceded by a plus sign does not affect any term within the group.
(2)   To remove a grouping symbol preceded by a minus sign, change the sign of each term within the group.

REMOVING GROUPING
SYMBOLS

(1)   Multiplication-division property: If the same number or expression, other than zero, is multiplied times (or divided into) both sides of an equation, then the resulting equation is equivalent to the original.
(2)   Addition-subtraction property: If the same number or expression is added to (or subtracted from) both sides of an equation, then the resulting equation is equivalent to the original.

PROPERTIES OF
EQUIVALENT
EQUATIONS

(1)   On each side of the equation, perform indicated multiplications and, if possible, combine like terms.
(2)   Use the addition-subtraction property, if necessary, to obtain an

SOLVING LINEAR
EQUATIONS

equivalent form where the variable is on one side and the number is on the other side.

(3)   Use the multiplication-division property, if necessary, to obtain an equivalent equation of the form $x =$ a number.

(4)   Check the solution by substituting it for the variable in the original equation.

SOLVING WORD PROBLEMS

(1)   Let a letter represent one of the unknown numbers.

(2)   Translate the word phrases into variable expressions and form an equation.

(3)   Solve the equation.

(4)   Interpret the solution and check by using the wording of the original problem.

Each of the following equations illustrates one of the properties of exponents. Name the property.

1. $x^2 \cdot x^5 = x^7$

2. $(a^2)^4 = a^8$

3. $\dfrac{m^8}{m^2} = m^6$ where $m \neq 0$

4. $(3x^3y^2)^2 = 9x^6y^4$

5. $7^0 = 1$

6. $\left(\dfrac{m^4}{2}\right)^2 = \dfrac{m^8}{4}$

Evaluate the following expressions.

7. $(-10)^0$

8. $3 \cdot 9^0$

9. $3^2 \cdot 2^3 - 5^3$

10. $3(2^3 - 4^2)$

Multiply or divide as indicated.

11. $2(5a)$

12. $(-3a)(-4a)^2$

13. $(2x)(3x^4)(5x)$

14. $(2a^3b^2c)(3ab^3)(-4ab)$

15. $\dfrac{-12m^5n^2}{3m^3n^2}$ where $m, n \neq 0$

16. $5x(x + y)$

17. $3ab(2a^3 + 3b^3)$

18. $5m^2n(2mn^2 + m^3n - 3n^3)$

Combine the following expressions, using addition or subtraction.

19. $3x - x$

20. $2a^2 + 3a - a^2 + 4a$

21. $3a^2b + 2ab^2 - 5a^2b - ab^2$

22. $(2a^2 + 3a - 5) + (3a^2 - 5a - 1)$

23. $(6m^2 - m + 2) - (3m^2 + 4m - 6)$

24. $(5x^2 - 2x + 1) + (3x^2 - x + 2) - (7x^2 + 2x - 5)$

Remove grouping symbols and combine like terms.

25. $-(2a + 3b)$

26. $3x - (5x + 2)$

27. $(3a - b) - (4a + 3b)$

28. $5m - [3m + (2m - 1) - 3]$

Solve these equations.

29. $\dfrac{1}{2}x = 7$

30. $-\dfrac{1}{5}x = 10$

31. $\dfrac{p}{3} = 5$

32. $\dfrac{-n}{6} = 1$

33. $-7y = 21$

34. $7a - 9a = 4$

35. $5n - 6n = 0$

36. $x - 5 = 7$

37. $p - 1 = 4$

38. $7 - 3h = 1$

39. $3x + 5 = -10$

40. $7 - t = 0$

41. $-7y = 28 - 3y$

42. $3x + 5 = -10 - 2x$

43. $3(x + 1) = 15$

44. $5x + 3 = 6x + 5$

45. $2m + 3 = 3$

46. $2(3a - 1) = 3(a + 5) + 1$

Solve the following word problems. Give both the equation and the solution.

47. The sum of two numbers is 16. The larger number is 4 less than 3 times the smaller number. Find both numbers.

48. A rope 48 inches long is cut into two parts. One piece is 12 inches longer than the other. Find the length of each piece.

49. The sum of two consecutive integers is 65. Find both integers.

Solutions to Self-Checking Exercise

1. Addition property of exponents

2. Power to a power property of exponents

3. Subtraction property of exponents

4. Distributive property of exponents

5. Zero power property

6. Distributive property of exponents

7. $(-10)^0 = 1$

8. $3 \cdot 9^0 = 3 \cdot 1 = 3$

9. $3^2 \cdot 2^3 - 5^3 = 9 \cdot 8 - 125$
$$= 72 - 125$$
$$= -53$$

10. $3(2^3 - 4^2) = 3(8 - 16)$
$$= 3(-8)$$
$$= -24$$

11. $10a$

12. $48a^3$

13. $30x^6$

14. $-24a^5b^6c$

15. $-4m^2$

16. $5x^2 + 5xy$

17. $6a^4b + 9ab^4$

18. $10m^3n^3 + 5m^5n^2 - 15m^2n^4$

19. $2x$

20. $a^2 + 7a$

21. $-2a^2b + ab^2$

22. $5a^2 - 2a - 6$

23. $3m^2 - 5m + 8$

24. $x^2 - 5x + 8$

25. $-2a - 3b$

26. $-2x - 2$

27. $-a - 4b$

28. 4

29. $\dfrac{1}{2}x = 7$

$2 \cdot \dfrac{1}{2}x = 2 \cdot 7$

$x = 14$

30. $-\dfrac{1}{5}x = 10$

$-5\left(-\dfrac{1}{5}x\right) = -5(10)$

$x = -50$

31. $\dfrac{p}{3} = 5$

$3 \cdot \dfrac{p}{3} = 3 \cdot 5$

$p = 15$

32. $\dfrac{-n}{6} = 1$

$-6\left(\dfrac{-n}{6}\right) = -6(1)$

$n = -6$

33. $-7y = 21$

$$\frac{-7y}{-7} = \frac{21}{-7}$$

$$y = -3$$

34. $7a - 9a = 4$

$$-2a = 4$$

$$\frac{-2a}{-2} = \frac{4}{-2}$$

$$a = -2$$

35. $5n - 6n = 0$

$$-n = 0$$

$$\frac{-n}{-1} = \frac{0}{-1}$$

$$n = 0$$

36. $x - 5 = 7$

$$x - 5 + 5 = 7 + 5$$

$$x = 12$$

37. $p - 1 = 4$

$$p - 1 + 1 = 4 + 1$$

$$p = 5$$

38. $7 - 3h = 1$

$$7 - 3h - 7 = 1 - 7$$

$$-3h = -6$$

$$\frac{-3h}{-3} = \frac{-6}{-3}$$

$$h = 2$$

39. $3x + 5 = -10$

$$3x + 5 - 5 = -10 - 5$$

$$3x = -15$$

$$\frac{3x}{3} = \frac{-15}{3}$$

$$x = -5$$

40. $7 - t = 0$

$$7 - t - 7 = 0 - 7$$

$$-t = -7$$

$$\frac{-t}{-1} = \frac{-7}{-1}$$

$$t = 7$$

41. $-7y = 28 - 3y$

$$-7y + 3y = 28 - 3y + 3y$$

$$-4y = 28$$

$$\frac{-4y}{-4} = \frac{28}{-4}$$

$$y = -7$$

42. $3x + 5 = -10 - 2x$

$$3x + 5 - 5 = -10 - 2x - 5$$

$$3x = -15 - 2x$$

$$3x + 2x = 15 - 2x + 2x$$

$$5x = -15$$

$$\frac{5x}{5} = \frac{-15}{5}$$

$$x = -3$$

43. $3(x + 1) = 15$

$$3x + 3 = 15$$

$$3x + 3 - 3 = 15 - 3$$

$$3x = 12$$

$$\frac{3x}{3} = \frac{12}{3}$$

$$x = 4$$

44. $5x + 3 = 6x + 5$

$$5x + 3 - 3 = 6x + 5 - 3$$

$$5x = 6x + 2$$

$$5x - 6x = 6x + 2 - 6x$$

$$-x = 2$$

$$\frac{-x}{-1} = \frac{2}{-1}$$

$$x = -2$$

45.
$$2m + 3 = 3$$
$$2m + 3 - 3 = 3 - 3$$
$$2m = 0$$
$$\frac{2m}{2} = \frac{0}{2}$$
$$m = 0$$

46.
$$2(3a - 1) = 3(a + 5) + 1$$
$$6a - 2 = 3a + 15 + 1$$
$$6a - 2 = 3a + 16$$
$$6a - 2 + 2 = 3a + 16 + 2$$
$$6a = 3a + 18$$
$$6a - 3a = 3a + 18 - 3a$$
$$3a = 18$$
$$\frac{3a}{3} = \frac{18}{3}$$
$$a = 6$$

47.
$$x + (3x - 4) = 16$$
$$4x - 4 = 16$$
$$4x = 20$$
$$x = 5$$
The smaller number is 5.
The larger number is 11.

48.
$$x + (x + 12) = 48$$
$$2x + 12 = 48$$
$$2x = 36$$
$$x = 18$$
One piece is 18 inches.
The other is 30 inches.

49.
$$x + (x + 1) = 65$$
$$2x + 1 = 65$$
$$2x = 64$$
$$x = 32$$
One integer is 32. The other is 33.

FRACTIONAL EXPRESSIONS

3

FRACTIONAL EXPRESSIONS AND THEIR BASIC PROPERTIES

3.1

In this section we begin a study of fractions. We will develop their basic properties in order to help us understand the operations of addition, subtraction, multiplication, and division.

You know from previous experience that the symbol for a fraction has a numerator, fraction bar, and a denominator.

$$\frac{3}{5} \begin{matrix} \leftarrow \text{numerator} \\ \leftarrow \text{fraction bar} \\ \leftarrow \text{denominator} \end{matrix}$$

A fraction has two meanings: it can either represent a division or a multiplication. For example, $\frac{3}{5} = 3 \div 5$ and $\frac{3}{5} = 3 \cdot \frac{1}{5}$. Thus, we have the following definition.

Definition of a Fraction

Since a fraction can be written as either a basic quotient or a basic product, we can state the definition:

$$\frac{a}{b} = a \div b \quad \text{or} \quad \frac{a}{b} = a \cdot \frac{1}{b} \qquad \text{where } b \neq 0$$

As you know, division by zero is impossible. Therefore, given any fraction, we will always assume that the denominator does not represent zero.

The fraction bar is considered a grouping symbol. It groups the expression above as well as the expression below. Thus, $\frac{x + 3}{x + 5}$ means $(x + 3) \div (x + 5)$ or $(x + 3) \cdot \frac{1}{x + 5}$. Using the definition of a fraction and remembering that the fraction bar is a grouping symbol, we can convert back and forth between basic quotients, basic products, and the fractional form.

Example 1

Convert $\frac{2}{x + 1}$ to a basic quotient.

    Solution: $\frac{2}{x + 1} = 2 \div (x + 1)$

Example 2

Convert $\frac{3r^2 - 8}{r + 1}$ to a basic product.

Solution: $\dfrac{3r^2 - 8}{r + 1} = (3r^2 - 8) \cdot \dfrac{1}{r + 1}$

Convert $(2m + 1) \div (5m - 7)$ to a fraction.   **Example 3**

Solution: $(2m + 1) \div (5m - 7) = \dfrac{2m + 1}{5m - 7}$

A fraction can indicate division. This gives rise to four basic properties of fractions. You can see that each property is true by dividing the numerator by the denominator.

Basic Properties of Fractions

(1)   The denominator of a fraction can never be zero.

   Example: $\dfrac{x + 1}{0}$ does not represent a number.

(2)   If the denominator of a fraction is 1, then the fraction is equal to the numerator.

   Example: $\dfrac{2m + 3n}{1} = 2m + 3n$

(3)   If the numerator of a fraction is zero and the denominator is not zero, then the fraction is equal to zero.

   Example: $\dfrac{0}{5m} = 0,$ where $m \neq 0$

(4)   If the numerator and denominator of a fraction are equal but not zero, then the fraction is equal to 1.

   Example: $\dfrac{6x}{6x} = 1,$ where $x \neq 0$

A fraction has three places to attach a minus sign or a plus sign: in front of the numerator, in front of the denominator, or in front of the fraction bar. Remembering that division of unlike signs produces a negative number, we can see that the following equation is true:

$$\frac{-2}{3} = \frac{2}{-3} = -\frac{2}{3}$$

Generalizing, we obtain this property.

The Sign Property of Fractions

$$\frac{-a}{b} = \frac{a}{-b} = -\frac{a}{b} \qquad \text{where } b \neq 0$$

FRACTIONAL EXPRESSIONS

This property states that any two signs may be changed without affecting the value of the fraction. Note that this is actually a restatement of the opposite law of division from section 2.2

Example 4

These equations illustrate the sign property of fractions.

(a) $\dfrac{-1}{2} = -\dfrac{1}{2}$

(b) $\dfrac{2x}{-3} = \dfrac{-2x}{3}$

(c) $\dfrac{-2m}{-3n} = \dfrac{2m}{3n}$

(d) $\dfrac{-5}{8} \neq \dfrac{-5}{-8}$

(e) $\dfrac{3}{x} \neq -\dfrac{-3}{-x}$

We now examine the cross products of two fractions to derive an interesting test that will tell whether two fractions are equal or unequal.

cross products

You know from previous experience that $\dfrac{1}{2} = \dfrac{5}{10}$. Observe that the **cross products** are equal.

$$\dfrac{1}{2} = \dfrac{5}{10}$$

$$1 \cdot 10 = 2 \cdot 5$$

Also, you know that $\dfrac{1}{2} \neq \dfrac{3}{4}$. You should see that the cross products are not equal.

$$\dfrac{1}{2} \neq \dfrac{3}{4}$$

$$1 \cdot 4 \neq 2 \cdot 3$$

Thus, we have the following test for equality.

The Equality Test for Fractions

Two fractions are said to be equal (or equivalent) if their cross products are equal. That is,

$$\frac{a}{b} = \frac{c}{d} \quad \text{if and only if} \quad a \cdot d = b \cdot c \text{ where } b, d \neq 0$$

These statements show the equality test for fractions. **Example 5**

(a) $\dfrac{2}{3} = \dfrac{6}{9}$ because $2 \cdot 9 = 3 \cdot 6$

(b) $\dfrac{2}{5} \neq \dfrac{1}{3}$ because $2 \cdot 3 \neq 5 \cdot 1$

(c) $\dfrac{2x}{y} = \dfrac{6x}{3y}$ because $(2x)(3y) = (y)(6x)$

(d) $\dfrac{2a}{b} \neq \dfrac{a}{2b}$ because $(2a)(2b) \neq (b)(a)$

The equality test can also be used to solve equations when both the left and right members are fractions.

Solve $\dfrac{x}{5} = \dfrac{12}{20}$ **Example 6**

Solution: Use the equality test to set the cross products equal.

$$\frac{x}{5} = \frac{12}{20}$$
$$20x = (5)(12)$$
$$20x = 60$$
$$x = \frac{60}{20}$$
$$x = 3$$

Solve $\dfrac{2y}{5} = \dfrac{7}{3}$ **Example 7**

Solution: Set the cross products equal.

$$\frac{2y}{5} = \frac{7}{3}$$
$$6y = 35$$
$$y = \frac{35}{6}$$

Example 8

Solve $\dfrac{m + 2}{5} = \dfrac{-2}{3}$

Solution: Set the cross products equal.

$$\frac{m + 2}{5} = \frac{-2}{3}$$

$$3(m + 2) = -10$$

$$3m + 6 = -10$$

$$3m = -10 - 6$$

$$3m = -16$$

$$m = \frac{-16}{3}$$

Exercise 3.1

Indicate the numerator and denominator for each of these fractions.

1. $\dfrac{3m}{2}$

2. $\dfrac{5x}{6y}$

3. $\dfrac{x + 7}{x - 5}$

4. $\dfrac{2a - 9}{-5a}$

Using the definition of a fraction, convert each of these fractions to a basic quotient and a basic product.

5. $\dfrac{3}{7}$

6. $\dfrac{x + 2}{3}$

7. $\dfrac{2a}{a - 5}$

8. $\dfrac{3x - 5}{4x + 7}$

Using the definition of a fraction, convert each of the following expressions to fractional form.

9. $2a \div 3$

10. $3y \div (y + 4)$

11. $(m - 3) \div (m + 5)$

12. $(5x) \cdot \dfrac{1}{6}$

13. $(2b + 1) \cdot \dfrac{1}{7}$

14. $(3x^2 + 4) \cdot \dfrac{1}{x + 7}$

Simplify these fractions according to the four basic properties.

15. $\dfrac{a}{1}$

16. $\dfrac{0}{x}$, where $x \neq 0$

17. $\dfrac{7m + 3}{1}$

18. $\dfrac{m + 2}{m + 2}$,   where $m + 2 \neq 0$

19. $\dfrac{0}{2y + 1}$,   where $2y + 1 \neq 0$

20. $\dfrac{5b}{5b}$,   where $5b \neq 0$

True or false.

21. $\dfrac{-7}{9} = \dfrac{7}{-9}$

22. $\dfrac{-3}{5} = -\dfrac{3}{5}$

23. $\dfrac{-2}{5b} = \dfrac{2}{5b}$

24. $\dfrac{7x}{5y} = \dfrac{-7x}{-5y}$

25. $-\dfrac{2m}{n} = -\dfrac{-2m}{-n}$

26. $\dfrac{7}{-3x} = \dfrac{-7}{-3x}$

Use the equality test for fractions to determine which of the following are equal. (Answer yes or no.)

27. $\dfrac{5}{8} = \dfrac{15}{24}$

28. $\dfrac{-3}{7} = \dfrac{6}{-14}$

29. $\dfrac{36}{41} = \dfrac{37}{42}$

30. $\dfrac{3m}{7n} = \dfrac{7m}{3n}$

31. $\dfrac{a}{b} = \dfrac{ax}{bx}$

32. $\dfrac{m + 3}{4} = \dfrac{2m + 6}{8}$

Use the equality test for fractions to solve these equations.

33. $\dfrac{x}{2} = \dfrac{6}{4}$

34. $\dfrac{m}{3} = \dfrac{6}{9}$

35. $\dfrac{y}{7} = \dfrac{-6}{42}$

36. $\dfrac{12}{n} = \dfrac{6}{5}$

37. $\dfrac{3y}{2} = \dfrac{4}{5}$

38. $\dfrac{3x}{5} = \dfrac{1}{4}$

39. $\dfrac{6}{x} = \dfrac{3}{2}$

40. $\dfrac{2m}{3} = \dfrac{-3}{5}$

41. $\dfrac{x + 3}{7} = \dfrac{1}{10}$

42. $\dfrac{m + 2}{5} = \dfrac{2}{3}$

43. $\dfrac{3y - 2}{3} = \dfrac{5}{4}$

44. $\dfrac{5x + 1}{3} = \dfrac{2x}{4}$

45. $\dfrac{3m}{2} = \dfrac{m + 1}{3}$

46. $\dfrac{2y + 1}{3y - 2} = \dfrac{4}{5}$

47. $\dfrac{n + 1}{n - 1} = \dfrac{2}{-3}$

48. $\dfrac{2(p + 1)}{p} = \dfrac{3}{2}$

49. $\dfrac{3(h - 1)}{2(h + 3)} = \dfrac{1}{2}$

50. $\dfrac{2(2a - 3)}{3} = \dfrac{a - 6}{3}$

THE FUNDAMENTAL PRINCIPLE OF FRACTIONS

3.2

In this section we will study one of the most important properties concerning fractions. Since it is so basic to all work with fractions, it is called the *fundamental principle of fractions*. This property states that we can multiply or divide both the numerator and denominator by the same nonzero number without affecting the value of the fraction. Using symbols, the property is written as follows.

The Fundamental Principle of Fractions

$$\frac{a}{b} = \frac{a \cdot n}{b \cdot n} \quad \text{and} \quad \frac{a}{b} = \frac{a \div n}{b \div n}, \quad \text{where } b, n \neq 0$$

The fundamental principle can be proved by using the equality test for fractions and showing that the cross products are equal. Thus,

$$\frac{a}{b} = \frac{a \cdot n}{b \cdot n} \quad \text{because} \quad a \cdot b \cdot n = b \cdot a \cdot n$$

and

$$\frac{a}{b} = \frac{a \div n}{b \div n} \quad \text{because} \quad a \cdot b \div n = b \cdot a \div n$$

Example 1

These illustrations use the fundamental principle of fractions.

(a) $\dfrac{1}{2} = \dfrac{1 \cdot 5}{2 \cdot 5}$ or $\dfrac{5}{10}$

(b) $\dfrac{x}{y} = \dfrac{x \cdot 2}{y \cdot 2}$ or $\dfrac{2x}{2y}$

(c) $\dfrac{8}{20} = \dfrac{8 \div 4}{20 \div 4}$ or $\dfrac{2}{5}$

(d) $\dfrac{4x}{6y} = \dfrac{4x \div 2}{6y \div 2}$ or $\dfrac{2x}{3y}$

expanded to higher terms

When applying the fundamental principle results in a fraction having a larger numerator and denominator, we say the fraction has been "expanded to higher terms." This occurs when the numerator and denominator are multiplied by a number greater than one.

Example 2

Expand $\dfrac{3}{5}$ so that its denominator becomes 20

Solution: Since 5 times 4 is 20, multiply numerator and denominator by 4

$$\frac{3}{5} = \frac{3 \cdot 4}{5 \cdot 4} = \frac{12}{20}$$

Expand $\dfrac{2x}{5y}$ so that its denominator becomes $15y^2$

Example 3

Solution: Since $5y$ times $3y$ is $15y^2$, multiply numerator and denominator by $3y$

$$\frac{2x}{5y} = \frac{(2x)\cdot(3y)}{(5y)\cdot(3y)} = \frac{6xy}{15y^2}$$

Expand $\dfrac{3b}{2a^2}$ so that its denominator becomes $10a^2b^2$

Example 4

Solution: Since $2a^2$ times $5b^2$ is $10a^2b^2$, multiply numerator and denominator by $5b^2$

$$\frac{3b}{2a^2} = \frac{(3b)\cdot(5b^2)}{(2a^2)\cdot(5b^2)} = \frac{15b^3}{10a^2b^2}$$

Expand $\dfrac{2n}{3}$ so that its denominator becomes $3(n+1)$

Example 5

Solution: Since 3 times $(n+1)$ is $3(n+1)$, multiply numerator and denominator by $(n+1)$

$$\frac{2n}{3} = \frac{2n\,(n+1)}{3\,(n+1)} \quad \text{or} \quad \frac{2n^2+2n}{3n+3}$$

When applying the fundamental principle results in a fraction having the smallest possible numerator and denominator, we say the fraction has been **"reduced to lowest terms."** This occurs when the numerator and denominator are divided by their largest common factor.

Reduce $\dfrac{4}{8}$ to lowest terms.

Example 6

Solution: 4 is the largest factor common to both the numerator and denominator because $\dfrac{4}{8} = \dfrac{1\cdot4}{2\cdot4}$. Therefore, divide numerator and denominator by 4

$$\frac{4}{8} = \frac{4\div4}{8\div4} = \frac{1}{2}$$

Example 7

Reduce $\dfrac{2a}{3a^2}$ to lowest terms.

Solution: a is the largest factor common to both the numerator and denominator because $\dfrac{2a}{3a^2} = \dfrac{2 \cdot a}{(3a) \cdot a}$. Therefore, divide numerator and denominator by a

$$\frac{2a}{3a^2} = \frac{2a \div a}{3a^2 \div a} = \frac{2}{3a}$$

Example 8

Reduce $\dfrac{-2mn}{4m^2n}$ to lowest terms.

Solution: $2mn$ is the largest factor common to both the numerator and denominator because $\dfrac{-2mn}{4m^2n} = \dfrac{-1 \cdot (2mn)}{2m \cdot (2mn)}$. Therefore, divide numerator and denominator by $2mn$

$$\frac{-2mn}{4m^2n} = \frac{-2mn \div (2mn)}{4m^2n \div (2mn)} = \frac{-1}{2m}$$

Example 9

Reduce $\dfrac{2x + 2y}{3x + 3y}$ to lowest terms.

Solution: $(x + y)$ is the largest factor common to both the numerator and denominator because $\dfrac{2x + 2y}{3x + 3y} = \dfrac{2(x + y)}{3(x + y)}$. Therefore, divide numerator and denominator by $(x + y)$

$$\frac{2x + 2y}{3x + 3y} = \frac{2(x + y)}{3(x + y)}$$

$$= \frac{2(x + y) \div (x + y)}{3(x + y) \div (x + y)}$$

$$= \frac{2}{3}$$

Exercise 3.2

Use the fundamental principle to expand the following fractions as directed.

1. Expand $\dfrac{1}{3}$ so that its denominator becomes 12

2. Expand $\dfrac{2}{7}$ so that its denominator becomes 21

3. Expand $\dfrac{-3}{5}$ so that its denominator becomes 10

4. Expand $\dfrac{5}{-6}$ so that its denominator becomes 18

5. Expand $\dfrac{9}{2}$ so that its denominator becomes 14

6. Expand $\dfrac{5}{3x}$ so that its denominator becomes $12x$

7. Expand $\dfrac{5}{3x}$ so that its denominator becomes $12x^2$

8. Expand $\dfrac{7}{2m}$ so that its denominator becomes $6m^2$

9. Expand $\dfrac{3}{5b}$ so that its denominator becomes $10b^2$

10. Expand $\dfrac{4m}{3n}$ so that its denominator becomes $6mn^2$

11. Expand $\dfrac{2a}{-5b^2}$ so that its denominator becomes $25a^2b^3$

12. Expand $\dfrac{2c^3}{7a^2b^3}$ so that its denominator becomes $21a^5b^4c$

13. Expand $\dfrac{3x}{2}$ so that its denominator becomes $2(x+y)$

14. Expand $\dfrac{5m}{3n}$ so that its denominator becomes $3n(m+2)$

15. Expand $\dfrac{3}{a+4}$ so that its denominator becomes $(a+4)^2$

Use the fundamental principle to reduce the following fractions to lowest terms.

16. $\dfrac{3}{6}$ **17.** $\dfrac{15}{20}$

18. $\dfrac{18}{24}$ **19.** $\dfrac{12}{15}$

20. $\dfrac{9}{6}$ **21.** $\dfrac{21}{-14}$

22. $\dfrac{5x}{6x^2}$ **23.** $\dfrac{7m}{m}$

24. $\dfrac{5a}{2a}$ **25.** $\dfrac{3n}{2n}$

26. $\dfrac{-3a^2}{2a}$ **27.** $\dfrac{12xy}{9x^2y}$

28. $\dfrac{35m^2n}{7mn}$

29. $\dfrac{16a^2b}{-24b}$

30. $\dfrac{-6x^2y^3}{15x^2y^4}$

31. $\dfrac{100r^3t^4}{120r^2t^2}$

32. $\dfrac{ab^2c}{-b^2c^3}$

33. $\dfrac{42hk^2}{-7hk}$

34. $\dfrac{-pt^2}{-p^2}$

35. $\dfrac{15ab^3c^2}{25abc^4}$

36. $\dfrac{6ab}{9a^2b}$

37. $\dfrac{6x^3}{36x^2y}$

38. $\dfrac{3m^2n}{6m^2n}$

39. $\dfrac{5(a + b)}{7(a + b)}$

40. $\dfrac{3m + 3n}{4m + 4n}$

41. $\dfrac{5a + 10}{5a + 15}$

42. $\dfrac{m + 2n}{3(m + 2n)}$

43. $\dfrac{a - 5b}{2(a - 5b)}$

44. $\dfrac{2a - 6}{2a}$

45. $\dfrac{3m - 15}{3m}$

46. $\dfrac{3a + 3b}{6a + 6b}$

47. $\dfrac{7a - 14}{a - 2}$

48. $\dfrac{5(a - b + c)}{6(a - b + c)}$

49. $\dfrac{2(x + y - z)}{8(x + y - z)}$

50. $\dfrac{2a + 2b - 2c}{3a + 3b - 3c}$

51. $\dfrac{7x - 7y + 7z}{14x - 14y + 14z}$

52. $\dfrac{2a + 4b + 6c}{3a + 6b + 9c}$

53. $\dfrac{(x - 3)^2}{x - 3}$

54. $\dfrac{(y + 5)^2}{3y + 15}$

55. $\dfrac{12xy^2}{3xy^2 + 6y}$

56. $\dfrac{2a + 6b}{2(a + 3b)}$

57. $\dfrac{-3m(2n + 5)}{9m^2}$

58. $\dfrac{-2x^2(y + 3)}{4xy + 12x}$

REDUCING FRACTIONAL EXPRESSIONS

3.3

As seen in the previous section, the fundamental principle of fractions states that $\dfrac{a \cdot n}{b \cdot n} = \dfrac{a}{b}$ where $b, n \neq 0$. Thus we can write

$$\frac{a \cdot \cancel{n}}{b \cdot \cancel{n}} = \frac{a}{b}$$

where the slashes are used to indicate a shortcut for dividing numerator and denominator by the same non-zero number. This reduction process is accomplished by using these two steps:

Step (1) Find the largest factor common to the numerator and denominator.

Step (2) Use the slashes to show that the largest factor common to both numerator and denominator has been "divided out."

Reduce $\dfrac{12}{15}$ to lowest terms.

Example 1

Solution:

Step (1) 3 is the largest factor common to both numerator and denominator. Thus $\dfrac{12}{15} = \dfrac{4 \cdot 3}{5 \cdot 3}$

Step (2) Divide out the largest common factor of 3.

$$\frac{12}{15} = \frac{4 \cdot \cancel{3}}{5 \cdot \cancel{3}} = \frac{4}{5}$$

Reduce $\dfrac{2a}{3a^2}$ to lowest terms.

Example 2

Solution:

Step (1) a is the largest factor common to both numerator and denominator. Thus $\dfrac{2a}{3a^2} = \dfrac{2 \cdot a}{(3a) \cdot a}$

Step (2) Divide out the largest common factor of a.

$$\frac{2a}{3a^2} = \frac{2 \cdot \cancel{a}}{(3a) \cdot \cancel{a}} = \frac{2}{3a}$$

Reduce $\dfrac{2mn}{4m^2n}$ to lowest terms.

Example 3

Solution: $\dfrac{2mn}{4m^2n} = \dfrac{1 \cdot (\cancel{2mn})}{2m \cdot (\cancel{2mn})} = \dfrac{1}{2m}$

Reduce $\dfrac{10a^2b}{5ab}$ to lowest terms.

Example 4

Solution: $\dfrac{10a^2b}{5ab} = \dfrac{2a \cdot \cancel{(5ab)}}{1 \cdot \cancel{(5ab)}} = \dfrac{2a}{1}$ or $2a$

[note that when the entire numerator or denominator is divided out a factor of 1 is left in its place]

Example 5

Reduce $\dfrac{5(a + 3)}{6(a + 3)}$ to lowest terms.

Solution: $\dfrac{5\cancel{(a + 3)}}{6\cancel{(a + 3)}} = \dfrac{5}{6}$

It is extremely important for you to remember that this process is a short-cut for dividing numerator and denominator by the same number. You may do this only under the following two conditions:

(1) Both the numerator and the denominator must be basic products.
(2) The expressions divided out must be factors.

Examples 6 and 7 have been done *incorrectly*.

Example 6

Reduce $\dfrac{2 \cdot a}{a + b}$ to lowest terms.

Incorrect Solution: $\dfrac{2 \cdot \cancel{a}}{\cancel{a} + b} = \dfrac{2}{b}$

This is incorrect because the denominator is a basic sum, not a basic product. Also, a in the denominator is not a factor. The original expression cannot be reduced; it is in lowest terms.

Example 7

Reduce $\dfrac{m + 2}{m + 3}$ *to lowest terms.*

Incorrect Solution: $\dfrac{\cancel{m} + 2}{\cancel{m} + 3} = \dfrac{2}{3}$

This is incorrect because the numerator and denominator are not basic products. Also, neither m is a factor. The original expression cannot be reduced; it is in lowest terms.

Many times we will need to reduce fractions containing a basic sum or a basic difference. This is done by first using the distributive property to factor out every number or expression common to each term of the numerator and denominator.

Example 8

Reduce $\dfrac{2x + 2y}{3x + 3y}$ to lowest terms.

Solution:

Step (1) Factor the numerator and the denominator.

$$\frac{2x + 2y}{3x + 3y} = \frac{2(x + y)}{3(x + y)}$$

Step (2) Divide out the common factor of $(x + y)$

$$\frac{2\cancel{(x + y)}}{3\cancel{(x + y)}} = \frac{2}{3}$$

Reduce $\dfrac{5y^2 - 15}{10y}$ to lowest terms.   **Example 9**

Solution:

Step (1) Factor the numerator and denominator.

$$\frac{5y^2 - 15}{10y} = \frac{5(y^2 - 3)}{5 \cdot (2y)}$$

Step (2) Divide out the common factor of 5.

$$\frac{\cancel{5}(y^2 - 3)}{\cancel{5} \cdot (2y)} = \frac{y^2 - 3}{2y}$$

This procedure of reducing fractions can be shortened as shown by the following three examples.

Reduce $\dfrac{2m^2 + 4}{6m}$ to lowest terms.   **Example 10**

Solution: $\dfrac{2m^2 + 4}{6m} = \dfrac{\cancel{2}(m^2 + 2)}{\cancel{2} \cdot (3m)} = \dfrac{m^2 + 2}{3m}$

Reduce $\dfrac{3a - 3b}{3}$ to lowest terms.   **Example 11**

Solution: $\dfrac{3a - 3b}{3} = \dfrac{\cancel{3}(a - b)}{\cancel{3} \cdot 1} = a - b$

Reduce $\dfrac{x - 2}{6x - 12}$ to lowest terms.   **Example 12**

Solution: $\dfrac{x - 2}{6x - 12} = \dfrac{1 \cdot \cancel{(x - 2)}}{6 \cdot \cancel{(x - 2)}} = \dfrac{1}{6}$

Exercise 3.3   Reduction has been performed in each of the following examples. Indicate whether it has been done correctly. (Answer yes or no.)

1. $\dfrac{6}{8} = \dfrac{3 \cdot \cancel{2}}{4 \cdot \cancel{2}} = \dfrac{3}{4}$

2. $\dfrac{\cancel{a} + 3}{5\cancel{a}} = \dfrac{3}{5}$

3. $\dfrac{\cancel{m} + 3}{\cancel{m} + 7} = \dfrac{3}{7}$

4. $\dfrac{\cancel{y} + 2}{3\cancel{y}} = \dfrac{2}{3}$

5. $\dfrac{5\cancel{(x-3)}}{7\cancel{(x-3)}} = \dfrac{5}{7}$

6. $\dfrac{2\cancel{(a+1)}}{3\cancel{(a+1)}} = \dfrac{2}{3}$

7. $\dfrac{\cancel{m}}{3\cancel{m}} = \dfrac{0}{3}$

8. $\dfrac{\cancel{m}}{3\cancel{m}} = \dfrac{1}{3}$

9. $\dfrac{\cancel{y} + 8}{\cancel{y} + 3} = \dfrac{8}{3}$

10. $\dfrac{6}{7} = \dfrac{4 + \cancel{2}}{5 + \cancel{2}} = \dfrac{4}{5}$

11. $\dfrac{2m\cancel{(m+1)}}{3\cancel{(m+1)}} = \dfrac{2m}{3}$

12. $\dfrac{2\cancel{a}b}{\cancel{a} + 1} = 5b$

Reduce each of the following fractions to lowest terms.

13. $\dfrac{10}{25}$

14. $\dfrac{-6}{15}$

15. $\dfrac{5}{20}$

16. $\dfrac{8}{24}$

17. $\dfrac{15}{6}$

18. $\dfrac{-21}{3}$

19. $\dfrac{7a}{14b}$

20. $\dfrac{2m}{4}$

21. $\dfrac{6a}{12a}$

22. $\dfrac{-2a}{3a^2}$

23. $\dfrac{-7a}{9a}$

24. $\dfrac{3m}{-5m^2}$

25. $\dfrac{6m^2}{2m}$

26. $\dfrac{8ab}{10a^2b}$

27. $\dfrac{12bc}{15b^2c}$

28. $\dfrac{5s^3}{25s^2t}$

29. $\dfrac{50x^4}{100x^4}$

30. $\dfrac{14b^5}{42b^2}$

31. $\dfrac{3m^7}{12m^2}$

32. $\dfrac{-8a^5b^3}{12a^7b^6}$

33. $\dfrac{-4m^6n^9}{-12m^6n^4}$

34. $\dfrac{16a^2bc^5}{4abc^4}$

35. $\dfrac{rs^2t}{4rt^3}$

36. $\dfrac{5(a + 3)}{6(a + 3)}$

37. $\dfrac{2x(y + 1)}{4x^2(y + 1)}$

38. $\dfrac{3(m + 2)^2}{4(m + 2)}$

39. $\dfrac{6(m + 2)}{2m}$

40. $\dfrac{3x(x - 2)}{x - 2}$

41. $\dfrac{2(a + b)}{-3(a + b)}$

42. $\dfrac{3a + 3b}{5a + 5b}$

43. $\dfrac{5x + 10y}{5}$

44. $\dfrac{2x + 2y}{7x + 7y}$

45. $\dfrac{3a + 6}{2a + 4}$

46. $\dfrac{5y + 10}{5y}$

47. $\dfrac{m - 3}{4m - 12}$

48. $\dfrac{5y^2 - 15}{10y}$

49. $\dfrac{m + 3}{7m + 21}$

50. $\dfrac{2a^2 - 4a}{2a(a - 2)}$

51. $\dfrac{3x - 6}{3x + 6}$

52. $\dfrac{3a^2 - 9a}{3a(a - 3)}$

53. $\dfrac{2y + 2}{y + 1}$

54. $\dfrac{3ab^2 + 6b}{9ab^2}$

MULTIPLICATION AND DIVISON OF FRACTIONAL EXPRESSIONS

In arithmetic you learned to multiply fractions by multiplying their numerators and multiplying their denominators. Here is the rule, stated in symbols,

3.4

Multiplication of Fractions

$$\frac{a}{b} \cdot \frac{c}{d} = \frac{a \cdot c}{b \cdot d}, \quad \text{where } b, d \neq 0$$

Multiply $\dfrac{3}{5} \cdot \dfrac{2}{7}$

Example 1

Solution: $\dfrac{3}{5} \cdot \dfrac{2}{7} = \dfrac{3 \cdot 2}{5 \cdot 7} = \dfrac{6}{35}$

Multiply $\dfrac{3x}{5y^2} \cdot \dfrac{2x^2}{7yz}$

Example 2

Solution: $\dfrac{3x}{5y^2} \cdot \dfrac{2x^2}{7yz} = \dfrac{(3x)(2x^2)}{(5y^2)(7yz)} = \dfrac{6x^3}{35y^3z}$

When multiplying fractions, the numerators will often share common factors with the denominators, requiring us to reduce the answer. There are two methods of multiplying fractions and obtaining the answer in lowest terms. The first method is to multiply before reducing. The second method is the reverse: reduce first by dividing out the common factors, then multiply the remaining fractions. You will see that the second method is preferred, because it usually lets us work with simpler expressions. However, the first method gives meaning to the second.

Method I Multiplying Before Reducing

Example

Multiply $\dfrac{4}{5} \cdot \dfrac{7}{6}$

Solution:

Step (1)   Multiply the fractions.

$$\frac{4}{5} \cdot \frac{7}{6} = \frac{28}{30}$$

Step (2)   Reduce the answer.

$$\frac{28}{30} = \frac{14 \cdot \cancel{2}}{15 \cdot \cancel{2}} = \frac{14}{15}$$

Method II Reducing Before Multiplying (Preferred)

Example

Multiply $\dfrac{4}{5} \cdot \dfrac{7}{6}$

Solution:

Step (1)   Write each fraction in factored form and divide out the common factors.

$$\frac{4}{5} \cdot \frac{7}{6} = \frac{2 \cdot \cancel{2}}{5} \cdot \frac{7}{\cancel{2} \cdot 3}$$

Step (2)   Multiply the remaining factors.

$$\frac{4}{5} \cdot \frac{7}{6} = \frac{2 \cdot \cancel{2}}{5} \cdot \frac{7}{\cancel{2} \cdot 3} = \frac{14}{15}$$

Again, to use the preferred method, write each fraction in factored form and divide out the common factors. Then multiply the remaining factors. This method automatically gives the answer in lowest terms. It is correct to work from any numerator to any denominator. Remember, however, this method is merely a shortcut process of applying the fundamental principle to divide numerator and denominator by the same number. Therefore, you may divide out only identical factors. Do not do this if either the numerator or denominator is a basic sum or a basic difference. Also, never divide out numbers that are added or subtracted.

Multiply $\dfrac{5a}{9} \cdot \dfrac{6}{7a}$

Example 3

Solution: $\dfrac{5a}{9} \cdot \dfrac{6}{7a} = \dfrac{5 \cdot \cancel{a}}{3 \cdot \cancel{3}} \cdot \dfrac{2 \cdot \cancel{3}}{7 \cdot \cancel{a}} = \dfrac{10}{21}$

Multiply $\dfrac{2m^2}{3n^2} \cdot \dfrac{n^3}{6m} \cdot \dfrac{5}{n}$

Example 4

Solution: $\dfrac{2m^2}{3n^2} \cdot \dfrac{n^3}{6m} \cdot \dfrac{5}{n} = \dfrac{\cancel{2} \cdot m \cdot \cancel{m}}{3 \cdot \cancel{n^2}} \cdot \dfrac{\cancel{n^2} \cdot \cancel{n}}{\cancel{2} \cdot 3 \cdot \cancel{m}} \cdot \dfrac{5}{\cancel{n}} = \dfrac{5m}{9}$

Multiply $\dfrac{2(a + 3)}{b} \cdot \dfrac{b^2}{3(a + 3)}$

Example 5

Solution: $\dfrac{2(a + 3)}{b} \cdot \dfrac{b^2}{3(a + 3)} = \dfrac{2 \cdot \cancel{(a + 3)}}{1 \cdot \cancel{b}} \cdot \dfrac{\cancel{b} \cdot b}{3 \cdot \cancel{(a + 3)}} = \dfrac{2b}{3}$

When multiplying fractions containing basic sums or differences, we must be especially careful to factor the numerators and denominators first.

Multiply $\dfrac{3m + 6}{5} \cdot \dfrac{2m}{m + 2}$

Example 6

Solution: $\begin{aligned} \dfrac{3m + 6}{5} \cdot \dfrac{2m}{m + 2} &= \dfrac{3 \cdot (m + 2)}{5} \cdot \dfrac{2m}{(m + 2)} \\ &= \dfrac{3 \cdot \cancel{(m + 2)}}{5} \cdot \dfrac{2m}{1 \cdot \cancel{(m + 2)}} \\ &= \dfrac{6m}{5} \end{aligned}$

Multiply $\dfrac{2n}{2m - 4n} \cdot \dfrac{3m - 6n}{3n^3}$

Example 7

Solution: $\begin{aligned} \dfrac{2n}{2m - 4n} \cdot \dfrac{3m - 6n}{3n^3} &= \dfrac{2 \cdot n}{2 \cdot (m - 2n)} \cdot \dfrac{3 \cdot (m - 2n)}{3 \cdot n \cdot n^2} \\ &= \dfrac{2 \cdot \cancel{n}}{2 \cdot \cancel{(m - 2n)}} \cdot \dfrac{3 \cdot \cancel{(m - 2n)}}{3 \cdot \cancel{n} \cdot n^2} \\ &= \dfrac{1}{n^2} \end{aligned}$

Next we use the properties of fractions, together with the fundamental principle, to obtain a method for dividing two fractions. Consider the following proof:

$$\frac{a}{b} \div \frac{c}{d} = \frac{\dfrac{a}{b}}{\dfrac{c}{d}} \quad \text{[definition of a fraction]}$$

$$= \frac{\dfrac{a}{b} \cdot \dfrac{d}{c}}{\dfrac{c}{d} \cdot \dfrac{d}{c}} \quad \text{[fundamental principle]}$$

$$= \frac{\dfrac{a}{b} \cdot \dfrac{d}{c}}{1} \quad \text{[multiplication of fractions]}$$

$$= \frac{a}{b} \cdot \frac{d}{c} \quad \text{[dividing by 1]}$$

This proof gives the following property for dividing fractions.

Division of Fractions

$$\frac{a}{b} \div \frac{c}{d} = \frac{a}{b} \cdot \frac{d}{c}, \quad \text{where } b, c, d \neq 0$$

reciprocals

Observe that $\dfrac{c}{d}$ and $\dfrac{d}{c}$ are multiplicative inverses (reciprocals) because their product, $\dfrac{c}{d} \cdot \dfrac{d}{c}$, equals 1. This property states that to divide fractions, multiply by the multiplicative inverse (or reciprocal) of the divisor. Sometimes this is stated as "Invert the divisor and multiply."

Example 8

Divide $\dfrac{2a^2}{3b^4} \div \dfrac{6a}{7b^5}$

Solution:
$$\frac{2a^2}{3b^4} \div \frac{6a}{7b^5} = \frac{2a^2}{3b^4} \cdot \frac{7b^5}{6a}$$

$$= \frac{\cancel{2} \cdot \cancel{a} \cdot a}{3 \cdot \cancel{b^4}} \cdot \frac{7 \cdot \cancel{b^4} \cdot b}{\cancel{2} \cdot 3 \cdot \cancel{a}}$$

$$= \frac{7ab}{9}$$

Divide $\dfrac{8rs^2}{2r} \div (4s^2)$

Example 9

Solution: $\dfrac{8rs^2}{2r} \div (4s^2) = \dfrac{8rs^2}{2r} \div \dfrac{4s^2}{1}$

$$= \dfrac{8rs^2}{2r} \cdot \dfrac{1}{4s^2}$$

$$= \dfrac{2 \cdot 4 \cdot r \cdot s^2}{2 \cdot r} \cdot \dfrac{1}{4 \cdot s^2}$$

$$= 1$$

Divide $\dfrac{5m + 10n}{3m^2} \div \dfrac{2m + 4n}{m}$

Example 10

Solution: $\dfrac{5m + 10n}{3m^2} \div \dfrac{2m + 4n}{m} = \dfrac{5m + 10n}{3m^2} \cdot \dfrac{m}{2m + 4n}$

$$= \dfrac{5 \cdot (m + 2n)}{3 \cdot m \cdot m} \cdot \dfrac{m}{2(m + 2n)}$$

$$= \dfrac{5}{6m}$$

Multiply the following expressions. All answers should be in lowest terms.

Exercise 3.4

1. $\dfrac{1}{3} \cdot \dfrac{1}{5}$

2. $\dfrac{1}{2} \cdot \dfrac{1}{-7}$

3. $\dfrac{1}{2} \cdot \dfrac{1}{-3} \cdot \dfrac{1}{-4}$

4. $\dfrac{1}{2x} \cdot \dfrac{1}{x}$

5. $\dfrac{2}{3} \cdot \dfrac{7}{3}$

6. $\dfrac{2}{3} \cdot \dfrac{4}{7} \cdot \dfrac{2}{9}$

7. $\dfrac{3}{7} \cdot \dfrac{21}{15}$

8. $7 \cdot \dfrac{5}{7}$

9. $\dfrac{6}{14} \cdot \dfrac{21}{2}$

10. $\dfrac{5m}{2} \cdot \dfrac{6}{7m}$

11. $\dfrac{2x}{y} \cdot \dfrac{3x^2}{y^2}$

12. $\dfrac{5m}{2n^2} \cdot \dfrac{4n^3}{3m^2}$

13. $\dfrac{3a^2}{b^2} \cdot \dfrac{5b}{6a^3}$

14. $\dfrac{2a^2b}{c} \cdot \dfrac{3ab^3}{c^3}$

15. $\dfrac{3x^2y}{2z^2} \cdot \dfrac{z}{12xy}$

16. $\dfrac{4x^2y}{16x^3} \cdot \dfrac{4x}{3y^2}$

17. $\dfrac{a}{3(b+2)} \cdot \dfrac{4(b+2)}{2a^3}$

18. $\dfrac{x+3}{4y} \cdot \dfrac{8y^2}{2(x+3)}$

19. $\dfrac{7(m+n)}{3m} \cdot \dfrac{6m^2}{(m+n)^2}$

20. $\dfrac{3x+3y}{x} \cdot \dfrac{5}{4x+4y}$

21. $\dfrac{3a}{a-2} \cdot \dfrac{2a-4}{9}$

22. $\dfrac{m}{m+6} \cdot \dfrac{2m+12}{2m}$

23. $\dfrac{x^2-xy}{3x^3} \cdot \dfrac{xy}{x-y}$

24. $\dfrac{2a-6b}{5b^2} \cdot \dfrac{10b^4}{3a-9b}$

25. $\dfrac{3m+3n}{6m} \cdot \dfrac{2m}{5m+5n}$

26. $\dfrac{3a-3b}{4} \cdot \dfrac{20}{5a-5b}$

27. $\dfrac{2m^2+4}{m} \cdot \dfrac{3m^3}{4m^2+16}$

28. $\dfrac{3x+12}{3x-12} \cdot \dfrac{2x-8}{6x+24}$

29. $\dfrac{2a+2b}{3} \cdot \dfrac{12}{(a+b)^2}$

30. $\dfrac{3y+6}{15y-30} \cdot \dfrac{9y-18}{6y+12}$

Divide the following expressions. All answers should be in lowest terms.

31. $\dfrac{-2}{5} \div \dfrac{4}{25}$

32. $\dfrac{3}{5} \div \dfrac{7}{2}$

33. $1 \div \dfrac{2}{3}$

34. $\dfrac{5}{6} \div 15$

35. $\dfrac{10x^2}{3y} \div \dfrac{9y^2}{6xy}$

36. $\dfrac{6m}{n} \div (12mn)$

37. $4x^2y \div \dfrac{2x}{y}$

38. $\dfrac{3a}{5b} \div \dfrac{6a^3}{25b}$

39. $1 \div \dfrac{3r}{2s}$

40. $\dfrac{3x^2}{y} \div \dfrac{2x}{y^3}$

41. $\dfrac{3m^2n}{2a^2} \div \dfrac{15m^4n^2}{6a^3}$

42. $\dfrac{ab^2}{c} \div \dfrac{b^2}{c^2}$

43. $\dfrac{2m^3n}{3} \div \dfrac{9m}{n}$

44. $\dfrac{12x^2}{3y} \div \dfrac{9y^2}{4xy}$

45. $1 \div \dfrac{m}{m+n}$

46. $\dfrac{3x+3y}{4x^2} \div \dfrac{5x+5y}{x}$

47. $\dfrac{4a^2+8ab}{2a} \div \dfrac{6a^2+12ab}{6a^2}$

48. $\dfrac{2x+6y}{x^2} \div \dfrac{x+3y}{x^4}$

49. $\dfrac{2ab-2a}{2} \div \dfrac{b-1}{3a^2}$

50. $\dfrac{2m+2n}{5n} \div \dfrac{3m+3n}{15n^2}$

51. $\dfrac{2m+n}{rs} \div \dfrac{6m+3n}{r^2}$

ADDITION AND SUBTRACTION OF FRACTIONS HAVING THE SAME DENOMINATORS

3.5

To determine a method of adding fractions having the same denominators, we use the definition of a fraction, together with our ability to factor. Consider the following steps:

$$\frac{a}{c} + \frac{b}{c} = a \cdot \frac{1}{c} + b \cdot \frac{1}{c} \qquad \text{[definition of a fraction]}$$

$$= \frac{1}{c} \cdot (a + b) \qquad \text{[factoring]}$$

$$= \frac{a + b}{c} \qquad \text{[definition of a fraction]}$$

We now have the following property.

Addition of Fractions Having the Same Denominators
$$\frac{a}{c} + \frac{b}{c} = \frac{a + b}{c}, \quad \text{where } c \neq 0$$

This property tells us to add the numerators and place the sum over the same (common) denominator. Also, you should examine the answer to see if it can be reduced to lowest terms.

Add $\dfrac{4}{11} + \dfrac{3}{11}$

Example 1

Solution: $\quad \dfrac{4}{11} + \dfrac{3}{11} = \dfrac{4 + 3}{11} = \dfrac{7}{11}$

Add $\dfrac{3x}{8} + \dfrac{x}{8}$

Example 2

Solution: $\quad \dfrac{3x}{8} + \dfrac{x}{8} = \dfrac{3x + x}{8}$

$$= \frac{4x}{8}$$

Reduce to lowest terms.

$$= \frac{\cancel{4}x}{\cancel{4} \cdot 2}$$

$$= \frac{x}{2} \text{ or } \frac{1}{2}x$$

Example 3

Add $\dfrac{4a}{3b} + \dfrac{2a}{3b} + \dfrac{a}{3b}$

Solution:

$$\dfrac{4a}{3b} + \dfrac{2a}{3b} + \dfrac{a}{3b} = \dfrac{4a + 2a + a}{3b}$$

$$= \dfrac{7a}{3b}$$

Example 4

Add $\dfrac{2x^2 + x - 3}{x + 5} + \dfrac{3x^2 + 2x - 5}{x + 5}$

Solution: $\dfrac{2x^2 + x - 3}{x + 5} + \dfrac{3x^2 + 2x - 5}{x + 5} = \dfrac{(2x^2 + x - 3) + (3x^2 + 2x - 5)}{x + 5}$

$$= \dfrac{2x^2 + x - 3 + 3x^2 + 2x - 5}{x + 5}$$

$$= \dfrac{5x^2 + 3x - 8}{x + 5}$$

To determine a method for subtracting fractions having the same denominators, we follow the same steps we used for addition.

$$\dfrac{a}{c} - \dfrac{b}{c} = a \cdot \dfrac{1}{c} - b \cdot \dfrac{1}{c} \qquad \text{[definition of a fraction]}$$

$$= \dfrac{1}{c} \cdot (a - b) \qquad \text{[factoring]}$$

$$= \dfrac{a - b}{c} \qquad \text{[definition of a fraction]}$$

This gives yet another property.

Subtraction of Fractions Having the Same Denominators

$$\dfrac{a}{c} - \dfrac{b}{c} = \dfrac{a - b}{c}, \quad \text{where } c \neq 0$$

The property says to subtract the numerators and place the difference over the common denominator. As in addition, you should examine the answer to see if it can be reduced to lower terms.

Example 5

Subtract $\dfrac{5}{7} - \dfrac{2}{7}$

Solution: $\dfrac{5}{7} - \dfrac{2}{7} = \dfrac{5 - 2}{7} = \dfrac{3}{7}$

Subtract $\dfrac{8a}{3b} - \dfrac{2a}{3b}$

Example 6

Solution: $\dfrac{8a}{3b} - \dfrac{2a}{3b} = \dfrac{8a - 2a}{3b}$

$$= \dfrac{6a}{3b}$$

Reduce to lowest terms.

$$= \dfrac{\cancel{3} \cdot 2 \cdot a}{\cancel{3} \cdot b}$$

$$= \dfrac{2a}{b}$$

Subtract $\dfrac{5x}{y} - \dfrac{x}{y} - \dfrac{7x}{y}$

Example 7

Solution: $\dfrac{5x}{y} - \dfrac{x}{y} - \dfrac{7x}{y} = \dfrac{5x - x - 7x}{y}$

$$= \dfrac{-3x}{y}$$

Subtract $\dfrac{2m + 3}{4n} - \dfrac{5m - 4}{4n}$

Example 8

Solution: $\dfrac{2m + 3}{4n} - \dfrac{5m - 4}{4n} = \dfrac{(2m + 3) - (5m - 4)}{4n}$

$$= \dfrac{2m + 3 - 5m + 4}{4n}$$

$$= \dfrac{-3m + 7}{4n}$$

As illustrated in Examples 9 and 10, addition and subtraction may both appear in the same problem.

Combine $\dfrac{5x}{2y} + \dfrac{3x}{2y} - \dfrac{x + 1}{2y}$

Example 9

Solution: $\dfrac{5x}{2y} + \dfrac{3x}{2y} - \dfrac{x + 1}{2y} = \dfrac{5x + 3x - (x + 1)}{2y}$

$$= \dfrac{5x + 3x - x - 1}{2y}$$

$$= \dfrac{7x - 1}{2y}$$

Example 10

Combine $\dfrac{3m - 2n}{m - n} + \dfrac{2m + n}{m - n} - \dfrac{m + 3n}{m - n}$

Solution: $\dfrac{3m - 2n}{m - n} + \dfrac{2m + n}{m - n} - \dfrac{m + 3n}{m - n}$

$$= \frac{(3m - 2n) + (2m + n) - (m + 3n)}{m - n}$$

$$= \frac{3m - 2n + 2m + n - m - 3n}{m - n}$$

$$= \frac{4m - 4n}{m - n}$$

Reduce to lowest terms.

$$= \frac{4\cancel{(m - n)}}{\cancel{(m - n)}}$$

$$= 4$$

Exercise 3.5

Combine the following fractions, using addition or subtraction.

1. $\dfrac{2}{7} + \dfrac{1}{7}$

2. $\dfrac{4}{5} + \dfrac{1}{5}$

3. $\dfrac{5}{12} + \dfrac{6}{12}$

4. $\dfrac{1}{9} + \dfrac{3}{9} + \dfrac{2}{9}$

5. $\dfrac{1}{7} - \dfrac{5}{7}$

6. $\dfrac{3}{5} - \dfrac{2}{5}$

7. $\dfrac{2}{11} - \dfrac{1}{11} - \dfrac{5}{11}$

8. $\dfrac{1}{9} + \dfrac{2}{9} - \dfrac{1}{9}$

9. $\dfrac{3}{5} - \dfrac{4}{5} + \dfrac{1}{5}$

10. $\dfrac{x}{y} + \dfrac{3x}{y}$

11. $\dfrac{m}{n} + \dfrac{3m}{n}$

12. $\dfrac{2m}{n} - \dfrac{7m}{n}$

13. $\dfrac{2x}{3} + \dfrac{4}{3} - \dfrac{5}{3}$

14. $\dfrac{6}{5y} - \dfrac{2}{5y} + \dfrac{11}{5y}$

15. $\dfrac{8}{7a} + \dfrac{9}{7a} - \dfrac{3}{7a}$

16. $\dfrac{3a}{2b} - \dfrac{a + 7}{2b}$

17. $\dfrac{3a}{2b} + \dfrac{a + 7}{2b}$

18. $\dfrac{m + 1}{3} + \dfrac{4}{3}$

19. $\dfrac{a - 3b}{4b} + \dfrac{a + 5b}{4b}$

20. $\dfrac{2x - y}{4x} - \dfrac{2x + 2y}{4x}$

21. $\dfrac{y^2 - y}{3} + \dfrac{y^2}{3} + \dfrac{2y}{3}$

22. $\dfrac{x^2 + 2x - 1}{x + 1} + \dfrac{3x^2 - 5x + 2}{x + 1}$

23. $\dfrac{5a^2 + 2a - 3}{a - 2} - \dfrac{2a^2 + 3a - 1}{a - 2}$

24. $\dfrac{3}{10a^2} - \dfrac{3 - a}{10a^2}$

25. $\dfrac{2m - 3n}{2m + 2n} + \dfrac{2m - n}{2m + 2n}$

26. $\dfrac{2 - r}{2r + 4} - \dfrac{r}{2r + 4}$

27. $\dfrac{x - 2}{6} - \dfrac{x + 1}{6}$

28. $\dfrac{2}{a + b} + \dfrac{5}{a + b} - \dfrac{3}{a + b}$

29. $\dfrac{x + 2}{x - 1} - \dfrac{2x - 3}{x - 1} + \dfrac{x - 4}{x - 1}$

30. $\dfrac{3m + 2n}{5} - \dfrac{4m - n}{5} + \dfrac{n}{5}$

31. $\dfrac{2a + 3}{2} - \dfrac{a - 5}{2} + \dfrac{1}{2}$

32. $\dfrac{m - 1}{n} - \dfrac{m + 1}{n} - \dfrac{m - 3}{n}$

33. $\dfrac{6r - 6s}{5} - \dfrac{r - s}{5}$

34. $\dfrac{4a + b}{a + 3} - \dfrac{6a - b}{a + 3} + \dfrac{2a - b}{a + 3}$

35. $\dfrac{m + 2}{m - 3} + \dfrac{m + 3}{m - 3} - \dfrac{1}{m - 3}$

36. $\dfrac{3x - 2y}{x - y} - \dfrac{2x - y}{x - y}$

37. $\dfrac{3a^2 - 2a + 1}{a - 3} + \dfrac{2a^2 + a - 5}{a - 3}$

38. $\dfrac{m^2 + m - 2}{m + 4} + \dfrac{2m^2 - m + 3}{m + 4}$

39. $\dfrac{x^2 + 3x - 1}{x + 2} - \dfrac{x^2 + 2x - 3}{x + 2}$

40. $\dfrac{2n^2 - 3}{n} + \dfrac{3n^2 + 1}{n} + \dfrac{n^2 + 2}{n}$

41. $\dfrac{3y^2 - 2y + 3}{3y - 1} + \dfrac{y^2 + y + 1}{3y - 1} - \dfrac{4y^2 - 4y + 5}{3y - 1}$

42. $\dfrac{x}{x + 3} - \dfrac{5}{x + 3} - \dfrac{x^2}{x + 3}$

43. $\dfrac{2(m + 1)}{m - 1} + \dfrac{3(m - 4)}{m - 1} - \dfrac{4(m + 2)}{m - 1}$

44. $\dfrac{2p - 1}{2p + 1} - \dfrac{3p - 4}{2p + 1} - \dfrac{p}{2p + 1}$

45. $\dfrac{x^2 + 3x - 5}{x - 2} + \dfrac{3x^2 - 5x + 1}{x - 2} - \dfrac{x^2 + 2x + 7}{x - 2}$

46. $\dfrac{x + 1}{x^2 + 1} - \dfrac{x^2 + 2}{x^2 + 1} - \dfrac{3 - x}{x^2 + 1} + \dfrac{3x^2 - x + 4}{x^2 + 1}$

47. $\dfrac{a - 2b}{3ab} + \dfrac{4a - 5b}{3ab} + \dfrac{2(a - b)}{3ab} - \dfrac{a + b}{3ab}$

48. $\dfrac{2(m + 3n)}{m - n} - \dfrac{3(m - 2n)}{m - n} + \dfrac{4(2m + n)}{m - n} - \dfrac{m + n}{m - n}$

49. $\dfrac{2(x + y)}{2x - y} + \dfrac{3(x - 2y)}{2x - y} - \dfrac{5(y - x)}{2x - y} + \dfrac{1}{2x - y}$

50. $\dfrac{7(r - 2s)}{r + s} - \dfrac{6(r + 3s)}{r + s} - \dfrac{2(3r - s)}{r + s} - \dfrac{s}{r + s}$

PRIME NUMBERS, COMPOSITE NUMBERS, AND LOWEST COMMON DENOMINATOR

3.6

In this section we will develop a method for finding the lowest common denominator of a group of fractions. To do this, we must discuss two types of numbers: prime numbers and composite numbers.

prime number
evenly divisible

A **prime number** is any natural number which is evenly divisible by exactly two different divisors, itself and 1. The phrase "evenly divisible" means the remainder is zero.

Example 1

Which of the following are prime numbers?
(a)   2 is a prime number because it is evenly divisible only by 2 and 1
(b)   13 is a prime number because it is evenly divisible only by 13 and 1
(c)   4 is not a prime number because it is evenly divisible by 1, 2, and 4
(d)   15 is not a prime number because it is evenly divisible by 1, 3, 5, and 15

The number 1 is not considered to be a prime number. It is a special case because it has only one divisor, itself. The prime numbers begin with 2 and continue on without end. Thus, it is impossible to know all the prime numbers. For our purposes, you will find it helpful to memorize the prime numbers which are less than 30:

$$\{2, 3, 5, 7, 11, 13, 17, 19, 23, 29\}$$

composite number

A **composite number** is any natural number other than 1 which is not prime. Therefore, a composite number is evenly divisible by some natural number other than itself or 1.

Example 2

Here are some illustrations of composite numbers.
(a)   6 is composite because it is evenly divisible by 2 and 3
(b)   21 is composite because it is evenly divisible by 3 and 7
(c)   45 is composite because it is evenly divisible by 3, 5, 9, and 15

Every composite number can be written as a product of prime factors. For example, $15 = 3 \cdot 5$, where 3 and 5 are the prime factors. In order to find the prime factors of a composite number, a process is used where we attempt to divide by each prime number as many times as possible. Each time a prime number divides evenly, it is recorded as a factor. When performing this process, it is important to begin with the first prime number, 2. Then, continue on trying each succeeding prime in order of succession. The process ends whenever a division produces a prime number for a quotient. This process is called **prime factorization** and is illustrated in the next two examples.

prime factorization

Find the prime factors of 60. Write them in exponential form.   **Example 3**

Solution: Begin testing with 2, the first prime number. Continue on
              until the quotient is a prime number.

$$
\begin{array}{r|r}
2 & 60 \\
2 & 30 \\
3 & 15 \\
\hline
& 5
\end{array}
$$

              The prime factors are $2 \cdot 2 \cdot 3 \cdot 5$. Thus in exponential form,
              $60 = 2^2 \cdot 3 \cdot 5$

Find the prime factors of 525. Write them in exponential form.   **Example 4**

Solution: 525 is not evenly divisible by 2, so begin with the next prime
              number, 3

$$
\begin{array}{r|r}
3 & 525 \\
5 & 175 \\
5 & 35 \\
\hline
& 7
\end{array}
$$

              The prime factors are $3 \cdot 5 \cdot 5 \cdot 7$. Thus, in exponential form,
              $525 = 3 \cdot 5^2 \cdot 7$

The **lowest common denominator** of a group of fractions is the smallest   **lowest common**
number which is evenly divisible by each denominator. Lowest common   **denominator**
denominator is abbreviated LCD.

Here are illustrations of lowest common denominator (LCD).   **Example 5**

(a)   The LCD of $\dfrac{1}{2}$ and $\dfrac{1}{5}$ is 10

(b)   The LCD of $\dfrac{2}{3}, \dfrac{1}{2},$ and $\dfrac{5}{6}$ is 6

(c)   The LCD of $\dfrac{3}{4a}$ and $\dfrac{5}{6a}$ is $12a$

The LCD must be large enough to contain all factors of each denominator, but it should not contain more factors than necessary. Consider
Example 6. We will use it as a model to develop a method for finding
an LCD.

Example 6

Find the LCD of $\frac{1}{6}$ and $\frac{1}{9}$

Solution: Write 6 and 9 in prime factored exponential form. Then, select the least number of factors necessary for the LCD to be evenly divisible by 6 and 9

$$6 = 2 \cdot 3$$

$$9 = 3^2$$

The LCD must contain $2 \cdot 3$ to be divisible by 6, and it must contain 3^2 to be divisible by 9. It is not necessary for it to have more than two factors of 3. Thus, the LCD $= 2 \cdot 3^2$ or   18

We now state a four-step procedure for finding the LCD.

Step (1) Write each denominator in prime factored exponential form.

Step (2) Write each prime factor used as a base, but write it only once.

Step (3) From Step (1) obtain the largest exponent used on each base.

Step (4) Evaluate the exponential expression.

Example 7

Find the LCD of $\frac{3}{8}$ and $\frac{1}{6}$

Solution:

Step (1) Write each denominator in prime factored exponential form.

$$8 = 2^3$$

$$6 = 2 \cdot 3$$

Step (2) Write each prime factor used as a base, but write it only once.

$$2 \cdot 3$$

Step (3) From Step (1) obtain the largest exponent used on each base.

$$LCD = 2^3 \cdot 3$$

Step (4) Evaluate the exponential expression.

$$LCD = 2^3 \cdot 3 = 24$$

Find the LCD of $\frac{5}{12}$, $\frac{7}{45}$, and $\frac{5}{18}$ **Example 8**

Solution:

    Step (1)   Factor each denominator.

            $12 = 2^2 \cdot 3$

            $45 = 3^2 \cdot 5$

            $18 = 2 \cdot 3^2$

    Step (2)   Write the bases.

            $2 \cdot 3 \cdot 5$

    Step (3)   Apply the largest exponents.

            $LCD = 2^2 \cdot 3^2 \cdot 5$

    Step (4)   Evaluate.

            $LCD = 180$

As illustrated by the final two examples, this four-step procedure also works for denominators containing algebraic expressions.

Find the LCD of $\frac{7}{15a^3b}$ and $\frac{5}{12ab^2}$ **Example 9**

Solution:

    Step (1)   Factor each denominator.

            $15a^3b = 3 \cdot 5 \cdot a^3 \cdot b$

            $12ab^2 = 2^2 \cdot 3 \cdot a \cdot b^2$

    Step (2)   Write the bases.

            $2 \cdot 3 \cdot 5 \cdot a \cdot b$

    Step (3)   Apply the largest exponents.

            $LCD = 2^2 \cdot 3 \cdot 5 \cdot a^3 \cdot b^2$

    Step (4) Evaluate.

            $LCD = 60a^3b^2$

Example 10

Find the LCD of $\dfrac{1}{9x}$ and $\dfrac{7}{3x + 6}$

Solution:

Step (1)   Factor each denominator.

$$9x = 3^2 \cdot x$$
$$3x + 6 = 3 \cdot (x + 2)$$

Step (2)   Write the bases.

$$3 \cdot x \cdot (x + 2)$$

Step (3)   Apply the largest exponents.

$$LCD = 3^2 \cdot x \cdot (x + 2)$$

Step (4)   Evaluate.

$$LCD = 9x(x + 2)$$

Many times the denominators will be simple enough that you will not need to use this four-step procedure, or you will be able to apply it mentally. However, don't hesitate to write it out when necessary.

Exercise 3.6

Classify each of the following numbers as either prime or composite.

1. 2	**2.** 5
3. 10	**4.** 20
5. 21	**6.** 25
7. 29	**8.** 36
9. 42	**10.** 100

Give the prime factorization for each of the following numbers. Put your answers in exponential form.

11. 4	**12.** 9
13. 15	**14.** 24
15. 28	**16.** 40
17. 60	**18.** 108
19. 180	**20.** 720

Find the LCD for each of the following groups of fractions. Write out the four-step procedure whenever you need it.

21. $\frac{1}{2}$ and $\frac{1}{3}$

22. $\frac{1}{4}$ and $\frac{1}{5}$

23. $\frac{1}{4}$ and $\frac{1}{6}$

24. $\frac{5}{12}$ and $\frac{1}{6}$

25. $\frac{5}{12}$ and $\frac{7}{18}$

26. $\frac{1}{2}, \frac{1}{3}, \frac{1}{4}$

27. $\frac{17}{40}, \frac{7}{20}$, and $\frac{5}{28}$

28. $\frac{5}{18}, \frac{3}{20}$, and $\frac{7}{24}$

29. $\frac{5}{12}, \frac{7}{36}, \frac{3}{5}$

30. $\frac{5}{12}, \frac{11}{36}$, and $\frac{3}{10}$

31. $\frac{5}{12x^2}$ and $\frac{7}{10xy}$

32. $\frac{a}{2xy}, \frac{b}{3x^2y}$, and $\frac{c}{6x^2y^3}$

33. $\frac{7}{x}$ and $\frac{5}{x^2}$

34. $\frac{3}{10a^3b}$ and $\frac{7}{8ab^2}$

35. $\frac{1}{a^2}$ and $\frac{2}{ab}$

36. $\frac{1}{a^2}, \frac{2}{ab}$, and, $\frac{1}{b^2}$

37. $\frac{7}{5m}$ and $\frac{3}{10m^2}$

38. $\frac{5a}{6b^2}, \frac{3a}{4b}$, and $\frac{7a}{10b}$

39. $\frac{3}{r^2}, \frac{2}{rs}$, and $\frac{5}{s^2}$

40. $\frac{5}{14m^2n}, \frac{6}{21mn^2}$, and $\frac{3}{28mn}$

41. $\frac{3}{2m + 2n}$ and $\frac{1}{m + n}$

42. $\frac{3}{2m + 2n}$ and $\frac{1}{(m + n)^2}$

43. $\frac{5}{a + b}$ and $\frac{6}{a - b}$

44. $\frac{3}{2x}$ and $\frac{1}{x + 2}$

45. $\frac{x}{y}$ and $\frac{3}{y + 1}$

46. $\frac{3}{m + 2}$ and $\frac{5}{m + 3}$

47. $\frac{3}{xy^2}$ and $\frac{5}{x^2(y + 1)}$

48. $\frac{3a}{2b + 2}, \frac{a}{3b + 3}$, and $\frac{2a}{9b + 9}$

49. $\frac{7}{r + s}$ and $\frac{9}{r - s}$

50. $\frac{4}{15m(m + 1)}$ and $\frac{3}{20m^2}$

ADDITION AND SUBTRACTION OF FRACTIONS HAVING DIFFERENT DENOMINATORS

3.7

In Section 3.5 you learned to add and subtract fractions having the same denominators. This was done by combining numerators and placing the result over the common denominator.

Example 1

Combine $\dfrac{3x}{2y} + \dfrac{5x}{2y} - \dfrac{x}{2y}$

Solution: $\dfrac{3x}{2y} + \dfrac{5x}{2y} - \dfrac{x}{2y} = \dfrac{3x + 5x - x}{2y} = \dfrac{7x}{2y}$

To add or subtract fractions having different denominators, we will first find their lowest common denominator. Then, using the fundamental principle, each fraction will be expanded to an equivalent form having the LCD. Finally, the numerators will be combined and the result placed over the LCD. This process can be summarized in three steps.

Addition and Subtraction of Fractions Having Different Denominators

Step (1)   Find the lowest common denominator (LCD).

Step (2)   Expand each fraction so it has the LCD.

Step (3)   Combine the numerators and place the result over the LCD. Reduce if possible.

Example 2

Combine $\dfrac{3}{8} + \dfrac{1}{6} - \dfrac{5}{12}$

Solution:

Step (1)   Find the LCD.

$$\left.\begin{array}{l} 8 = 2^3 \\ 6 = 2\cdot 3 \\ 12 = 2^2\cdot 3 \end{array}\right\} \qquad \text{LCD} = 2^3\cdot 3 = 24$$

Step (2)   Expand each fraction so it has the LCD.

$$\dfrac{3}{8} + \dfrac{1}{6} - \dfrac{5}{12} = \dfrac{3\cdot 3}{8\cdot 3} + \dfrac{1\cdot 4}{6\cdot 4} - \dfrac{5\cdot 2}{12\cdot 2}$$

$$= \dfrac{9}{24} + \dfrac{4}{24} - \dfrac{10}{24}$$

Step (3)   Combine the numerators and reduce if possible.

$$\dfrac{9}{24} + \dfrac{4}{24} - \dfrac{10}{24} = \dfrac{3}{24}$$

$$= \frac{1 \cdot \cancel{3}}{8 \cdot \cancel{3}}$$

$$= \frac{1}{8}$$

Therefore, $\frac{3}{8} + \frac{1}{6} - \frac{5}{12} = \frac{1}{8}$

Add $\frac{7}{8ab^2} + \frac{3}{10a^3b}$ **Example 3**

Solution:

Step (1) Find the LCD.

$$8ab^2 = 2^3 \cdot a \cdot b^2$$
$$10a^3b = 2 \cdot 5 \cdot a^3 \cdot b$$

$$LCD = 2^3 \cdot 5 \cdot a^3 \cdot b^2 = 40a^3b^2$$

Step (2) Expand each fraction so it has the LCD.

$$\frac{7}{8ab^2} + \frac{3}{10a^3b} = \frac{7 \cdot (5a^2)}{8ab^2 \cdot (5a^2)} + \frac{3 \cdot (4b)}{10a^3b \cdot (4b)}$$

$$= \frac{35a^2}{40a^3b^2} + \frac{12b}{40a^3b^2}$$

Step (3) Combine the numerators and reduce if possible.

$$\frac{35a^2}{40a^3b^2} + \frac{12b}{40a^3b^2} = \frac{35a^2 + 12b}{40a^3b^2}$$

Therefore, $\frac{7}{8ab^2} + \frac{3}{10a^3b} = \frac{35a^2 + 12b}{40a^3b^2}$

Subtract $\frac{7}{5x} - \frac{2+x}{10x^2}$ **Example 4**

Solution:

Step (1) Find the LCD.

$$5x = 5 \cdot x$$
$$10x^2 = 2 \cdot 5 \cdot x^2$$

$$LCD = 2 \cdot 5 \cdot x^2 = 10x^2$$

Step (2) Expand each fraction so it has the LCD.

$$\frac{7}{5x} - \frac{2+x}{10x^2} = \frac{7 \cdot (2x)}{(5x) \cdot (2x)} - \frac{2+x}{10x^2}$$

$$= \frac{14x}{10x^2} - \frac{2+x}{10x^2}$$

Step (3) Combine the numerators and reduce if possible.

$$\frac{14x}{10x^2} - \frac{2+x}{10x^2} = \frac{14x - (2+x)}{10x^2}$$

$$= \frac{14x - 2 - x}{10x^2}$$

$$= \frac{13x - 2}{10x^2}$$

Therefore, $\dfrac{7}{5x} - \dfrac{2+x}{10x^2} = \dfrac{13x-2}{10x^2}$

Example 5

Subtract $\dfrac{5}{a+b} - \dfrac{2}{a-b}$

Step (1) Find the LCD.

$$\left.\begin{array}{l} a + b = (a + b) \\ a - b = (a - b) \end{array}\right\} \quad LCD = (a+b)(a-b)$$

Step (2) Expand each fraction so it has the LCD.

$$\frac{5}{a+b} - \frac{2}{a-b} = \frac{5 \cdot (a-b)}{(a+b) \cdot (a-b)} - \frac{2 \cdot (a+b)}{(a-b) \cdot (a+b)}$$

Step (3) Combine the numerators and reduce if possible.

$$\frac{5 \cdot (a-b)}{(a+b)(a-b)} - \frac{2 \cdot (a+b)}{(a-b)(a+b)} = \frac{5 \cdot (a-b) - 2 \cdot (a+b)}{(a+b)(a-b)}$$

$$= \frac{5a - 5b - 2a - 2b}{(a+b)(a-b)}$$

$$= \frac{3a - 7b}{(a+b)(a-b)}$$

Therefore, $\dfrac{5}{a+b} - \dfrac{2}{a-b} = \dfrac{3a-7b}{(a+b)(a-b)}$

As you gain experience combining fractions you will be able to shorten the process as shown in the next two examples.

Combine $\dfrac{1}{a^2} - \dfrac{2}{ab} + \dfrac{1}{b^2}$

Example 6

Solution: The LCD is a^2b^2

$$\frac{1}{a^2} - \frac{2}{ab} + \frac{1}{b^2} = \frac{1 \cdot (b^2)}{a^2 \cdot (b^2)} - \frac{2 \cdot (ab)}{(ab) \cdot (ab)} + \frac{1 \cdot (a^2)}{b^2 \cdot (a^2)}$$

$$= \frac{b^2}{a^2b^2} - \frac{2ab}{a^2b^2} + \frac{a^2}{a^2b^2}$$

$$= \frac{b^2 - 2ab + a^2}{a^2b^2}$$

Combine $m + \dfrac{5}{7m} - \dfrac{m+1}{14m^2}$

Example 7

Solution: Since $m = \dfrac{m}{1}$, the LCD is $14m^2$

$$\frac{m}{1} + \frac{5}{7m} - \frac{m+1}{14m^2} = \frac{m \cdot (14m^2)}{1 \cdot (14m^2)} + \frac{5 \cdot (2m)}{7m \cdot (2m)} - \frac{m+1}{14m^2}$$

$$= \frac{14m^3}{14m^2} + \frac{10m}{14m^2} - \frac{m+1}{14m^2}$$

[Remember, the fraction bar is an automatic grouping symbol. See section 3.1]

$$= \frac{14m^3 + 10m - (m+1)}{14m^2}$$

$$= \frac{14m^3 + 10m - m - 1}{14m^2}$$

$$= \frac{14m^3 + 9m - 1}{14m^2}$$

Often in algebra and arithmetic we encounter mixed numerals. A **mixed numeral** is a whole number mixed with a fraction. For example, $5\frac{2}{3}$ is a mixed numeral. It means $5 + \frac{2}{3}$. We find in algebra that it is inconvenient to work with mixed numerals, so we usually convert them to a single improper fraction as follows (an improper fraction has a numerator greater than or equal to its denominator).

mixed numeral

Example 8

Convert $5\frac{2}{3}$ to a single fraction.

Solution: $\quad 5\frac{2}{3} = 5 + \frac{2}{3}$

$$= \frac{5}{1} + \frac{2}{3}$$

$$= \frac{5 \cdot 3}{1 \cdot 3} + \frac{2}{3}$$

$$= \frac{15}{3} + \frac{2}{3}$$

$$= \frac{17}{3}$$

Example 9

Convert $-3\frac{2}{5}$ to a single fraction.

Solution: $\quad -3\frac{2}{5} = -\left(3\frac{2}{5}\right)$

$$= -\left(3 + \frac{2}{5}\right)$$

$$= -\left(\frac{3}{1} + \frac{2}{5}\right)$$

$$= -\left(\frac{15}{5} + \frac{2}{5}\right)$$

$$= -\frac{17}{5}$$

You may remember from arithmetic that an improper fraction is converted to a mixed numeral by dividing the denominator into the numerator and placing the remainder over the denominator.

Example 10

Convert $\frac{13}{5}$ to a mixed numeral.

Solution: $\quad 13 \div 5 = 2$ with a remainder of 3

$$= 2 + \frac{3}{5}$$

$$= 2\frac{3}{5}$$

Convert $-\frac{4}{3}$ to a mixed numeral

Example 11

Solution: $\quad -\frac{4}{3} = -\left(\frac{4}{3}\right)$

$$= -\left(1\frac{1}{3}\right)$$

$$= -1\frac{1}{3}$$

Combine the following fractions. Reduce the answer if possible.

1. $\frac{3}{4} + \frac{1}{3}$

2. $\frac{3}{4} - \frac{1}{12}$

3. $\frac{1}{3} + \frac{9}{10}$

4. $\frac{5}{8} - \frac{3}{4}$

5. $\frac{3}{10} - \frac{5}{6}$

6. $\frac{11}{4} - \frac{11}{3}$

7. $\frac{2}{3} + \frac{4}{9}$

8. $\frac{1}{15} - \frac{3}{5}$

9. $3 + \frac{1}{5}$

10. $6 - \frac{3}{4}$

11. $\frac{3}{4} + \frac{1}{3} - \frac{1}{12}$

12. $\frac{3}{5} - \frac{7}{20} + \frac{1}{2}$

13. $\frac{1}{3} + \frac{1}{5} + \frac{1}{10}$

14. $\frac{1}{4} - \frac{5}{7} + \frac{1}{2}$

15. $\frac{2}{3} - \frac{1}{6} + \frac{1}{2}$

16. $\frac{5}{18} + \frac{1}{3} - \frac{1}{2} + \frac{2}{9}$

17. $\frac{2}{a} + \frac{3}{a^2}$

18. $\frac{3}{x} + \frac{5}{y}$

19. $\frac{c}{d} + 1$

20. $\frac{4}{a} + \frac{3}{b}$

21. $\frac{2}{mn} + \frac{2}{n}$

22. $\frac{3}{m} + \frac{4}{m^2}$

23. $1 + \frac{x}{y}$

24. $\frac{5a}{3b^2} + \frac{3}{6b}$

25. $\frac{3a}{2b^2} + \frac{3}{4b}$

26. $\frac{5}{6mn^2} - \frac{3}{4m^2n}$

27. $\frac{5}{7x} - \frac{1+x}{14x^2}$

28. $\frac{7}{3x} - \frac{1-x}{6x^2}$

29. $\dfrac{7}{2a} + \dfrac{4 + a}{4a^2}$

30. $\dfrac{7}{2a} - \dfrac{4 + a}{4a^2}$

31. $\dfrac{2}{5a} - \dfrac{3 - a}{10a^2}$

32. $\dfrac{7}{x + 2} - \dfrac{3}{x - 2}$

33. $\dfrac{x}{x + 2} - \dfrac{2x}{x + 3}$

34. $\dfrac{r + 2}{3} + \dfrac{r - 3}{9}$

35. $\dfrac{x - 2}{6} - \dfrac{x + 1}{3}$

36. $\dfrac{2a - b}{2} - \dfrac{a + b}{3}$

37. $\dfrac{3}{x + y} + \dfrac{1}{x - y}$

38. $\dfrac{5}{m + n} + \dfrac{2}{(m + n)^2}$

39. $\dfrac{x}{2(x - 2)} + \dfrac{2x - 1}{x - 2}$

40. $\dfrac{3a}{a - b} + \dfrac{2b}{a + b}$

41. $\dfrac{4r - 3}{8} - \dfrac{3r + 1}{6} + \dfrac{r + 2}{2}$

42. $\dfrac{3x + y}{3} + \dfrac{x + 2y}{6} - \dfrac{x + y}{2}$

Convert these mixed numerals to single improper fractions.

43. $1\dfrac{1}{4}$

44. $2\dfrac{1}{2}$

45. $8\dfrac{2}{3}$

46. $7\dfrac{1}{5}$

47. $-1\dfrac{1}{3}$

48. $-4\dfrac{5}{6}$

49. $-10\dfrac{1}{2}$

50. $-13\dfrac{5}{9}$

Convert these improper fractions to mixed numerals.

51. $\dfrac{6}{5}$

52. $\dfrac{15}{2}$

53. $\dfrac{10}{3}$

54. $\dfrac{22}{3}$

55. $-\dfrac{5}{2}$

56. $-\dfrac{19}{6}$

57. $-\dfrac{14}{3}$

58. $-\dfrac{37}{5}$

RATIOS AND PROPORTIONS

3.8   ratio

An important application of fractions is to describe ratios. A **ratio** is a comparison of two quantities by division. The ratio of a to b can be written three ways:

(1) a to b

(2) $a : b$ (read a is to b)

(3) $\dfrac{a}{b}$

In algebra the fractional form is preferred and is usually reduced to lowest terms.

Describe the ratio of 5 to 10

Example 1

Solution: Write a fraction and reduce to lowest terms.

$$\frac{5}{10} = \frac{1}{2}$$

Thus, the ratio of 5 to 10 is equivalent to the ratio of 1 to 2.

When a ratio compares two numbers with the same unit of measurement, the unit of measurement is not written.

Joe's weekly salary is $400, while Jan's weekly salary is $600. What is the ratio of Joe's salary to Jan's?

Example 2

Solution: Write a fraction, but do not include the unit of measurement (dollars); then reduce to lowest terms.

$$\frac{400}{600} = \frac{2}{3}$$

The ratio of Joe's salary to Jan's is 2 to 3.

A ratio comparing two numbers with different units of measurement is called a **rate**. Units of measurement must be included in a rate. Also, when working with rates, it is helpful to remember that the word "per" means "divided by." Thus, "miles per hour" means "miles divided by hours."

rate

A jet airliner travels a distance of 1,500 miles in 3 hours. Find its rate of miles to hours.

Example 3

Solution: Write a fraction including the different units of measurement, then reduce to lowest terms.

$$\frac{1500 \text{ miles}}{3 \text{ hours}} = \frac{500 \text{ miles}}{1 \text{ hour}}$$

Thus, its rate is 500 miles per hour.

On a trip of 250 miles, Cheri's car used 10 gallons of gasoline. Find the rate of miles to gallons.

Example 4

Solution: $$\frac{250 \text{ miles}}{10 \text{ gallons}} = \frac{25 \text{ miles}}{1 \text{ gallon}}$$

Thus, Cheri's car averaged 25 miles per gallon.

proportion

An equation showing two ratios to be equal is called a **proportion**. Thus, the equation

$$\frac{a}{b} = \frac{c}{d} \text{ where } b, d \neq 0$$

is a proportion. It is read, "*a* is to *b* as *c* is to *d*."

To determine whether two ratios are proportional, we use the equality test for fractions and see if the cross products are equal.

Example 5

The equality test for fractions has been used to test the proportionality of the following ratios.

(a) $\frac{3}{5}$ and $\frac{6}{10}$ are proportional because $3 \cdot 10 = 5 \cdot 6$

(b) $\frac{5}{6}$ and $\frac{6}{7}$ are not proportional because $5 \cdot 7 \neq 6 \cdot 6$

A proportion involves four numbers. If any three of these numbers are known, we can solve for the fourth by setting the cross products equal as first discussed in Section 3.1.

Example 6

Solve the proportion $\frac{5}{x} = \frac{12}{7}$

Solution: Set the cross products equal and solve for x.

$$\frac{5}{x} = \frac{12}{7}$$
$$12x = 35$$
$$x = \frac{35}{12}$$

Example 7

Solve the proportion $\frac{n + 3}{2n - 1} = \frac{6}{5}$

Solution: Set the cross products equal and solve for *n*.

$$\frac{n + 3}{2n - 1} = \frac{6}{5}$$
$$5(n + 3) = 6(2n - 1)$$
$$5n + 15 = 12n - 6$$

$$5n = 12n - 21 \quad \text{[subtract 15]}$$
$$-7n = -21 \quad \text{[subtract 12}n\text{]}$$
$$n = 3$$

As illustrated by the next example, proportions occur in many word problems.

If it takes 4 gallons of paint to cover 1200 square feet, how many gallons are needed to cover 2700 square feet?   **Example 8**

Solution: Let x represent the number of gallons needed to cover 2700 square feet. Set up a proportion where one ratio relates gallons of paint and the other relates square feet. Corresponding numbers must appear in the numerators and in the denominators. It is helpful to think, "x gallons is to 4 gallons as 2700 square feet is to 1200 square feet." Thus,

x gal. ⟶ $\dfrac{x}{4}$ = $\dfrac{2700}{1200}$ ⟵ 2700 sq. ft.
⟵ 1200 sq. ft.

Solve the proportion for x.

$$\frac{x}{4} = \frac{2700}{1200}$$
$$1200x = 4(2700) \quad \text{[cross products]}$$
$$1200x = 10800$$
$$x = \frac{10800}{1200}$$
$$x = 9$$

9 gallons of paint will be needed to cover 2700 square feet.

Describe each of these ratios by writing a fraction in lowest terms. Be sure both numerator and denominator have the same unit of measurement.   **Exercise 3.8**

Describe the ratio of 3 days to 1 week.   **Example**

Solution: Convert 1 week to 7 days and form the ratio of 3 to 7, producing $\dfrac{3}{7}$.

1. 50 miles to 100 miles   **2.** 10 miles to 5 miles
3. 10 feet to 15 feet   **4.** 30 dollars to 90 dollars
5. 50 cents to 2 dollars   **6.** 60 people to 100 people

7. 121 people to 11 people   **8.** 4 feet to 2 yards
9. 3 pints to 5 quarts   **10.** 4 inches to 1 foot
11. 50 inches to 5 yards   **12.** 5 hours to 1 day
13. 15 minutes to 1 hour   **14.** 12 days to 2 weeks
15. 4 pounds to 4 ounces   **16.** 2 gallons to 2 quarts

Describe each of these rates by writing a fraction in lowest terms. Be sure to include the unit of measurement.

17. A car travels 250 miles in 5 hours. Find the rate in miles per hour.
18. A jet airliner travels 2000 miles in 4 hours. Find its rate in miles per hour.
19. Harry's car travels 240 miles using 8 gallons of gasoline. Find the rate of fuel consumption in miles per gallon.
20. A 10 ounce bar of pure gold is worth $3500. Find its rate in dollars per ounce.
21. A 1000 ounce bar of pure silver is worth $6750. Find its rate of worth in dollars per ounce.
22. A quart of milk costs 64¢. Find its rate of cost in cents per ounce.
23. A quart of orange juice costs $1.28. Find its rate of cost in cents per ounce.
24. Jim earns an annual salary of $18,000. Find his monthly rate.
25. Joyce earns an annual salary of $18,250. Find her daily rate (1 year = 365 days).

Use the equality test for fractions to determine whether the following ratios are proportional. Answer yes or no.

26. $\frac{4}{5}$ and $\frac{8}{10}$ **27.** $\frac{2}{3}$ and $\frac{3}{5}$

28. $\frac{1}{6}$ and $\frac{3}{18}$ **29.** $\frac{12}{15}$ and $\frac{16}{20}$

30. $\frac{8}{6}$ and $\frac{12}{9}$ **31.** $\frac{7}{5}$ and $\frac{20}{14}$

32. $\frac{1.2}{3.6}$ and $\frac{3.4}{10.2}$ **33.** $\frac{10.4}{12.6}$ and $\frac{9.3}{11.2}$

Solve the following proportions.

34. $\frac{a}{10} = \frac{2}{5}$ **35.** $\frac{6}{x} = \frac{24}{20}$

36. $\frac{45}{2} = \frac{90}{k}$ **37.** $\frac{3}{n} = \frac{24}{32}$

38. $\frac{20}{x} = \frac{100}{80}$ **39.** $\frac{4}{n} = \frac{20}{35}$

40. $\dfrac{1}{3} = \dfrac{n}{21}$

41. $\dfrac{9}{15} = \dfrac{5}{z}$

42. $\dfrac{7}{8} = \dfrac{z}{5}$

43. $\dfrac{76}{9} = \dfrac{n}{36}$

44. $\dfrac{4}{11} = \dfrac{100}{y}$

45. $\dfrac{3}{2n} = \dfrac{3}{8}$

46. $\dfrac{3x}{5} = \dfrac{7}{8}$

47. $\dfrac{2}{5y} = \dfrac{3}{10}$

48. $\dfrac{x + 1}{x - 1} = \dfrac{1}{3}$

49. $\dfrac{h - 1}{h + 7} = \dfrac{4}{3}$

50. $\dfrac{3}{x - 9} = \dfrac{7}{x + 8}$

51. $\dfrac{6x - 5}{x} = \dfrac{11}{5}$

52. $\dfrac{3}{a + 1} = \dfrac{2}{3a}$

53. $\dfrac{3y - 2}{5} = \dfrac{2y - 1}{6}$

Use proportions to solve these word problems.

54. If your heart beats 17 times in 15 seconds, how many times will it beat in 60 seconds?

55. If 6 cans of orange juice cost $1.62, what would 10 cans cost?

56. If a car uses 15 gallons of gas for a trip of 471 miles, how many gallons are needed for a trip of 628 miles?

57. If 4 pounds of fertilizer will cover 1000 square feet of lawn, how many pounds will be needed to cover 1750 square feet?

58. On a map a 3-inch segment represents 51 miles. How many miles does a 7-inch segment represent?

59. The scale on a map reads "1 inch = 40 miles." How many inches are needed to show a distance of 200 miles?

60. If the sales tax on a $50 item is $3, find the tax on a $90 item.

61. At a rate of 2 defective television sets for each 75 manufactured, how many defective sets will there be for 6000 manufactured?

62. Jane can type 11 pages in 2 hours. At this rate, how long will it take her to type a 132-page report?

63. If a car uses 14 gallons of gasoline to travel 308 miles, how far can it travel on 20 gallons?

Summary

		SYMBOLS
LCD	Lowest common denominator	
$a : b$	Ratio of a to b	

$$\dfrac{a}{b} = a \div b \quad \text{or} \quad \dfrac{a}{b} = a \cdot \dfrac{1}{b} \text{ where } b \neq 0$$

DEFINITION OF A FRACTION

BASIC PROPERTIES OF FRACTIONS

(1) The denominator of a fraction can never be zero; $\dfrac{b}{0}$ does not represent a number.

(2) If the denominator is 1, then the fraction is equal to the numerator: $\dfrac{5a}{1} = 5a$

(3) If the numerator is zero and the denominator is not zero, then the fraction is equal to zero: $\dfrac{0}{6} = 0$

(4) If the numerator and denominator are equal, but not zero, then the fraction is equal to 1: $\dfrac{7}{7} = 1$

THE SIGN PROPERTY OF FRACTIONS

$$\frac{-a}{b} = \frac{a}{-b} = -\frac{a}{b} \text{ where } b \neq 0$$

THE EQUALITY TEST FOR FRACTIONS

$$\frac{a}{b} = \frac{c}{d} \text{ if and only if } ad = bc, \text{ where } b, d \neq 0$$

THE FUNDAMENTAL PRINCIPLE OF FRACTIONS

$$\frac{a}{b} = \frac{a \cdot n}{b \cdot n} \text{ and } \frac{a}{b} = \frac{a \div n}{b \div n} \text{ where } b, n \neq 0$$

MULTIPLICATION OF FRACTIONS

$$\frac{a}{b} \cdot \frac{c}{d} = \frac{a \cdot c}{b \cdot d} \text{ where } b, d \neq 0$$

DIVISION OF FRACTIONS

$$\frac{a}{b} \div \frac{c}{d} = \frac{a}{b} \cdot \frac{d}{c} \text{ where } b, c, d \neq 0$$

DEFINITION OF A PRIME NUMBER

A prime number is any natural number that is evenly divisible by exactly two different divisors, itself and 1. The number 1 is not a prime number.

DEFINITION OF A COMPOSITE NUMBER

A composite number is any natural number other than 1 which is not prime.

LOWEST COMMON DENOMINATOR (LCD)

The lowest common denominator of a group of fractions is the smallest number which is evenly divisible by each denominator.

ADDITION AND SUBTRACTION OF FRACTIONS

Step (1) Find the LCD.
Step (2) Expand each fraction so it has the LCD.
Step (3) Combine the numerators and place the result over the LCD. Reduce if possible.

DEFINITION OF A PROPORTION

A proportion is an equation showing two ratios to be equal.

Convert each of these fractions to a basic product and a basic quotient.

1. $\dfrac{x}{y}$

2. $\dfrac{3m}{m + 2}$

3. $\dfrac{x + 5}{x + 7}$

4. $\dfrac{2y + 7}{5y}$

Simplify these fractional expressions.

5. $\dfrac{a}{1}$

6. $\dfrac{2n + 5}{1}$

7. $\dfrac{5x}{5x}$, where $x \neq 0$

8. $\dfrac{x + 7}{x + 7}$, where $x \neq -7$

9. $\dfrac{0}{a + 3}$, where $a \neq -3$

10. $\dfrac{0}{2a - 6}$, where $a \neq 3$

True or false. (Use the sign property of fractions.)

11. $\dfrac{-5}{8} = \dfrac{5}{-8}$

12. $\dfrac{-14}{15} = -\dfrac{14}{15}$

13. $\dfrac{3a}{5b} = \dfrac{-3a}{-5b}$, where $b \neq 0$

14. $\dfrac{n}{2} = -\dfrac{-n}{-2}$

15. $\dfrac{7m}{n} = -\dfrac{7m}{n}$, where $b \neq 0$

16. $\dfrac{x - 3}{5} = -\dfrac{x - 3}{-5}$

Expand these fractions as directed.

17. $\dfrac{3}{5y}$ so that its denominator becomes $15y^2$

18. $\dfrac{2a}{3bc^2}$ so that its denominator becomes $6b^2c^3$

Reduce to lowest terms. (Assume that denominators are not zero.)

19. $\dfrac{3x}{6x^2}$

20. $\dfrac{-12m^2n}{4mn}$

21. $\dfrac{10a^2b^3}{15a^3b^5}$

22. $\dfrac{3(b + 2)}{5a(b + 2)}$

23. $\dfrac{2x + 6}{5x + 15}$

24. $\dfrac{3m(n - 2)}{n - 2}$

Multiple or divide the following fractions and put the answers in reduced form. (Assume that denominators do not represent zero.)

25. $\dfrac{15}{21} \cdot \dfrac{7}{3}$

26. $\dfrac{2m}{n} \cdot \dfrac{5n^2}{m^2}$

27. $\dfrac{3x}{7} \cdot \dfrac{5}{x^2}$

28. $\dfrac{3a^2b}{2c^2} \cdot \dfrac{c}{9ab}$

29. $\dfrac{5a - 5b}{3} \cdot \dfrac{6}{7a - 7b}$

30. $\dfrac{3m + 3n}{4m^2} \div \dfrac{5m + 5n}{m}$

31. $\dfrac{15x - 30}{3x + 6} \div \dfrac{9x - 18}{6x + 12}$ **32.** $\dfrac{2a + 6}{2a^3} \div (a + 3)$

Use the equality test to solve these equations.

33. $\dfrac{x}{3} = \dfrac{4}{6}$ **34.** $\dfrac{2}{y} = \dfrac{3}{5}$

35. $\dfrac{3m}{5} = \dfrac{1}{2}$ **36.** $\dfrac{x - 3}{2} = 1$

37. $\dfrac{2}{n + 1} = \dfrac{3}{5}$ **38.** $\dfrac{a - 3}{2} = \dfrac{a}{4}$

Classify the following natural numbers as being either prime or composite.

39. 10   **40.** 21

41. 13   **42.** 23

43. 27   **44.** 51

Give the prime factorization for each of these numbers. Put your answers in exponential form.

45. 24

46. 108

Combine these fractions by addition or subtraction. Be sure your answers are in reduced form. (Assume that denominators are not zero.)

41. $\dfrac{3}{x} + \dfrac{5}{x^2}$ **48.** $\dfrac{3}{10x^3y} + \dfrac{7}{8x^2y^2}$

49. $\dfrac{3}{2x} - \dfrac{x + 2}{4x^2}$ **50.** $x + \dfrac{x + 2}{3} - \dfrac{x - 1}{9}$

51. $\dfrac{5}{m + 3} - \dfrac{7}{m - 3}$

Do as directed.

52. Convert $-2\dfrac{5}{8}$ to fractional form.

53. Convert $\dfrac{17}{3}$ to a mixed numeral.

Use proportions to solve these word problems.

54. If a car uses 12 gallons of gasoline to travel 312 miles, how many gallons are needed for a trip of 572 miles?

55. If 8 cans of frozen lemon concentrate cost $5.20, what would 12 cans cost?

1. $\dfrac{x}{y} = x \cdot \dfrac{1}{y}$ and $\dfrac{x}{y} = x \div y$

2. $\dfrac{3m}{m+2} = (3m) \cdot \dfrac{1}{m+2}$ and $\dfrac{3m}{m+2} = (3m) \div (m+2)$

3. $\dfrac{x+5}{x+7} = (x+5) \cdot \dfrac{1}{x+7}$ and $\dfrac{x+5}{x+7} = (x+5) \div (x+7)$

4. $\dfrac{2y+7}{5y} = (2y+7) \cdot \dfrac{1}{5y}$ and $\dfrac{2y+7}{5y} = (2y+7) \div (5y)$

5. $\dfrac{a}{1} = a$

6. $\dfrac{2n+5}{1} = 2n+5$

7. $\dfrac{5x}{5x} = 1$

8. $\dfrac{x+7}{x+7} = 1$

9. $\dfrac{0}{a+3} = 0$

10. $\dfrac{0}{2a-6} = 0$

11. True

12. True

13. True

14. False

15. False

16. True

17. $\dfrac{3}{5y} = \dfrac{3 \cdot (3y)}{5y \cdot (3y)} = \dfrac{9y}{15y^2}$

18. $\dfrac{2a}{3bc^2} = \dfrac{2a \cdot (2bc)}{3bc^2 \cdot (2bc)} = \dfrac{4abc}{6b^2c^3}$

19. $\dfrac{3x}{6x^2} = \dfrac{\cancel{3} \cdot \cancel{x}}{2 \cdot \cancel{3} \cdot \cancel{x} \cdot x} = \dfrac{1}{2x}$

20. $\dfrac{-12m^2n}{4mn} = \dfrac{-3 \cdot \cancel{4} \cdot \cancel{m} \cdot m \cdot \cancel{n}}{\cancel{4} \cdot \cancel{m} \cdot \cancel{n}} = -3m$

21. $\dfrac{10a^2b^3}{15a^3b^5} = \dfrac{2 \cdot \cancel{5} \cdot \cancel{a^2} \cdot \cancel{b^3}}{3 \cdot \cancel{5} \cdot \cancel{a^2} \cdot a \cdot \cancel{b^3} \cdot b^2} = \dfrac{2}{3ab^2}$

22. $\dfrac{3(b+2)}{5a(b+2)} = \dfrac{3 \cdot \cancel{(b+2)}}{5 \cdot a \cdot \cancel{(b+2)}} = \dfrac{3}{5a}$

23. $\dfrac{2x+6}{5x+15} = \dfrac{2 \cdot \cancel{(x+3)}}{5 \cdot \cancel{(x+3)}} = \dfrac{2}{5}$

24. $\dfrac{3m(n-2)}{n-2} = \dfrac{3m\cancel{(n-2)}}{\cancel{(n-2)}} = 3m$

25. $\dfrac{15}{21} \cdot \dfrac{7}{3} = \dfrac{\cancel{3} \cdot 5}{\cancel{3} \cdot \cancel{7}} \cdot \dfrac{\cancel{7}}{3} = \dfrac{5}{3}$

26. $\dfrac{2m}{n} \cdot \dfrac{5n^2}{m^2} = \dfrac{2 \cdot \not{m}}{\not{n}} \cdot \dfrac{5 \cdot \not{n} \cdot n}{\not{m} \cdot m} = \dfrac{10n}{m}$

27. $\dfrac{3x}{7} \cdot \dfrac{5}{x^2} = \dfrac{3 \cdot \not{x}}{7} \cdot \dfrac{5}{\not{x} \cdot x} = \dfrac{15}{7x}$

28. $\dfrac{3a^2b}{2c^2} \cdot \dfrac{c}{9ab} = \dfrac{\not{3} \cdot \not{a} \cdot a \cdot \not{b}}{2 \cdot c \cdot \not{c}} \cdot \dfrac{\not{c}}{\not{3} \cdot 3 \cdot \not{a} \cdot \not{b}} = \dfrac{a}{6c}$

29. $\dfrac{5a - 5b}{3} \cdot \dfrac{6}{7a - 7b} = \dfrac{5 \cdot (a - b)}{\not{3}} \cdot \dfrac{2 \cdot \not{3}}{7 \cdot (a - b)} = \dfrac{10}{7}$

30. $\dfrac{3m + 3n}{4m^2} \div \dfrac{5m + 5n}{m} = \dfrac{3 \cdot (m + n)}{4 \cdot m \cdot \not{m}} \cdot \dfrac{\not{m}}{5 \cdot (m + n)} = \dfrac{3}{20m}$

31. $\dfrac{15x - 30}{3x + 6} \div \dfrac{9x - 18}{6x + 12} = \dfrac{\not{3} \cdot 5 \cdot (x - 2)}{\not{3} \cdot (x + 2)} \cdot \dfrac{2 \cdot \not{3} \cdot (x + 2)}{3 \cdot \not{3} \cdot (x - 2)} = \dfrac{10}{3}$

32. $\dfrac{2a + 6}{2a^3} \div (a + 3) = \dfrac{\not{2} \cdot (a + 3)}{\not{2} \cdot a^3} \cdot \dfrac{1}{(a + 3)} = \dfrac{1}{a^3}$

33. $\dfrac{x}{3} = \dfrac{4}{6}$

$6x = 12$

$x = 2$

34. $\dfrac{2}{y} = \dfrac{3}{5}$

$3y = 10$

$y = \dfrac{10}{3}$

35. $\dfrac{3m}{5} = \dfrac{1}{2}$

$6m = 5$

$m = \dfrac{5}{6}$

36. $\dfrac{x - 3}{2} = \dfrac{1}{1}$

$x - 3 = 2$

$x = 5$

37. $\dfrac{2}{n + 1} = \dfrac{3}{5}$

$3(n + 1) = 10$

$3n + 3 = 10$

$3n = 7$

$n = \dfrac{7}{3}$

38. $\dfrac{a - 3}{2} = \dfrac{a}{4}$

$4(a - 3) = 2a$

$4a - 12 = 2a$

$2a = 12$

$a = 6$

39. Composite

40. Composite

41. Prime

42. Prime

43. Composite

44. Composite

45. $24 = 2^3 \cdot 3$

46. $108 = 2^2 \cdot 3^3$

47. $\dfrac{3}{x} + \dfrac{5}{x^2} = \dfrac{3x}{x^2} + \dfrac{5}{x^2} = \dfrac{3x + 5}{x^2}$

48. $\dfrac{3}{10x^3y} + \dfrac{7}{8x^2y^2} = \dfrac{12y}{40x^3y^2} + \dfrac{35x}{40x^3y^2} = \dfrac{12y + 35x}{40x^3y^2}$

49. $\dfrac{3}{2x} - \dfrac{x + 2}{4x^2} = \dfrac{6x}{4x^2} - \dfrac{x + 2}{4x^2}$

$= \dfrac{6x - (x + 2)}{4x^2}$

$= \dfrac{6x - x - 2}{4x^2}$

$= \dfrac{5x - 2}{4x^2}$

50. $x + \dfrac{x + 2}{3} - \dfrac{x - 1}{9} = \dfrac{9x}{9} + \dfrac{3(x + 2)}{9} - \dfrac{x - 1}{9}$

$$= \dfrac{9x + 3(x + 2) - (x - 1)}{9}$$

$$= \dfrac{9x + 3x + 6 - x + 1}{9}$$

$$= \dfrac{11x + 7}{9}$$

51. $\dfrac{5}{m + 3} - \dfrac{7}{m - 3} = \dfrac{5(m - 3)}{(m + 3)(m - 3)} - \dfrac{7(m + 3)}{(m - 3)(m + 3)}$

$$= \dfrac{5(m - 3) - 7(m + 3)}{(m + 3)(m - 3)}$$

$$= \dfrac{5m - 15 - 7m - 21}{(m + 3)(m - 3)}$$

$$= \dfrac{-2m - 36}{(m + 3)(m - 3)} \ \text{ or } \ \dfrac{-2(m + 18)}{(m + 3)(m - 3)}$$

52. $-2\dfrac{5}{8} = \dfrac{-21}{8}$ **53.** $\dfrac{17}{3} = 5\dfrac{2}{3}$

54. $\quad \dfrac{12}{x} = \dfrac{312}{572}$

$312x = (12)(572)$

$312x = 6864$

$\quad\ x = \dfrac{6864}{312}$

$\quad\ x = 22$

22 gallons of gasoline are needed.

55. $\dfrac{8}{12} = \dfrac{5.20}{x}$

$8x = (12)(5.20)$

$8x = 62.4$

$\ x = \dfrac{62.4}{8}$

$\ x = 7.80$

The cost for 12 cans is $7.80.

OPERATIONS WITH POLYNOMIAL EXPRESSIONS

4

POLYNOMIAL EXPRESSIONS

4.1

terms

polynomial

In an expression such as $4x^3 + 2x^2 - 3x + 1$, the quantities which are added or subtracted are called **terms**. Thus, $4x^3$, $2x^2$, $3x$, and 1 are the terms of that expression. In mathematics we are particularly interested in expressions where each term is a product of a number and variable raised to a whole number power, such as $5x^6$. Expressions with terms of this type are called polynomials. That is, a **polynomial** is defined to be an expression containing a finite number of terms where each term fits the pattern ax^n. a represents any real number, x represents a variable, and n represents any whole number.

Example 1

Each of these expressions is a polynomial.
(a) $6x^4 + 5x^3 - 4x^2 - 2x$ $[2x = 2x^1]$
(b) $3y^2 - 5y + 2$ $[2 = 2y^0]$
(c) $m^2 - 3m + 5$ $[m^2 = 1m^2]$
(d)   6
(e)   0

Example 2

None of these expressions is a polynomial.

(a) $\dfrac{3}{x^2}$

(b) $\dfrac{1}{m}$

(c) $2y^3 + \dfrac{1}{y^2}$

(d) $\dfrac{x^2 + 5}{x - 1}$

Polynomials may also contain more than one variable. However, each term must still be a product where the variables are raised to whole number powers.

Example 3

Each of these expressions is a polynomial.
(a) $3x^2 + 2y^3$
(b) $5x^2y^3 - 2x^3y^4$
(c) $3a + 2b - 4c$

monomial

Three types of polynomials are so frequently seen in algebra that they are given special names. They are called monomials, binomials, and trinomials. A **monomial** is a polynomial consisting of one term. (The prefix *mon* means "one," as in *monorail*.)

Example 4

Each of these polynomials is a monomial.
(a)   0
(b) -7

(c) $5x$
(d) $-7m^4n^3$

A **binomial** is a polynomial consisting of two terms. (The prefix *bi* means "two," as in *bicycle*.)

binomial

Each of these polynomials is a binomial.
(a) $9x^2 - 1$
(b) $m^2 - n^2$
(c) $y + 3$

Example 5

A **trinomial** is a polynomial consisting of three terms. (The prefix *tri* means "three," as in *tricycle*.)

trinomial

Each of these polynomials is a trinomial.
(a) $x^2 - 5x + 6$
(b) $2m^2 - 3mn + n^2$
(c) $-2y^4 + y^2 - 1$

Example 6

Polynomials of one variable are often symbolized by an uppercase letter together with the variable. For example, the polynomial $2x^3 - x + 4$ might be symbolized by $P(x)$, read "P of x." Thus, $P(x) = 2x^3 - x + 4$. This notation allows us to indicate the value of the polynomial for specific values of the variable. $P(1)$ would represent the value of the polynomial when $x = 1$.

If $P(x) = 2x^3 - x + 4$, find $P(1)$

Example 7

Solution: Let $x = 1$ and evaluate the polynomial.

$$P(x) = 2x^3 - x + 4$$
$$P(1) = 2(1)^3 - (1) + 4$$
$$P(1) = \quad 2 \quad - 1 + 4$$
$$P(1) = 5$$

If $P(m) = 3m^2 + 2m - 1$, find $P(-3)$

Example 8

Solution: Let $m = -3$ and evaluate the polynomial.

$$P(m) \ = 3m^2 + 2m - 1$$
$$P(-3) = 3(-3)^2 + 2(-3) - 1$$
$$P(-3) = \quad 27 \quad - \quad 6 \quad - 1$$
$$P(-3) = 20$$

Polynomial expressions can be added or subtracted by combining the numerical coefficients of like terms. As you learned in section 2.4, this can be done either horizontally or vertically.

Example 9

Add $(3x^2 + 2x - 5) + (x^2 - 3x + 7)$

Horizontal Solution:

$$
\begin{aligned}
(3x^2 + 2x - 5) + (x^2 - 3x + 7) &= 3x^2 + 2x - 5 + x^2 - 3x + 7 \\
&= 3x^2 + x^2 + 2x - 3x - 5 + 7 \\
&= 4x^2 - x + 2
\end{aligned}
$$

Vertical Solution:

$$
\begin{array}{r}
3x^2 + 2x - 5 \\
x^2 - 3x + 7 \\
\hline
4x^2 - x + 2
\end{array}
$$

Example 10

Subtract $(5a^2 - 3a + 2) - (2a^2 - 4a + 1)$

Horizontal Solution:

$$
\begin{aligned}
(5a^2 - 3a + 2) - (2a^2 - 4a + 1) &= 5a^2 - 3a + 2 - 2a^2 + 4a - 1 \\
&= 5a^2 - 2a^2 - 3a + 4a + 2 - 1 \\
&= 3a^2 + a + 1
\end{aligned}
$$

Vertical Solution:

$$
\begin{array}{r}
5a^2 - 3a + 2 \\
2a^2 - 4a + 1 \\
\hline
\end{array}
\quad \xrightarrow{\text{change signs}} \quad
\begin{array}{r}
5a^2 - 3a + 2 \\
-2a^2 + 4a - 1 \\
\hline
3a^2 + a + 1
\end{array}
$$

Exercise 4.1

Which of the following expressions are polynomials?

1. $5x^3 + 2x^2 - 3x + 4$

2. $\dfrac{m^2 + 1}{m^2 - 1}$

3. $6y + 2$

4. $\dfrac{1}{x}$

5. $3y^2 + \dfrac{2}{y}$

6. 0

7. -1

8. $3a^2b^3 - 5ab + 2$

Classify each of these polynomials as being a monomial, binomial, or trinomial.

9. $x^2 + 5x + 6$

10. $m - 3$

11. $4x$

12. $x^2 - y^2$

13. $3a^4b^2$

14. $a + b - c$

15. -6

16. 0

17. $m^2 - 9$

If $P(x) = 2x^3 - 3x^2 + x - 5$, find each of the following values.

18. $P(0)$

19. $P(1)$

20. $P(3)$

21. $P(-1)$

22. $P(-2)$

23. $P(-3)$

If $P(x) = 3x^2 - x + 4$, find each of the following values.

24. $P(0)$

25. $P(1)$

26. $P(2)$

27. $P(-1)$

28. $P(-2)$

29. $P(-3)$

Add or subtract the following polynomials as indicated. You may use either the horizontal or vertical method, whichever you prefer.

30. $(4x^2 + 3x - 5) + (2x^2 - x + 7)$

31. $(2m^2 - 3m - 1) + (m^2 + m - 6)$

32. $(3y^2 + 2y) + (5y^2 - 6y)$

33. $(2a^2 + 3a) + (a^2 - 5)$

34. $(7x^2 - 3x + 1) - (5x^2 + 4x + 3)$

35. $(m^2 - m + 1) - (m^2 + 2m + 2)$

36. $(4y^2 - y) - (3y^2 + 4y)$

37. $(3a^2 + 1) - (a^2 + a)$

38. $(2x^2 + 3x - 2) + (x^2 - 5x + 1) + (3x^2 - x + 4)$

39. $(m^2 + 4m - 3) + (3m^2 - 5m + 1) + (2m^2 - 3m + 2)$

40. $(y^2 + 2y - 3) - (3y^2 + 3y - 4) + (y^2 + 5y + 7)$

41. $(a^2 - 5a - 3) + (3a^2 - 2a + 1) - (4a^2 + 3a - 5)$

42. $(m^3 + 2m^2 + m - 5) + (2m^3 - 3m^2 + m - 6) - (m^3 + m^2 - 3m + 1)$

43. $(2y^4 - 3y^2 + 4) - (y^4 + 2y^2 - 1) - (y^4 + 3y^2 + 2)$

44. $(5x^3 - 3x + 1) + (2x^3 + 2x - 5) - (3x^3 - x + 4)$

45. $(2x^2 + 3) + (5x - 7) - (4x^3 + x) - (2x + 1)$

46. $(3a^2 + 2a - 5) - (4a^2 + 6a - 1) + (2a^2 - a) - (3a + 5)$

47. $(p^3 - p^2 + p - 4) + (p^3 - 2p^2 - p + 1) - (2p^3 + 3p^2) - (p^2 + p - 4)$

48. $(h^3 - 1) + (h^2 + 1) + (3h - 5) - (2h^3 + h) - (3h^2 + 2h - 3)$

49. $(3t^5 - 4t^3 + 7t) + (t^4 - 3t^2 + 5) - (2t^2 - t) - (t - 1)$

50. $(4x^2 - 3x + 1) - (2x^2 - 5x + 2) + (x^2 - 5x - 3) - (3x^2 + 7x - 4)$

MULTIPLICATION OF POLYNOMIALS

4.2

To multiply polynomials we will discuss three cases. First, we will use the addition property of exponents to multiply monomials. Secondly, we will use the distributive properties of multiplication to multiply monomials times polynomials having more than one term. And third, we will use vertical multiplication to find the product of any two polynomials.

Case (1)   Multiplying Monomials

Recall from Section 2.2 that the commutative and associative properties allow us to multiply the numerical coefficients. And the addition property of exponents permits us to add the exponents that appear on identical bases.

Example 1

Multiply $(-3x^2)(4x^3)$

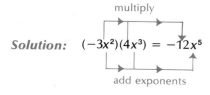

Solution: $(-3x^2)(4x^3) = -12x^5$

Example 2

Multiply $(-4a)(2a^3b^2)(-3abc)$

Solution: $(-4a)(2a^3b^2)(-3abc) = 24a^5b^3c$

Case (2)   Multiplying a Monomial Times a Polynomial Having More Than One Term

As studied in section 2.2, the distributive properties of multiplication are used to multiply each term of the polynomial by the monomial.

Example 3

Multiply $3m^2(2m^2 - 4m + 1)$

Solution: $3m^2(2m^2 - 4m + 1) = 6m^4 - 12m^3 + 3m^2$

Multiply $-2x^3y^2(5x^2y - 3xy^2 + 4)$ Example 4

Solution: $-2x^3y^2(5x^2y - 3xy^2 + 4) = -10x^5y^3 + 6x^4y^4 - 8x^3y^2$

Case (3) Multiplying Polynomials

In general, to multiply two polynomials we use a vertical method similar to ordinary multiplication of numbers. Each term of one polynomial is multiplied by each term of the other polynomial.

Multiply $3x(2x^2 - 5x + 2)$ Example 5

Solution: Arrange vertically and multiply the monomial times each term of the trinomial.

$$\begin{array}{r} 2x^2 - 5x + 2 \\ 3x \\ \hline 6x^3 - 15x^2 + 6x \end{array}$$

Multiply $(x + 2)(3x^2 - 5x + 4)$ Example 6

Solution: Arrange vertically and multiply each term of the binomial times each term of the trinomial.

$$\begin{array}{r} 3x^2 - 5x + 4 \\ x + 2 \\ \hline 6x^2 - 10x + 8 \\ 3x^3 - 5x^2 + 4x \\ \hline 3x^3 + x^2 - 6x + 8 \end{array}$$

arrange like terms in the same columns and add

Multiply $(2m^2 - 5m - 2)(3m^3 + 2m^2 - m + 3)$ Example 7

Solution: Arrange vertically and multiply each term of one polynomial times each term of the other.

$$\begin{array}{r} 3m^3 + 2m^2 - m + 3 \\ 2m^2 - 5m - 2 \\ \hline - 6m^3 - 4m^2 + 2m - 6 \\ - 15m^4 - 10m^3 + 5m^2 - 15m \\ 6m^5 + 4m^4 - 2m^3 + 6m^2 \\ \hline 6m^5 - 11m^4 - 18m^3 + 7m^2 - 13m - 6 \end{array}$$

Since addition and multiplication are commutative and associative, it makes no difference whether you multiply from right to left or left to right. Many mathematicians prefer the following format. However, you may take your choice.

Example 8   Multiply $(2x - 3)(3x + 4)$

Solution: Arrange vertically and multiply from left to right.

$$
\begin{array}{r}
3x + 4 \\
2x - 3 \\
\hline
6x^2 + 8x \\
- 9x - 12 \\
\hline
6x^2 - x - 12
\end{array}
$$

Exercise 4.2   Multiply the following polynomials.

1. $-4(5x)$
2. $(-2a)(-3a)$
3. $(-5x)(-7x)$
4. $(-3a^2)(2a^3)$
5. $(3m^2)(-2m^5)$
6. $(-4x^2y)(-3xy^2)$
7. $(3a^2b)(-4ab^2)$
8. $(5m^2n)(-2mn^4)$
9. $(2m^2n)(-6mn)(3mn^3)$
10. $(-4x^2y^3)(-2xyz)(-3y^3z^2)$
11. $(-2ab^2c)(-3ab)(4a^2c^2)$
12. $(xy^2)(x^2y)(3y^2z)(-2xy)$
13. $3(2x - 4)$
14. $-3x(4x^2 - 5)$
15. $2(m - 5)$
16. $-3a(7a^2 + 2)$
17. $-2a(3a^2 - 4b^2)$
18. $-5xy(2x^2 - 3y^2)$
19. $2xy(3x - 4y + 5)$
20. $3a^2(4a^2b - 3ab + 2)$
21. $-5r^2t(3r^2 - t^2 + 2r)$
22. $-2mn(4m^2n - 5m + 2n)$
23. $3x^2y(-2xy + 3x^2 - 4y^2)$
24. $7a^2bc^3(2a^3b - 3b^2c + 4ac^2)$
25. $(x + 4)(x + 2)$
26. $(m - 3)(m + 2)$
27. $(2x + 3)(3x - 5)$
28. $(x - 2)(x^2 + 2x - 1)$
29. $(a + 4)(5a^2 - 3a + 2)$
30. $(2m - 3)(3m^2 - 5m + 2)$
31. $(r + 3)(2r^2 - 7r + 3)$
32. $(2a + 1)(a^2 - a + 2)$
33. $(2x^2 - 5x + 3)(3x^2 + x - 1)$
34. $(m^2 + 2m + 1)(m^2 - 3m - 2)$
35. $(3a^2 - 4a - 5)(2a^2 + a - 3)$
36. $(r^2 + 2rs - s^2)(3r^2 - 5rs + 2s^2)$
37. $(2x - 3)(5x^4 - 3x^3 + x^2 - 5x + 1)$
38. $(3a - 1)(2a^4 - 4a^3 + 3a^2 - a - 2)$
39. $(2m^2 + m - 3)(3m^4 - 4m^3 + m^2 - m + 2)$
40. $(3x^2 - 2x + 1)(x^4 - x^3 + 3x^2 - 5x + 2)$
41. $(2x - 3)(4x^2 + 6x + 9)$
42. $(3m + n)(9m^2 - 3mn + n^2)$
43. $(x - 1)(x^3 + x^2 + x + 1)$
44. $(x + y)(x^4 - x^3y + x^2y^2 - xy^3 + y^4)$
45. $(y + 2)(y + 3)(y + 4)$
46. $(2m + 3)(m - 1)(3m - 2)$
47. $(r + 2)(r - 2)(r + 3)$

48. $(3a + 2)(a - 1)(a + 2)$ **49.** $(x + 1)(2x + 3)(x - 4)(x + 2)$

50. $(3y + 2)(y - 1)(y + 2)(y - 3)$

SPECIAL PRODUCTS OF BINOMIALS

4.3

Vertical multiplication, discussed in the previous section, works for any two polynomials. Since, in algebra, many polynomials are binomials, it will be helpful to develop three short cuts. These shortcuts will enable us to multiply two binomials quickly and efficiently. They are called **special products** and fall into these categories:

special products

(1)   Product of Two Binomials (FOIL shortcut)

Example: $(2x + 3)(3x + 2)$

(2)   Product of a Sum Times a Difference

Example: $(3x + 5)(3x - 5)$

(3)   Squaring a Binomial

Example: $(3x - 4)^2$

First we consider the product of two binomials and develop the FOIL shortcut. To do this we find the product of two general binomials, $(a + b)$ and $(c + d)$, using the vertical method. We then analyze the answer to get a useful shortcut.

$$
\begin{array}{r}
c + d \\
a + b \\
\hline
ac + ad \quad\quad\quad\quad \\
+ \ bc + bd \\
\hline
ac + ad + bc + bd
\end{array}
$$

Comparing the indicated product with the answer, we find the following pattern, called the *FOIL* shortcut:

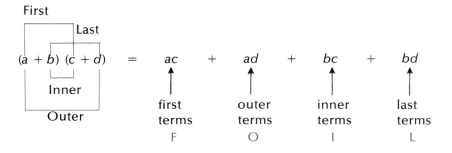

Multiplying Binomials

To use the FOIL shortcut:
(1)   Multiply the *first* terms of the two binomials.
(2)   Multiply their *outer* terms.
(3)   Multiply their *inner* terms.
(4)   Multiply their *last* terms.

Example 1

Multiply $(x + 2)(x + 3)$

Solution: Use the FOIL shortcut.
(1)   Multiply their first terms: $(x)(x) = x^2$
(2)   Multiply their outer terms: $(x)(3) = 3x$
(3)   Multiply their inner terms: $(2)(x) = 2x$
(4)   Multiply their last terms: $(2)(3) = 6$
Thus, $(x + 2)(x + 3) = x^2 + 3x + 2x + 6$
$$= x^2 + 5x + 6$$

To further shorten this process consider the following examples.

Example 2

Multiply $(a + 3)(2a + 1)$

Solution:
$$(a + 3)(2a + 1) = \overset{F}{a \cdot 2a} + \overset{O}{a \cdot 1} + \overset{I}{3 \cdot 2a} + \overset{L}{3 \cdot 1}$$
$$= 2a^2 + a + 6a + 3$$
$$= 2a^2 + 7a + 3$$

Example 3

Multiply $(2m - 1)(3m + 2)$

Solution:
$$(2m - 1)(3m + 2) = \overset{F}{2m \cdot 3m} + \overset{O}{2m \cdot 2} - \overset{I}{1 \cdot 3m} - \overset{L}{1 \cdot 2}$$
$$= 6m^2 + 4m - 3m - 2$$
$$= 6m^2 + m - 2$$

In each of the examples you should see that the outer product and the inner product were combined to give the middle term of the trinomial answer. Thus, you can think of the FOIL shortcut as $F + (O + I) + L$. You should learn to do this mentally, finding the answer in one step.

Example 4

Multiply $(2y + 3)(y + 5)$

Solution:
$$(2y + 3)(y + 5) = \overset{F}{2y^2} + \overset{(O + I)}{13y} + \overset{L}{15}$$

Multiply $(5x - 1)(2x - 3)$ **Example 5**

Solution:
$$(5x - 1)(2x - 3) = \overset{F}{10x^2} \overset{(O + I)}{- 17x} \overset{L}{+ 3}$$

The second type of special product is called a **sum times a difference,**   **sum times**
symbolized by $(a + b)(a - b)$. For products of this type, one binomial is   **a difference**
the *sum* of two terms while the other is the *difference* of exactly the
same two terms.

Each of these is classified as a sum times a difference.   **Example 6**
(a) $(x + 3)(x - 3)$
(b) $(2x + 5)(2x - 5)$
(c) $\left(3a - \dfrac{1}{2}\right)\left(3a + \dfrac{1}{2}\right)$

The shortcut for a sum times a difference is obtained by using the FOIL
method to multiply $(a + b)$ times $(a - b)$.

$$(a + b)(a - b) = a^2 - ab + ba - b^2$$
$$= a^2 + \quad 0 \quad - b^2$$
$$= a^2 - b^2$$

The result, $a^2 - b^2$, is called a **difference of two squares.**   **difference of**
                                                                               two squares

---

Multiplying a Sum Times a Difference
To multiply a sum times a difference, multiply the first terms and
then multiply the last terms. You will obtain a difference of two
squares. That is,
$$(a + b)(a - b) = a^2 - b^2$$

Multiply $(x + 3)(x - 3)$ **Example 7**

Solution: Multiply first terms and multiply last terms.

$$(x + 3)(x - 3) = \overset{F}{x \cdot x} - \overset{L}{3 \cdot 3}$$
$$= x^2 - 9$$

Multiply $(2x + 5)(2x - 5)$ **Example 8**

Solution: $(2x + 5)(2x - 5) = 4x^2 - 25$

Example 9

Multiply $\left(3a - \dfrac{1}{2}\right)\left(3a + \dfrac{1}{2}\right)$

Solution: $\left(3a - \dfrac{1}{2}\right)\left(3a + \dfrac{1}{2}\right) = 9a^2 - \dfrac{1}{4}$

squaring binomials

The third type of special product is a shortcut for **squaring binomials**. It is obtained by using the FOIL method to multiply a binomial times itself.

$$(a + b)^2 = (a + b)(a + b)$$
$$= a^2 + ab + ba + b^2$$
$$= a^2 + \underbrace{2ab} + b^2$$

This pattern tells us that the square of a binomial is a trinomial. It is obtained by squaring the first term, adding twice the product of the first and last terms, and squaring the last term. Carefully study the following procedure.

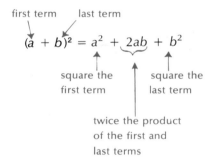

first term   last term

$(a + b)^2 = a^2 + 2ab + b^2$

square the
first term

square the
last term

twice the product
of the first and
last terms

Squaring a Binomial

To square a binomial:
(1)   Square its first term.
(2)   Multiply 2 times the product of its first and last terms.
(3)   Square its last term.

Example 10

Find $(x + 3)^2$

Solution: Use the shortcut for squaring a binomial.
    (1)   Square its first term: x^2
    (2)   Multiply 2 times the product of its first and last terms:
           $2(x)(3) = 6x$
    (3)   Square its last term: $3^2 = 9$

    Thus, $(x + 3)^2 = x^2 + 6x + 9$

Soon enough, as shown in the following examples, this procedure becomes automatic.

Find $(x + 5)^2$ **Example 11**

Solution: $(x + 5)^2 = x^2 + 2(5x) + 25$
                        $= x^2 + 10x + 25$

Find $(a - 4)^2$ **Example 12**

Solution: $(a - 4)^2 = a^2 + 2(-4a) + 16$
                        $= a^2 - 8a + 16$

Find $(3m - 2n)^2$ **Example 13**

Solution: $(3m - 2n)^2 = 9m^2 + 2(-6mn) + 4n^2$
                           $= 9m^2 - 12mn + 4n^2$

Multiply, using the FOIL shortcut.   **Exercise 4.3**

1. $(x + 2)(x + 3)$ **2.** $(a + 5)(a + 1)$

3. $(a - 4)(a + 5)$ **4.** $(m - 2)(m - 3)$

5. $(2x + 3)(x + 2)$ **6.** $(3y - 2)(y + 3)$

7. $(3m - n)(2m + 5n)$ **8.** $(2x - 3y)(x - 7y)$

9. $(3a + 5)(2a + 1)$ **10.** $(8y + 3)(6y + 1)$

Multiply, using the sum times a difference shortcut.

11. $(x + 2)(x - 2)$ **12.** $(m + 1)(m - 1)$

13. $(2y - 5)(2y + 5)$ **14.** $(3a + 5)(3a - 5)$

15. $(7x - y)(7x + y)$ **16.** $(3m - 2n)(3m + 2n)$

17. $(2a^2 + 3)(2a^2 - 3)$ **18.** $(3xy + 4)(3xy - 4)$

19. $\left(5x + \dfrac{2}{3}\right)\left(5x - \dfrac{2}{3}\right)$ **20.** $\left(\dfrac{1}{2}m + \dfrac{1}{3}\right)\left(\dfrac{1}{2}m - \dfrac{1}{3}\right)$

Square the following binomials, using the shortcut.

21. $(a + 3)^2$ **22.** $(a - 3)^2$

23. $(m + 1)^2$ **24.** $(m - 1)^2$

25. $(x + 7)^2$ **26.** $(x - 7)^2$

27. $(2m - 1)^2$ **28.** $(3a + 4)^2$

29. $(3x - 2y)^2$ **30.** $\left(2y + \dfrac{2}{5}\right)^2$

Multiply the following binomials, using any shortcut which applies.

31. $(a + 3)(a + 2)$　　　　　　　　**32.** $(a + 4)(a - 2)$

33. $(2x + 3)(3x - 2)$　　　　　　**34.** $(x - 3)(3x - 2)$

35. $(2x + 1)(x + 3)$　　　　　　　**36.** $(m + 4)(m - 4)$

37. $(a + 3)^2$　　　　　　　　　　**38.** $(2x - 3)(2x + 3)$

39. $(2x + 5)(2x - 5)$　　　　　　**40.** $(x + y)(x - y)$

41. $(5x + 2)(4x - 3)$　　　　　　**42.** $(2m^2 - 3n^2)(2m^2 + 3n^2)$

43. $(2a^3 + 5b^2)(2a^3 - 5b^2)$　　**44.** $(3m^2 + 2n^3)^2$

45. $(3x - 2)(x + 5)$　　　　　　　**46.** $(2a + 1)(3a - 4)$

47. $(x + 4)(x - 5)$　　　　　　　**48.** $(x + 5)(x - 5)$

49. $(2x + 5)(3x - 2)$　　　　　　**50.** $(3x + 7y)(3x - 7y)$

51. $(x + 6)^2$　　　　　　　　　　**52.** $(x - 6)^2$

53. $(3x + 4)^2$　　　　　　　　　**54.** $(5y + 4)(3y - 7)$

55. $(2x^2 + y^2)(2x^2 - y^2)$　　　**56.** $(4x - 7)(2x - 1)$

57. $\left(a + \dfrac{2}{3}\right)\left(a - \dfrac{2}{3}\right)$　　　　**58.** $(2a^2 - 3b^3)^2$

59. $(x - y)^2$　　　　　　　　　　**60.** $(2r + 7s)(5r - 6s)$

61. $(2 - x)(3 - 2x)$　　　　　　　**62.** $(x^2 - 4)^2$

63. $(a^2 + b^2)(a^2 - b^2)$　　　　**64.** $(4 - x)^2$

65. $(-x - 3)(-x + 2)$　　　　　　**66.** $(2 - x)(2 + x)$

67. $(a - 3)(2a + 3)$　　　　　　　**68.** $(5 - 2n)(5 + 2n)$

69. $(6m - n)^2$　　　　　　　　　**70.** $(4 - 3y)(5 - 2y)$

71. $(7 - 2z)^2$　　　　　　　　　**72.** $(3x^2 - 4y^3)^2$

73. $\left(\dfrac{2}{3}x - \dfrac{1}{2}\right)^2$　　　　　**74.** $\left(\dfrac{3}{5}m - \dfrac{2}{3}\right)\left(\dfrac{3}{5}m + \dfrac{2}{3}\right)$

75. $(0.3y + 0.2z)(0.3y - 0.2z)$

DIVISION OF POLYNOMIALS

4.4

We discuss division of polynomials by considering two cases. First, we divide by a monomial divisor. And second, we divide by a polynomial divisor of more than one term.

Case (1)　Dividing by a Monomial

Recall that fractions are reduced to lowest terms by factoring the numerator and denominator and dividing out the common factors. This same procedure can be used to divide a polynomial by a monomial where the monomial occurs as a factor in every term of the polynomial.

Divide $\dfrac{4m^3 + 6m^2 - 2m}{2m}$

Example 1

Solution: Factor the numerator and divide out the common factors.

$$\frac{4m^3 + 6m^2 - 2m}{2m} = \frac{\cancel{2m}(2m^2 + 3m - 1)}{\cancel{2m}}$$
$$= 2m^2 + 3m - 1$$

The same type of division can be done another way. We know that the sum or difference of several fractions having a common denominator can be expressed by a single fraction as follows.

$$\frac{a}{x} + \frac{b}{x} + \frac{c}{x} = \frac{a + b + c}{x}$$

This property can be reversed to show that a fraction whose numerator is a polynomial can be separated into several fractions whose numerators are the terms of the original polynomial.

Dividing a Polynomial by a Monomial
$$\frac{a + b + c}{x} = \frac{a}{x} + \frac{b}{x} + \frac{c}{x}$$
This property states that to divide a polynomial by a monomial, divide the monomial into each term of the polynomial.

Divide $\dfrac{12a^3 - 9a^2 + 3a}{3a}$

Example 2

Solution: $\dfrac{12a^3 - 9a^2 + 3a}{3a} = \dfrac{12a^3}{3a} - \dfrac{9a^2}{3a} + \dfrac{3a}{3a}$
$$= 4a^2 - 3a + 1$$

Divide $15a^3b^2 + 20a^2b^3$ over $-5a^2b^2$

Example 3

Solution: $\dfrac{15a^3b^2 + 20a^2b^3}{-5a^2b^2} = \dfrac{15a^3b^2}{-5a^2b^2} + \dfrac{20a^2b^3}{-5a^2b^2}$
$$= -3a - 4b$$

This method is particularly useful when the monomial divisor is not a factor of the numerator.

Example 4

Divide $\dfrac{5x^2 + 7x - 2}{x}$

Solution: $\dfrac{5x^2 + 7x - 2}{x} = \dfrac{5x^2}{x} + \dfrac{7x}{x} - \dfrac{2}{x}$

$$= 5x + 7 - \dfrac{2}{x}$$

Case (2)   Dividing by a Polynomial

To divide by a polynomial divisor containing more than one term, we follow a process similar to long division of two whole numbers. As an illustration, we will divide $2x^2 + 7x + 6$ by $x + 2$ and compare it to the long division process by simultaneously dividing 782 by 23.

Step (1)   Divide 782 by 23

$$23\overline{)782}$$

Step (1)   Divide $2x^2 + 7x + 6$ by $x + 2$

$$x + 2\overline{)2x^2 + 7x + 6}$$

Step (2)   Divide 2 into 7

$$\begin{array}{r} 3 \\ 23\overline{)782} \end{array}$$

Step (2)   Divide x into $2x^2$

$$\begin{array}{r} 2x \\ x + 2\overline{)2x^2 + 7x + 6} \end{array}$$

Step (3)   Subtract the product of (3)(23) from 78

$$\begin{array}{r} 3 \\ 23\overline{)782} \\ \underline{69} \\ 9 \end{array}$$

Step (3)   Subtract the product of $(2x)(x + 2)$ from $2x^2 + 7x$

$$\begin{array}{r} 2x \\ x + 2\overline{)2x^2 + 7x + 6} \\ \underline{2x^2 + 4x} \\ 3x \end{array}$$

Step (4)   Bring down the 2

$$\begin{array}{r} 3 \\ 23\overline{)782} \\ \underline{69\downarrow} \\ 92 \end{array}$$

Step (4)   Bring down the 6

$$\begin{array}{r} 2x \\ x + 2\overline{)2x^2 + 7x + 6} \\ \underline{2x^2 + 4x \quad \downarrow} \\ 3x + 6 \end{array}$$

Step (5) Divide 2 into 9   **Step (5)** Divide x into $3x$

$$\begin{array}{r} 34 \\ 23\overline{)782} \\ 69 \\ \hline 92 \end{array}$$

$$\begin{array}{r} 2x + 3 \\ x + 2\overline{)2x^2 + 7x + 6} \\ 2x^2 + 4x \\ \hline 3x + 6 \end{array}$$

Step (6) Subtract the product of   **Step (6)** Subtract the product of
(4)(23) from 92   (3)$(x + 2)$ from $3x + 6$

$$\begin{array}{r} 34 \\ 23\overline{)782} \\ 69 \\ \hline 92 \\ 92 \\ \hline 0 \end{array}$$

$$\begin{array}{r} 2x + 3 \\ x + 2\overline{)2x^2 + 7x + 6} \\ 2x^2 + 4x \\ \hline 3x + 6 \\ 3x + 6 \\ \hline 0 \end{array}$$

From the last step we obtain the answers. 782 divided by 23 is 34, and $2x^2 + 7x + 6$ divided by $x + 2$ is $2x + 3$. Since both problems had remainders of zero, the answers can be checked by showing that the quotient times the divisor gives the dividend. Thus, you should show that $(34)(23) = 782$ and $(2x + 3)(x + 2) = 2x^2 + 7x + 6$.

Divide $6m^3 + 3m^2 - m + 14$ by $2m + 3$ **Example 5**

Solution:

Step (1) $2m + 3\overline{)6m^3 + 3m^2 - m + 14}$

Step (2)
$$\begin{array}{r} 3m^2 \\ 2m + 3\overline{)6m^3 + 3m^2 - m + 14} \\ 6m^3 + 9m^2 \\ \hline -6m^2 - m \end{array}$$
$\left[\begin{array}{l} \text{subtract; change} \\ \text{signs and add} \end{array}\right]$

Step (3)
$$\begin{array}{r} 3m^2 - 3m \\ 2m + 3\overline{)6m^3 + 3m^2 - \;\; m + 14} \\ 6m^3 + 9m^2 \\ \hline -6m^2 - \;\; m \\ -6m^2 - 9m \\ \hline 8m + 14 \end{array}$$
[subtract]

$$
\begin{array}{r}
3m^2 - 3m + 4 \\
2m + 3\overline{)6m^3 + 3m^2 - m + 14}
\end{array}
$$

Step (4)

$$6m^3 + 9m^2$$
$$\overline{- 6m^2 - m}$$
$$- 6m^2 - 9m$$
$$\overline{8m + 14}$$
$$8m + 12 \longleftarrow \text{[subtract]}$$
$$\overline{2}$$

This problem has a remainder of 2. The answer is $3m^2 - 3m + 4$ R2, or the remainder may be placed over the divisor and added to the quotient. This gives an answer of $3m^2 - 3m + 4 + \dfrac{2}{2m + 3}$. The problem can be checked by multiplying the quotient times the divisor and adding the remainder. The result should be the dividend.

The check $(3m^2 - 3m + 4)(2m + 3) + 2$ should   produce $6m^3 + 3m^2 - m + 14$

Computation

$$
\begin{array}{r}
3m^2 - 3m + 4 \\
2m + 3 \\
\hline
6m^3 - 6m^2 + 8m \\
9m^2 - 9m + 12 \\
\hline
6m^3 + 3m^2 - m + 12 \\
+ 2 \longleftarrow \text{[add remainder]} \\
\hline
6m^3 + 3m^2 - m + 14 \longleftarrow \text{[the dividend]}
\end{array}
$$

Example 6   Divide $8y^3 + 1$ by $2y + 1$

Solution: The polynomial $8y^3 + 1$ is missing the y^2 term and the y term. When this happens, each missing term should be inserted with a coefficient of zero.

$$
\begin{array}{r}
4y^2 - 2y + 1 \\
2y + 1\overline{)8y^3 + 0y^2 + 0y + 1}
\end{array}
$$

$$8y^3 + 4y^2$$
$$\overline{- 4y^2 + 0y}$$
$$- 4y^2 - 2y$$
$$\overline{2y + 1}$$
$$2y + 1$$
$$\overline{0}$$

The answer is $4y^2 - 2y + 1$. Since the remainder is zero, you can check by showing that $(4y^2 - 2y + 1)(2y + 1)$ equals $8y^3 + 1$.

When using long division to divide polynomials, both divisor and dividend should be written in descending powers of the variable. Thus, $(5x + x^2 + 6) \div (3 + x)$ should be written as follows:

$$x + 3\overline{)x^2 + 5x + 6}$$

Divide the following expressions.   **Exercise 4.4**

1. $\dfrac{9a^2 + 6a}{3a}$

2. $\dfrac{9a^2 + 6a}{-3a}$

3. $\dfrac{4x^3 + 2x}{2x}$

4. $\dfrac{6m^2 + 12m - 3}{3}$

5. $\dfrac{10r^3 - 15r^2 + 5r}{5r}$

6. $\dfrac{15a^3 + 10a^2 - 5a}{-5a}$

7. $\dfrac{4m^5 - 12m^3 + 16m^2}{4m^2}$

8. $\dfrac{14r^3t^2 + 7r^2t^3}{7r^2t^3}$

9. $\dfrac{12m^3n^5 - 6m^4n^3 + 3m^2n}{3mn}$

10. $\dfrac{20x^2y^3 - 10xy^2 + 5xy}{-5xy}$

11. $\dfrac{x^2 + x + 1}{x}$

12. $\dfrac{3y^2 + 6y - 9}{3y}$

13. $\dfrac{4m^3 + 2m^2 - 6m}{-2m^2}$

14. $\dfrac{6a^4 + 3a^2 - 2}{3a^2}$

15. $\dfrac{m^2n + mn - n}{-mn}$

16. $\dfrac{x^3 - 3x^2 + x + 1}{x}$

17. $\dfrac{8y^3 - 6y^2 + 2y + 1}{-2y}$

18. $\dfrac{a^2b + ab^2 + a}{ab}$

19. $\dfrac{a^3b^2 - a^2b - b}{a^2}$

20. $\dfrac{10x^3 + 5x^2y - 5xy^2 + 10y^3}{5xy}$

Divide the following expressions, using long division. Check your answers.

21. $(x^2 - 7x + 12) \div (x - 3)$

22. $(x^2 - 8x + 15) \div (x - 5)$

23. $(3a^2 - 2a + 4) \div (a + 3)$

24. $(m^2 - 7m + 10) \div (m - 2)$

25. $(6y^2 - 5y + 5) \div (3y + 2)$

26. $(x^2 + 4x + 3) \div (x + 3)$

27. $(a^2 - 3a - 10) \div (a - 5)$

28. $(2m^2 - m - 6) \div (2m + 3)$

29. $(3x^2 + 4x - 3) \div (x + 2)$

30. $(2x^2 - 5x + 1) \div (2x + 1)$

31. $(x^2 - 1) \div (x - 1)$

32. $(4m^2 - 25) \div (2m + 5)$

33. $(9m^2 - 4) \div (3m - 2)$

34. $(x^3 + 1) \div (x + 1)$

35. $(x^3 - 1) \div (x + 1)$

36. $(n^4 + 1) \div (n + 1)$

37. $(4x^3 - 4x^2 + 5x - 6) \div (2x - 1)$

38. $(2x^3 + 3x - 1) \div (x - 2)$ **39.** $(2r^4 - r + 6) \div (r - 5)$

40. $(n^4 - 1) \div (n^2 - 1)$ **41.** $(3x^4 + x^2 + 1) \div (x^2 + 1)$

42. $(m^4 + 2m^2 + 1) \div (m^2 + 1)$ **43.** $\dfrac{3x^3 + 4x^2 - 14x + 3}{x + 3}$

44. $\dfrac{4n^2 + 6n + 9}{2n - 3}$ **45.** $\dfrac{a^4 - 3a^3 + 3a^2 + 2a - 3}{a - 1}$

46. $\dfrac{p^4 - 1}{p^2 + 1}$ **47.** $\dfrac{h^4 - 1}{h^2 + 1}$

48. $\dfrac{x^3 + 1}{x - 1}$ **49.** $\dfrac{p^6 - 1}{p - 1}$

50. $\dfrac{t^6 - 1}{t + 1}$

COMMON MONOMIAL FACTORING

4.5

Previously you learned techniques for multiplying polynomials. Now we reverse these techniques to take a single polynomial and write it as a product of two or more simpler polynomials. This is called factoring and was first discussed in sections 1.6 and 2.3.

To factor a polynomial, we begin by looking for its largest common monomial factor. This is the largest monomial that divides evenly into each term of the polynomial. For example, the polynomial $4m^2 + 8m$ has several common factors, including 2, 4, and m. However, the largest common monomial factor is $4m$.

Example 1

Find the largest common monomial factor for each of these polynomials.

        Polynomial   Largest Monomial Factor

(a) $3a + 3b$ 3

(b) $4x^2 - 6x$ $2x$

(c) $6m^2n^2 + 12mn^2 + 3mn^3$ $3mn^2$

(Note, the exponents included in the largest common factor are the smallest exponents appearing on the corresponding factors.)

You should recall that the distributive properties of multiplication can be used to factor out the expression common to each term. Thus,

$$ab + ac = a(b + c)$$

common monomial factoring

When applied to polynomials, this process is called **common monomial factoring**.

When factoring a polynomial, it is important to factor completely. That is, factor as far as possible. To help ensure this, we always remove (factor out) the largest common monomial factor.

Common Monomial Factoring

To perform common monomial factoring follow these thee steps:

Step (1)   Find the largest common monomial factor.

Step (2)   Rewrite the polynomial so the largest common monomial factor is seen in each term.

Step (3)   Use the distributive properties to remove the largest common monomial factor from each term.

Factor $10ab + 5ac$ Example 2

Solution:

Step (1)   The largest common monomial factor is $5a$

Step (2)   Rewrite the polynomial so that $5a$ appears in each term.

$$10ab + 5ac = (5a)\cdot 2b + (5a)\cdot c$$

Step (3)   Remove $5a$ from each term.

$$(5a)\cdot 2b + (5a)\cdot c = 5a(2b + c)$$

Therefore, in completely factored form, $10ab + 5ac = 5a(2b + c)$. The answer to a factoring problem can be checked by multiplying it out according to the distributive properties. The result should be the original expression. Thus, we check this example as follows:

$$5a(2b + c) = 10ab + 5ac$$

Factor $4x^3 - 6x^2 + 2x$ Example 3

Solution:

Step (1)   The largest common monomial factor is $2x$

Step (2)   Rewrite the polynomial so that $2x$ appears in each term.

$$4x^3 - 6x^2 + 2x = (2x)\cdot 2x^2 - (2x)\cdot 3x + (2x)\cdot 1$$

Step (3)   Factor out $2x$ from each term.

$$(2x)\cdot 2x^2 - (2x)\cdot 3x + (2x)\cdot 1 = 2x(2x^2 - 3x + 1)$$

Therefore, $4x^3 - 6x^2 + 2x = 2x(2x^2 - 3x + 1)$. It was necessary in this example to write $2x$ as $(2x) \cdot 1$. This allows us to factor $2x$ out of the third term, leaving behind a 1.

As you become more familiar with common monomial factoring, you will be able to shorten the process by combining the steps.

Example 4

Factor $9m^5 + 6m^4 - 3m^2$

Solution: The largest common monomial factor is $3m^2$

$$9m^5 + 6m^4 - 3m^2 = 3m^2(3m^3 + 2m^2 - 1)$$

Example 5

Factor $25m^3n^2 - 5m^2n^3 + 10mn^4$

Solution: The largest common monomial factor is $5mn^2$

$$25m^3n^2 - 5m^2n^3 + 10mn^4 = 5mn^2(5m^2 - mn + 2n^2)$$

Exercise 4.5

Factor the largest common monomial from each of the following.

1. $7a + 7b$
2. $3x + 9$
3. $2m + 2n$
4. $5y + 15$
5. $6x + 6$
6. $2r - 2$
7. $3ab + 3ac$
8. $5x^2 - 5x$
9. $5x^2 + 2x$
10. $3m - 18n$
11. $5a + 3ab$
12. $3r^2t - 9rt^2$
13. $4m^2n - 12mn$
14. $18ab^2 - 6a^2b$
15. $6x^2 - 18x$
16. $5x^3 + 15x^2$
17. $21m^2 + 14m$
18. $a^5 - a^4$
19. $11r^2 + 121$
20. $32a^3b^2 - 64ab^2$
21. $6b^2 + 3b - 12$
22. $4m^3 + 8m^2 - 2m$
23. $15x^3y + 5x^2y - 10xy$
24. $6a^2b^3 - 12ab^4 - 18b^5$
25. $3m^2 - 6m + 3$
26. $25r^4 - 20r^3 + 4r^2$
27. $a^3b^4 + a^2b^5 - ab^3$
28. $13x^7 - 39x^6 + 26x^5$
29. $25y^4 + 5y^3 - 20y^2$
30. $8n^3 + 16n^2 - 24n$
31. $8x^2 + 10y^2 - 4z^2$
32. $-15a^2b + 5ab$
33. $25x^3y^2 - 100x^2y^2 + 50xy^2$
34. $81m^7n^5 + 27m^8n^3 - 36m^9n^2$
35. $6x^5 - 3x^4 + 12x^3 - 15x^2 + 9x$
36. $9a^6b - 18a^5b^2 + 15a^4b^3 + 12a^3b^2$

37. $3x^3y^2 - 6x^2y^3 + 9x^2y^2$

38. $6m^3n^2 - 21m^2n + 9m$

39. $9ab^2 - 6b + 3a - 12$

40. $\pi R^2 + \pi r^2$

41. $y^{10} + y^8 - y^6 + y^4$

42. $-3r^4s^2 + 6r^3s^2 + 12r^2s^2 - 9rs^2$

43. $13p^6q^4 + 26p^3q^4 - 39p^4q^4$

44. $50m^6 + 100m^4 - 100m^2$

45. $20t^2 - 100$

46. $16y^4 - 100y^2$

47. $24a^5b^3c - 36a^4b^3c^2 + 20a^3b^5c^3 - 16a^2b^6c^4$

48. $x^7y^6 - 3x^6y + 10x^5y^2 - 7x^2y^3$

49. $y^6 - y^5 - y^4 - y^2$

50. $14h^3k^2 - 42h^2k^3 - 28h^2k^4 + 56h^3k^5$

FACTORING TRINOMIALS

4.6

In this section, you will learn how to factor trinomials where the coefficient of the squared term is 1, such as $x^2 + 5x + 6$.

You should recall from section 4.3 that taking the product of two binomials usually produces a trinomial, as follows:

$$(x + 2)(x + 3) = \overset{F}{x^2} + \overset{O}{3x} + \overset{I}{2x} + \overset{L}{6}$$
$$= x^2 + 5x + 6$$

To factor the trinomial $x^2 + 5x + 6$, we reverse the FOIL process by following this procedure.

Step (1) $x^2 + 5x + 6$ was produced from a product of two binomials. So, we write two grouping symbols.

$$(\qquad) \cdot (\qquad)$$

Step (2) The x^2 term came from the product of the first terms of the binomials, namely $x \cdot x$. Therefore, we write x as the first term of each binomial.

$$(x \qquad) \cdot (x \qquad)$$

Step (3) The third term of the trinomial is 6. It came from the product of the last terms of the two binomials. The choices for factors of 6 are given below.

$(x + 6) \cdot (x + 1)$

or $\qquad$ {factors of 6}

$(x + 3) \cdot (x + 2)$

Step (4) The correct solution is found by taking the sum of the outer and inner products. This sum must produce the middle term of the original trinomial, namely $5x$.

Wrong choice is $(x + 6) \cdot (x + 1)$ because $1x + 6x \neq 5x$

Correct choice is $(x + 3) \cdot (x + 2)$ because $3x + 2x = 5x$

Therefore, $x^2 + 5x + 6 = (x + 3) \cdot (x + 2)$

Factoring a trinomial into two binomials unfortunately involves a certain amount of trial and error. Keep in mind that the last terms of the binomial factors must be such that their product gives the third (or constant) term of the trinomial, and their sum gives the coefficient of the trinomial's middle term. Consider this diagram:

$$x^2 + 5x + 6 = (x + 2)(x + 3)$$

$2 \cdot 3$

$2 + 3$

Example 1   Factor $x^2 + 7x + 12$

Solution:

Step (1) Write two grouping symbols.

$$(\quad) \cdot (\quad)$$

Step (2) x^2 came from the first terms of the binomials.

$$(x\quad) \cdot (x\quad)$$

Step (3) The third term of the trinomial is 12. Write the factors.

$(x + 1) \cdot (x + 12)$

or

$(x + 2) \cdot (x + 6)$ ——[factors of 12]

or

$(x + 3) \cdot (x + 4)$

Step (4) The sum of the outer and inner products must produce the middle term of the trinomial, $7x$.

Correct choice is $(x + 3) \cdot (x + 4)$ because $4x + 3x = 7x$

Therefore, $x^2 + 7x + 12 = (x + 3)(x + 4)$. Note again the product and sum relationship of the last terms of the binomials:

$$x^2 + 7x + 12 = (x + 3)(x + 4)$$

This procedure for factoring trinomials can be shortened by using the following product-sum relationship.

Product-Sum Relationship for Factoring Trinomials

A trinomial where the coefficient of the squared term is 1 will factor into two binomials according to this relationship:

$$x^2 + Ax + B = (x + c)(x + d)$$

where $c \cdot d = B$ and $c + d = A$

Using this relationship, we will consider only integers. This is called **factoring over the set of integers**.

factoring over the set of integers

Factor $x^2 + 8x + 15$

Example 2

Solution: According to the product-sum relationship, we must find two integers whose product is 15 and whose sum is 8. Since $3 \cdot 5 = 15$, and $3 + 5 = 8$, the two integers are 3 and 5. Thus,

$$x^2 + 8x + 15 = (\qquad)(\qquad)$$
$$= (x \qquad)(x \qquad)$$
$$= (x + 3)(x + 5)$$

Check: The sum of the outer and inner products must give $8x$.

$$x^2 + 8x + 15 = (x + 3)(x + 5) \text{ because } 5x + 3x = 8x$$

Example 3

Factor $m^2 - 4m - 5$

Solution: Find two integers whose product is -5 and whose sum is -4. Since $(-5)(1) = -5$, and $-5 + 1 = -4$, the two integers are -5 and 1. Thus,

$$m^2 - 4m - 5 = (m - 5)(m + 1)$$

Check: $m^2 - 4m - 5 = (m - 5)(m + 1)$ because $1m - 5m = -4m$.

Example 4

Factor $a^2 - 8a + 16$

Solution: List all pairs of integers whose product is 16. Then select the pair whose sum is -8.

Products	Sums
$(-1)(-16)$	$(-1) + (-16) = -17$
$(-2)(-8)$	$(-2) + (-8) = -10$
$(-4)(-4)$	$(-4) + (-4) = -8$

The correct choice consists of -4 and -4.

$$a^2 - 8a + 16 = (a - 4)(a - 4)$$
$$= (a - 4)^2$$

Example 5

Factor $y^2 + 6y - 8$

Solution: List all pairs of integers whose product is -8. Then examine each pair to find a sum of 6.

Products	Sums
$(-1)(8)$	$-1 + 8 = 7$
$(-2)(4)$	$-2 + 4 = 2$
$(1)(-8)$	$1 + (-8) = -7$
$(2)(-4)$	$2 + (-4) = -2$

But none of the pairs has a sum of 6. Thus, the trinomial $y^2 + 6y - 8$ cannot be factored using integers. It is said to be prime.

When factoring trinomials, it is important always to remove the largest common factor first.

Example 6

Factor $2x^3 - 14x^2 + 12x$

Solution: First remove the largest common factor, $2x$.

$$2x^3 - 14x^2 + 12x = 2x(x^2 - 7x + 6)$$

Now factor the trinomial. The integers -1 and -6 have a product of 6 and a sum of -7. Thus, the completely factored form is

$$2x^3 - 14x^2 + 12x = 2x(x - 1)(x - 6)$$

Factor these trinomials completely over the set of integers. If a trinomial cannot be factored, write *prime*.

Exercise 4.6

1. $x^2 + 3x + 2$ **2.** $x^2 - 3x + 2$

3. $x^2 + x - 2$ **4.** $x^2 - x - 2$

5. $m^2 + 5m + 4$ **6.** $m^2 + 2m + 1$

7. $n^2 - 6n + 8$ **8.** $n^2 - 2n - 8$

9. $a^2 + 6a + 9$ **10.** $a^2 - 6a + 9$

11. $p^2 + 4p + 4$ **12.** $x^2 + 3x - 4$

13. $t^2 + 6t + 5$ **14.** $y^2 + 8y + 15$

15. $n^2 - 8n + 7$ **16.** $h^2 + 11h + 30$

17. $m^2 + 12m + 35$ **18.** $a^2 - 9a + 20$

19. $k^2 - 11k + 10$ **20.** $b^2 + 7b + 6$

21. $n^2 + 8n + 12$ **22.** $t^2 - 3t - 40$

23. $x^2 + 3x - 18$ **24.** $y^2 - 11y + 28$

25. $p^2 + 8p - 48$ **26.** $b^2 - 2b - 24$

27. $h^2 - h - 42$ **28.** $x^2 + x + 1$

29. $m^2 - m - 72$ **30.** $y^2 - 10y - 56$

31. $t^2 + 2t + 10$ **32.** $z^2 + 4z - 77$

33. $n^2 - 6n - 72$ **34.** $x^2 - 15x + 26$

35. $k^2 + 8k - 65$ **36.** $r^2 + 12r + 32$

37. $t^2 - 12t + 27$ **38.** $y^2 + 11y + 28$

39. $a^2 - 18a + 45$ **40.** $x^2 + 5xy + 6y^2$

41. $a^2 + 3ab + 2b^2$ **42.** $m^2 - mn - 6n^2$

43. $r^2 - 7rs + 12s^2$ **44.** $x^2 - 3xy - 10y^2$

45. $p^2 + 9pg + 14q^2$ **46.** $h^2 + hk - 42k^2$

47. $x^2 - 5ax - 14a^2$ **48.** $x^2 - 11xy + 18y^2$

Factor completely, being sure to remove the largest common factor first.

49. $2a^2 + 10a + 8$ **50.** $3x^2 + 15x + 18$

51. $4b^2 + 32b + 28$ **52.** $k^3 - 4k^2 + 3k$

53. $t^3 - 11t^2 + 30t$ **54.** $x^4 - 7x^3 + 6x^2$

55. $m^5 - 4m^4 + 3m^3$ **56.** $3x^3 + 12x^2 + 9x$

57. $3p^4 - 18p^3 + 15p^2$ **58.** $2h^4 - 8h^3 - 10h^2$

59. $2x^5 - 8x^4 - 42x^3$ **60.** $t^6 - 5t^5 - 14t^4$

61   $2x^4 + 14x^3 + 20x^2$

62   $4a^3 - 4a^2 - 24a$

63   $3y^3 + 6y^2 - 9y$

64   $2p^2 - 26p + 60$

61. $2x^4 + 14x^3 + 20x^2$ **62.** $4a^3 - 4a^2 - 24a$

63. $3y^3 + 6y^2 - 9y$ **64.** $2p^2 - 26p + 60$

MORE ON FACTORING

4.7

In this section we continue our discussion of factoring by learning to factor trinomials where the coefficient of the squared term is not 1, for example $3x^2 + 7x + 2$. Also, you will learn to factor certain binomials known as differences of two squares.

To factor a trinomial where the coefficient of the squared term is not 1, we must find the correct combination of first term factors and last term factors, so that the sum of the outer and inner binomial products gives the middle term of the trinomial. Unfortunately, this involves a certain amount of trial and error. To illustrate, we factor the trinomial, $3x^2 + 7x + 2$.

Step (1) $3x^2 + 7x + 2$ came from a product of two binomials. Thus, the factors are represented with two sets of grouping symbols. So,
$$3x^2 + 7x + 2 = (\qquad)(\qquad)$$

Step (2)   The first term of the trinomial, $3x^2$, is the product of the first terms of the two binomials. Since $3x^2 = (3x)(x)$, we place $3x$ as the first term of one binomial and x as the first term of the other. Thus,
$$3x^2 + 7x + 2 = (3x\qquad)(x\qquad)$$

Step (3)   2 is the third term of the trinomial and is the product of the last terms of the two binomials. Hence, factors of 2 must be placed for the last terms. In counting order, 2 has two sets of factors, $1 \cdot 2$ and $2 \cdot 1$. This gives the following two possibilities:
$$3x^2 + 7x + 2 = (3x + 2)(x + 1)$$
$$\text{or} \qquad \qquad \qquad \text{(factors of 2)}$$
$$3x^2 + 7x + 2 = (3x + 1)(x + 2)$$

Step (4)   The correct combination is found by taking the sum of the outer and inner products. This sum must produce the middle term of the trinomial, $7x$.

(1) $3x^2 + 7x + 2 = (3x + 2)(x + 1)$ is incorrect

because $3x + 2x \ne 7x$.

(2) $3x^2 + 7x + 2 = (3x + 1)(x + 2)$ is correct

since $6x + 1x = 7x.$

Therefore, $3x^2 + 7x + 2 = (3x + 1)(x + 2)$

Factor $2m^2 + m - 15$ **Example 1**

Solution:

Step (1) Write two sets of grouping symbols.

$($ $)($ $)$

Step (2) The first term of the trinomial, $2m^2$, came from the product of $2m$ and m. So, we write

$(2m$ $)(m$ $)$

Step (3) The third term of the trinomial is -15. Now form the last terms by considering all factors of -15. We have:

$(2m + 3)(m - 5)$

$(2m - 3)(m + 5)$ —(factors of -15)

$(2m + 5)(m - 3)$

$(2m - 5)(m + 3)$

Step (4) The sum of the outer and inner products must produce the middle term of the trinomial, m.

$(2m + 3)(m - 5)$ is incorrect, since $-10m + 3m \neq m.$

$(2m - 3)(m + 5)$ is incorrect, since $10m - 3m \neq m$

$(2m + 5)(m - 3)$ is also incorrect because

$$-6m + 5m \neq m$$

$(2m - 5)(m + 3)$ is correct because $6m - 5m = m$

Therefore, $2m^2 + m - 15 = (2m - 5)(m + 3)$

Combining steps enables us to shorten this procedure as illustrated in the next example.

Example 2

Factor $6y^2 + 11y + 5$

Solution: We form all possible binomial factors. The first term of the trinomial, $6y^2$, came from either the product of $2y$ and $3y$ or from the product of $6y$ and y. The third term, 5, came from the product of 5 and 1.

$(2y + 5)(3y + 1)$ is incorrect since $2y + 15y \neq 11y$

$(2y + 1)(3y + 5)$ is incorrect since $10y + 3y \neq 11y$

$(6y + 1)(y + 5)$ is incorrect since $30y + 1y \neq 11y$

$(6y + 5)(y + 1)$ is correct because $6y + 5y = 11y$

Therefore, $6y^2 + 11y + 5 = (6y + 5)(y + 1)$.

Example 3

Factor $3n^2 - 10n + 3$

Solution: The only possibilities are
(1) $(3n - 3)(n - 1)$ or
(2) $(3n - 1)(n - 3)$
The second possibility is correct as shown:

$$3n^2 - 10n + 3 = (3n - 1)(n - 3)$$

since $-9n - 1n = -10n$

We could have seen from the beginning that the first possibility, $(3n - 3)(n - 1)$, could not be correct because $3n - 3$ has a common factor of 3, while the original trinomial, $3n^2 - 10n + 3$, does not. This suggests the following rule concerning common factors.

Common Factor Rule

If a polynomial has no common factor, then neither will its factors.

Factor $2a^2 - 5a + 2$ **Example 4**

Solution: The only possibilities are

  (1) $(2a - 2)(a - 1)$ and
  (2) $(2a - 1)(a - 2)$

We need not consider the first possibility because $2a - 2$ has a common factor of 2, while the original trinomial does not. This leaves only the second possibility which we show to be correct.

$$2a^2 - 5a + 2 = (2a - 1)(a - 2) \text{ since } -4a - 1a = -5a$$

Factoring trinomials is much easier if you memorize the various sign combinations described by the following three cases. Each case assumes that the coefficient of the squared term is positive.

Case (1)   If the trinomial contains only plus signs, then the binomial factors contain only plus signs.

$$2a^2 + 5a + 3 = (2a + 3)(a + 1)$$
all plus   all plus

Case (2)   If the middle term of the trinomial is minus while the last term is plus, then the middle sign of each binomial is minus.

$$m^2 - 5m + 6 = (m - 3)(m - 2)$$
minus plus   both minus

Case (3)   If the last term of the trinomial is minus, then the binomials have opposite middle signs.

$$2x^2 - x - 3 = (2x - 3)(x + 1)$$
minus   opposite signs

Use the pattern of signs to help factor $3x^2 - 7x + 4$.   **Example 5**

Solution: The middle term is minus and the last term is plus. So, the middle sign of each binomial is minus.

$$3x^2 - 7x + 4 = (\quad - \quad)(\quad - \quad)$$
$$= (3x - \quad)(x - \quad)$$
$$= (3x - 4)(x - 1)$$

Check: $3x^2 - 7x + 4 = (3x - 4)(x - 1)$ since $-3x - 4x = -7x$

> When factoring, it is important to remember these points:
>
> (1) Watch for common monomial factors. Remove them before you do anything else.
> (2) Always factor completely.
> (3) When factoring trinomials, use the pattern of signs and remember that the sum of the outer and inner products must produce the middle term of the trinomial.

Example 6

Factor $6m^3 - 33m^2 + 36m$

Solution: First remove the largest common monomial factor, $3m$.

$$6m^3 - 33m^2 + 36m = 3m(2m^2 - 11m + 12)$$
$$= 3m(\quad - \quad)(\quad - \quad)$$
$$= 3m(2m - 3)(m - 4)$$

Example 7

Factor $2p^2 + 3p - 1$

Solution: The only possibilities are

(1) $(2p + 1)(p - 1)$ and
(2) $(2p - 1)(p + 1)$

However, as seen below, neither checks.

$$2p^2 + 3p - 1 \ne (2p + 1)(p - 1) \text{ since } -2p + 1p \ne 3p$$

$$2p^2 + 3p - 1 \ne (2p - 1)(p + 1) \text{ since } 2p - 1p \ne 3p$$

Thus, $2p^2 + 3p - 1$ is not factorable using integers. The trinomial is prime.

We conclude our study of factoring by discussing a difference of two squares, such as $x^2 - 25$. You should recall from section 4.3 that a sum times a difference produces a product that is a difference of two squares. Thus,

$$\underbrace{(a + b)}_{\text{sum}}\underbrace{(a - b)}_{\text{difference}} = \underbrace{a^2 - b^2}_{\substack{\text{difference} \\ \text{of} \\ \text{two squares}}}$$

Reversing this pattern shows that a difference of two squares factors into a sum times a difference.

Factoring a Difference of Two Squares

First, remove the largest common factor (if any), then follow this pattern: $a^2 - b^2 = (a + b)(a - b)$

Factor the following differences of two squares.

Example 8

(a) $x^2 - 25 = x^2 - 5^2$
              $= (x + 5)(x - 5)$

(b) $a^2 - 1 = a^2 - 1^2$
           $= (a + 1)(a - 1)$

(c) $9n^2 - 49 = (3n)^2 - 7^2$
             $= (3n + 7)(3n - 7)$

When factoring, remember to first remove the largest common factor. Then, factor as far as possible (factor completely.)

Factor $3x^2 - 27$

Example 9

Solution: Remove the largest common factor, 3. Then factor as a difference of two squares.

        $3x^2 - 27 = 3(x^2 - 9)$
                $= 3(x + 3)(x - 3)$

Factor $6a^5 - 6a$

Example 10

Solution: Remove the largest common factor, $6a$. Then factor completely.

        $6a^5 - 6a = 6a(a^4 - 1)$
                $= 6a(a^2 + 1)(a^2 - 1)$

Note that $a^2 - 1$ is a difference of two squares, so continue factoring.

                $= 6a(a^2 + 1)(a^2 - 1)$
                $= 6a(a^2 + 1)(a + 1)(a - 1)$

The binomial $a^2 + 1$ is an example of a sum of two squares. In general, a sum of two squares, fitting the pattern $a^2 + b^2$, is not factorable; it is prime.

Complete the following factoring.

Exercise 4.7

  1. $2x^2 + 9x + 4 = (2x + 1)($ $)$
  2. $3a^2 + 8a + 4 = (3a + 2)($ $)$
  3. $2m^2 + 3m - 2 = (2m - 1)($ $)$
  4. $3n^2 + 4n - 15 = (a + 3)($ $)$

complete the following factoring

5. $6p^2 + p - 1 = (2p + 1)(\quad)$

6. $4k^2 - 7k - 2 = (k - 2)(\quad)$

7. $6y^2 - 11y + 3 = (2y - 3)(\quad)$

8. $8a^2 - 22a + 5 = (2a - 5)(\quad)$

9. $4x^2 + 10x + 6 = 2(x + 1)(\quad)$

10. $6t^2 - 3t - 18 = 3(t - 2)(\quad)$

11. $30h^3 - 9h^2 - 3h = 3h(2h - 1)(\quad)$

12. $24x^4 - 28x^3 + 8x^2 = 4x^2(2x - 1)(\quad)$

13. $y^2 - 9 = (y + 3)(\quad)$

14. $a^2 - 100 = (a - 10)(\quad)$

15. $4m^2 - 9 = (2m + 3)(\quad)$

16. $2x^2 - 32 = 2(x + 4)(\quad)$

17. $18n^2 - 50 = 2(3n - 5)(\quad)$

18. $5t^3 - 5t = 5t(t + 1)(\quad)$

19. $m^4 - 1 = (m^2 + 1)(m + 1)(\quad)$

20. $3y^5 - 48y = 3y(y^2 + 4)(y + 2)(\quad)$

Factor each of these polynomials completely.

21. $x^2 + 5x + 6$

22. $x^2 - 5x + 6$

23. $m^2 - 6m + 8$

24. $m^2 - 2m - 8$

25. $a^2 + 6a + 9$

26. $a^2 - 6a + 9$

27. $x^2 + x - 6$

28. $x^2 + 4x + 4$

29. $x^2 + 3x - 4$

30. $2a^2 + 5a + 3$

31. $2a^2 - a - 3$

32. $2a^2 - 5a + 3$

33. $6r^2 + 5r - 6$

34. $10k^2 - 19k - 15$

35. $9x^2 + 12xy + 4y^2$

36. $3m^2 + 7mn + 2n^2$

37. $2a^2 + 5ab - 3b^2$

38. $9h^2 - 3hk - 2k^2$

39. $x^2 + x + 1$

40. $2x^2 + 10x + 12$

41. $6a^3 - 3a^2 - 63a$

42. $2m^4 - 7m^3 - 15m^2$

43. $15y - 5x^2y^2$

44. $4x^2y^2 - 49$

45. $3x^2 + 7x + 4$

46. $5m^2 - 9m + 4$

47. $12a^2 - 26a + 12$

48. $4a^2 + 9a + 5$

49. $16x^2 - 25y^2$

50. $3m^3 - 3m$

51. $x^4 - y^6$

52. $6m^3n^2 - 21m^2n + 9m$

53. $2ab - 4a^2b^2 - 8a^3b^3$

54. $18x^3 - 8xy^2$

55. $3r^3s^2 + 6r^2s^2 + 3rs^2$

56. $9x^2 - 4$

57. $9x^2 + 4$

58. $10x^3 + 20x^2 - 55x$

59. $\pi R^3 - 25\pi R$

60. $3a^4 + 21a^3 + 30a^2$

61. $21k^2 + 13k + 2$

62. $9x^2 + 14x + 25$

63. $6 - 28h + 16h^2$

64. $30x^2y^2 - 14xy^2 - 8y^2$

65. $24m^5 + 10m^4 - 4m^3$

66. $15r^2 + r - 6$

67. $8k^2 + 10k - 3$

68. $21x^2 - 13x + 2$

69. $3m - 3m^{13}$

70. $6x^3 - 3x^2 + 9x$

Each of the following is a "difference of two squares." Factor each completely.

71. $a^2 - 1$

72. $x^2 - 4$

73. $m^2 - 9$

74. $y^2 - 16$

75. $r^2 - 36$

76. $4x^2 - 1$

77. $16a^2 - 9$

78. $9a^2 - 25b^2$

79. $x^2 - y^2$

80. $25x^4 - 9y^2$

81. $36r^2 - 100$

82. $49m^2 - 64n^2$

83. $5x^2 - 20$

84. $27m^2 - 12$

85. $m^4 - n^4$

86. $64y^2 - 16$

87. $n^4 - 81$

88. $50 - 98x^2$

89. $144r^2t^2 - 121$

90. $x^8 - y^8$

SOLVING QUADRATIC EQUATIONS BY FACTORING

4.8

In section 2.8 you learned to solve linear equations. You should recall that linear equations can be expressed in the form $ax + b = c$ where a, b, and c represent constants, and a is not equal to zero. Also, the highest exponent for the variable is 1. Linear equations are solved by obtaining the variables and constants on opposite sides of the equation. Then, you divide each side by the coefficient of the variable. As a review, study Example 1.

Solve $2(3x - 7) = 3(4x + 2) + 10$

Example 1

Solution: $2(3x - 7) = 3(4x + 2) + 10$

$$6x - 14 = 12x + 6 + 10$$

$$6x - 14 = 12x + 16$$

$$6x = 12x + 30 \qquad \text{[add 14]}$$

$$-6x = 30 \qquad \text{[subtract 12x]}$$

$$\frac{-6x}{-6} = \frac{30}{-6}$$

$$x = -5$$

quadratic equations

We are now ready to solve **quadratic equations**. These are equations which can be expressed in this form:

$$ax^2 + bx + c = 0$$

where a, b, and c are constants and $a \neq 0$. This is called the standard form of a quadratic equation. (Note that the largest exponent is 2.)

Example 2

Each of these is a quadratic equation.
(a) $x^2 + 6x + 5 = 0$
(b) $2y^2 + y = 1$
(c) $3m^2 - m = 0$
(d) $x^2 - 16 = 0$

In this section we will learn to solve quadratic equations by factoring. An important aid will be the zero factor property.

Zero Factor Property
If a and b represent real numbers, and if $a \cdot b = 0$, then $a = 0$ or $b = 0$.

This property tells us that whenever the product of two factors is zero, either the first factor is zero, the second factor is zero, or both factors are zero.

Example 3

Suppose $(x + 3) \cdot (x - 4) = 0$. Then by the zero factor property, either $x + 3 = 0$ or $x - 4 = 0$. Solving each equation for x, we have $x = -3$ or $x = 4$.

Example 4

Solve $(y - 1) \cdot (y - 5) = 0$

Solution: Use the zero factor property to set each factor equal to zero.

$$(y - 1)(y - 5) = 0$$
$$y - 1 = 0 \quad \text{or} \quad y - 5 = 0$$
$$y = 1 \quad \text{or} \quad y = 5$$

To solve a quadratic equation such as $x^2 + 2x = 8$ using the zero factor property, we need to obtain a product which is equal to zero. Therefore, we add -8 to each side, getting $x^2 + 2x - 8 = 0$. Next, we factor the trinomial and set each factor equal to zero. This gives the two solutions as follows:

$$x^2 + 2x = 8$$
$$x^2 + 2x - 8 = 0$$
$$(x + 4)(x - 2) = 0$$
$$x + 4 = 0 \quad \text{or} \quad x - 2 = 0$$
$$x = -4 \quad \text{or} \quad x = 2$$

These solutions can be checked by substituting them, one at a time, into the original equation.

Check for $x = -4$

$$x^2 + 2x = 8$$
$$(-4)^2 + 2(-4) = 8$$
$$16 \quad - \quad 8 \quad = 8$$
$$8 = 8$$

Check for $x = 2$

$$x^2 + 2x = 8$$
$$2^2 + 2 \cdot 2 = 8$$
$$4 + \quad 4 \quad = 8$$
$$8 = 8$$

As illustrated by this example, quadratic equations usually have two solutions. Sometimes they will have less than two solutions, but never more than two.

The procedure for solving a quadratic equation can now be summarized by these four steps.

Step (1)   Obtain all terms on one side of the equation. (The quadratic equation must be placed in standard form.)

Step (2)   Factor the polynomial.

Step (3)   Use the zero factor property to set each factor equal to zero.

Step (4)   Solve the resulting first degree equations.

Solve $3x^2 - 2 = x$ Example 5

Solution:

Step (1)   Obtain all terms on one side of the equation. (Place the quadratic equation in standard form.)

$$3x^2 - 2 = x$$
$$3x^2 - x - 2 = 0 \qquad \text{[subtract x]}$$

Step (2)   Factor the polynomial.

$$(3x + 2)(x - 1) = 0$$

Step (3) Use the zero factor property to set each factor equal to zero.

$$3x + 2 = 0 \quad \text{or} \quad x - 1 = 0$$

Step (4) Solve the resulting first degree equations.

$$3x + 2 = 0 \qquad x - 1 = 0$$
$$3x = -2 \qquad x = 1$$
$$x = \frac{-2}{3}$$

Therefore, the solutions are $\frac{-2}{3}$ and 1. You should check these solutions by substituting them, one at a time, into the original equation.

Example 6

Solve $6m^2 = 3m$

Solution:

Step (1) Place the equation in standard form.
$$6m^2 = 3m$$
$$6m^2 - 3m = 0 \qquad \text{[subtract } 3m]$$

Step (2) Factor the polynomial.

$$3m(2m - 1) = 0$$

Step (3) Set each factor equal to zero.

$$3m = 0 \quad \text{or} \quad 2m - 1 = 0$$

Step (4) Solve the resulting first degree equations.

$$3m = 0 \qquad 2m - 1 = 0$$
$$m = \frac{0}{3} \qquad 2m = 1$$
$$m = 0 \qquad m = \frac{1}{2}$$

Therefore, the solutions are 0 and $\frac{1}{2}$. Be sure to check them in the original equation.

Example 7

Solve $y^2 - 6y = -9$

Solution:

Step (1) Place the equation in standard form.

$$y^2 - 6y = -9$$
$$y^2 - 6y + 9 = 0 \qquad \text{[add 9]}$$

Step (2) Factor the polynomial.

$$(y - 3)(y - 3) = 0$$

Step (3) Set each factor equal to zero. In this instance both factors are identical, so we only work with one.

$$y - 3 = 0$$

Step (4) Solve the resulting first degree equation.

$$y - 3 = 0$$
$$y = 3$$

This quadratic equation has only one solution, 3. You should verify this solution by checking it in the original equation. As you become more familiar with this procedure, you can use a shorter form, as shown in examples 8 and 9.

Solve $2x^2 + x = 15$ **Example 8**

Solution:
$$2x^2 + x = 15$$
$$2x^2 + x - 15 = 0 \qquad \text{[subtract 15]}$$
$$(2x - 5)(x + 3) = 0$$
$$2x - 5 = 0 \quad \text{or} \quad x + 3 = 0$$
$$2x = 5 \qquad\qquad x = -3$$
$$x = \frac{5}{2}$$

The solutions are $\frac{5}{2}$ and -3. You should check both answers.

Solve $9y^2 = 49$ **Example 9**

Solution:
$$9y^2 = 49$$
$$9y^2 - 49 = 0 \qquad \text{[subtract 49]}$$

$$(3y + 7)(3y - 7) = 0$$

$$3y + 7 = 0 \quad \text{or} \quad 3y - 7 = 0$$

$$3y = -7 \qquad\qquad 3y = 7$$

$$y = \frac{-7}{3} \qquad\qquad y = \frac{7}{3}$$

The solutions are $\frac{-7}{3}$ and $\frac{7}{3}$. Be sure to check both answers.

You have learned in this section to solve quadratic equations by factoring. However, some polynomials are prime and cannot be factored. Thus, in chapter 6 we will study a method which is not dependent upon factoring. It is called the *quadratic formula*.

Exercise 4.8

As a review solve the following first degree equations and check your answers.

1. $x - 2 = 0$
2. $x + 5 = 0$
3. $2y - 3 = 0$
4. $3m + 7 = 0$
5. $2x + 14 = 2$
6. $3y + 4 = -8$
7. $5a - 15 = 0$
8. $6t + 2 = -4$
9. $3(r + 1) = 0$
10. $3(2x - 5) = 2(5x + 1) - 1$
11. $3(1 - 2y) = y - 4$
12. $3(m - 1) + m = 2(m - 1) - 3$
13. $2(y - 5) = 5(y + 1)$
14. $5(2 + x) = 2(x + 6)$
15. $7(2r + 6) - 5 = 9(r + 3)$
16. $6(2t - 1) - 7 = 7(3t - 2) - 26$
17. $3(x - 4) - (2x - 3) = -4$
18. $5(2y - 1) - 7 = 4(2y + 1)$
19. $2(x - 3) - 9 = 5(x + 4)$
20. $4(2m + 3) - (m - 2) = 14$

Solve the following quadratic equations and check your answers.

21. $x^2 - 5x - 6 = 0$
22. $m^2 - 9m + 14 = 0$
23. $x^2 - 2x + 1 = 0$
24. $a^2 - 2a - 15 = 0$
25. $m^2 - 2m = 8$
26. $t^2 - 3t = -2$
27. $x^2 - x = 12$
28. $n^2 + 3n - 28 = 0$
29. $y^2 - 8y + 16 = 0$
30. $2x^2 + 3x - 20 = 0$
31. $2m^2 - m = 3$
32. $x^2 + 9 = 6x$
33. $r^2 - 9 = 0$
34. $2x^2 - x = 0$
35. $2y^2 - 9y = 0$
36. $5x^2 - 20 = 0$
37. $m^2 = 2m$
38. $p^2 = p + 12$
39. $3x^2 = 6 - 3x$
40. $5x^2 + 13x - 6 = 0$

41. $2x^2 - 25x + 72 = 0$ **42.** $72 = n^2 + 14n$

43. $3x^2 = 27$ **44.** $6y^2 - 7y = 5$

45. $9m^2 - 12m + 4 = 0$ **46.** $3y^2 - 11y + 6 = 0$

47. $3t^2 - 8t + 4 = 0$ **48.** $2a^2 - 5a + 2 = 0$

49. $3s^2 + 3 = 10s$ **50.** $3y^2 - 7y + 2 = 0$

51. $25y^2 + 20y + 4 = 0$ **52.** $9t^2 - 36 = 0$

53. $x(x + 7) = -10$ **54.** $y(3y + 7) = -4$

55. $x(x - 2) = 3$

56. $(x + 1)(3x + 1) = 2x(x + 1) + 4$

57. $(m - 1)(2m + 3) = 3m(m - 1)$

58. $5y^2 + 4y - 1 = 3y^2 + 7y + 1$

59. $3x^2 + 3(x - 1) = 2(x^2 + 11) + 3x$

60. $7t^2 - (t + 5) = 5(t + 1)(t - 1)$

APPLIED PROBLEMS INVOLVING LINEAR AND QUADRATIC EQUATIONS

4.9

As we discussed in section 2.8, word problems are solved by translating word phrases into variable expressions. Then the variable expressions are used to form an equation. It is helpful to follow these four steps.

Step (1)   Let some letter represent one of the unknown numbers. Then translate the word phrases into variable expressions.

Step (2)   Form the equation using information given in the problem.

Step (3)   Solve the equation.

Step (4)   Interpret the solution. There are times when the solution to the equation is not the answer to the written problem. Something else may have to be done. Check each answer using the wording of the given problem.

A 10-foot board is cut into 2 pieces. One piece is 2 feet shorter than the other. How long is each piece?

Example 1

Solution:

Step (1)   Let x represent the longer piece. Then $x - 2$ represents the shorter piece.

Step (2)   Form the equation, remembering that the sum of the 2 lengths is 10 feet.

$$x + (x - 2) = 10$$

Step (3)   Solve the equation.

$$x + (x - 2) = 10$$
$$x + x - 2 = 10$$
$$2x - 2 = 10$$
$$2x = 12$$
$$x = 6$$

Step (4)   Interpret the solution. If $x = 6$, then $x - 2 = 4$. Therefore, the longer piece is 6 feet and the shorter piece is 4 feet.

Example 2

Water tank A has a capacity of 75 gallons more than water tank B. Water tank C has a capacity 3 times greater than tank B. Together the tanks have a capacity of 1075 gallons. Find the capacity of each tank.

Solution:

Step (1)   Since the capacities of tank A and tank C are both given in terms of tank B, let x represent the capacity of tank B. Then, $x + 75$ represents the capacity of tank A and $3x$ represents the capacity of tank C.

Step (2)   Form the equation by noting that the sum of the capacities of the 3 tanks is 1075 gallons.

$$\underset{\begin{matrix}\text{tank}\\ B\\ \downarrow\end{matrix}}{x} + \underset{\begin{matrix}\text{tank}\\ A\\ \downarrow\end{matrix}}{(x + 75)} + \underset{\begin{matrix}\text{tank}\\ C\\ \downarrow\end{matrix}}{3x} = \underset{\begin{matrix}\text{total}\\ \text{capacity}\\ \downarrow\end{matrix}}{1075}$$

Step (3)   Solve the equation.

$$x + (x + 75) + 3x = 1075$$
$$x + x + 75 + 3x = 1075$$
$$5x + 75 = 1075$$
$$5x = 1000$$
$$x = 200$$

Step (4)   Interpret the solution. If $x = 200$, then $x + 75 = 275$ and $3x = 600$. Thus, the capacity of tank B is 200 gallons. The capacity of tank A is 275 gallons, and the capacity of tank C is 600 gallons.

Example 3

A person made a trip at an average speed of 50 miles per hour. A second person made the same trip in 1 hour less time at an average speed of 60 miles per hour. Find the distance of the trip.

Solution: Use the formula $d = rt$. (Distance equals rate multiplied by time.)

Step (1)   Let t represent the time necessary for the first person to make the trip. $t - 1$ would then represent the time necessary for the second person to make the same trip. Since distance equals rate multiplied by time, $50t$ would be the distance traveled by the first person and $60(t - 1)$ would be the distance traveled by the second person. Study the accompanying diagram representing the trip.

> 1st person: time is t hr
> rate is 50 mph
> distance is $50t$ mi

Trip ●————————————————————————→●

> 2nd person: time is $t - 1$ hr
> rate is 60 mph
> distance is $60(t - 1)$ mi

Step (2)   Form an equation, keeping in mind that the distance was the same for both persons.

$$50t = 60(t - 1)$$

Step (3)   Solve the equation.

$$50t = 60(t - 1)$$
$$50t = 60t - 60$$
$$-10t = -60$$
$$t = 6$$

Step (4)   Interpret the solution. We wish to find the distance, which is given by either $50t$ or $60(t - 1)$. Substitute 6 hours for t and evaluate either expression:

$$50t = 50 \cdot 6 = 300 \text{ miles}$$

$$60(t - 1) = 60(6 - 1) = 300 \text{ miles}$$

Thus, the distance of the trip was 300 miles.

Example 4   The length of a rectangle is 4 meters (m) more than the width. The area of the rectangle is 12 square meters. Find the length and width.

Solution: The area of a rectangle is found by multiplying the length times the width. $A = lw$

Step (1)   Let w represent the width of the rectangle. Then $w + 4$ would represent the length. Study the following diagram:

$$w \left\{ \boxed{\text{area is 12 } m^2} \right.$$
$$\underbrace{}_{w + 4}$$

Step (2)   Form the equation by remembering that the product of the length and width gives the area.

$$(w + 4) \cdot w = 12$$

Step (3)   Solve the equation, which is quadratic.

$$(w + 4) \cdot w = 12$$
$$w^2 + 4w = 12$$
$$w^2 + 4w - 12 = 0$$
$$(w + 6)(w - 2) = 0$$
$$w + 6 = 0 \quad \text{or} \quad w - 2 = 0$$
$$w = -6 \qquad\qquad w = 2$$

Step (4)   Interpret the solutions. The equation has two solutions, -6 and 2. However, a rectangle cannot have a negative width. The only solution which meets the physical situation is 2. Thus, the width of the rectangle is 2 meters. The length is $w + 4$, which would be 6 meters.

Example 5   The product of two consecutive even integers is twenty more than twice their sum. Find the integers.

Solution: Consecutive even integers differ by 2. For example, -4, -2, 0, 2, 4, and 6 are consecutive even integers.

Step (1)   Let n represent the smaller of the two integers; then $n + 2$ represents the larger. The product is $n(n + 2)$. Twenty more than twice their sum is $20 + 2[n + (n + 2)]$.

Step (2)   Form the equation.

the   more   their
product   is   20 than   twice   sum

$$n(n + 2) = 20 + 2[n + (n + 2)]$$

Step (3)   Solve the equation, which is quadratic.

$$n(n + 2) = 20 + 2[n + (n + 2)]$$

$$n^2 + 2n = 20 + 2[2n + 2]$$

$$n^2 + 2n = 20 + 4n + 4$$

$$n^2 - 2n - 24 = 0$$

$$(n - 6)(n + 4) = 0$$

$$n - 6 = 0 \quad \text{or} \quad n + 4 = 0$$

$$n = 6 \qquad\qquad n = -4$$

Step (4)   Interpret the solutions. The consecutive even integers are represented by n and $n + 2$. When $n = 6$, $n + 2 = 8$. And when $n = -4$, $n + 2 = -2$. This gives two sets of answers, 6 and 8 together with -4 and -2. Each of these check as follows.

Check for 6 and 8

$$6 \cdot 8 = 20 + 2(6 + 8)$$

$$48 = 20 + 28$$

$$48 = 48$$

Check for −4 and −2

$$(-4)(-2) = 20 + 2(-4 - 2)$$

$$8 = 20 - 12$$

$$8 = 8$$

Solve the following problems.   **Exercise 4.9**

1. A 100-ft rope is cut into two pieces. One piece is 16 ft longer than the other. How long is each piece?

2. Oil tank A has a capacity twice that of tank B. Tank C has a capacity of 100 gal more than tank B. Together the tanks have a capacity of 3700 gal. Find the capacity of each tank.

3. A jet airliner made a trip at an average speed of 400 mph. A second jet airliner made the same trip in 2 hr less time at an average speed of 600 mph. What was the distance of the trip?

4. The length of a rectangle is 10 m more than the width. The area of the rectangle is 24 m². Find the length and the width.

5. The product of two consecutive even integers is 4 more than twice their sum. Find the integers.

6. Six times a positive number when added to the square of the number will produce 16. Find the number.

7. A negative number subtracted from the square of the number produces 6. Find the number. (Hint: let x equal the negative number.)

8. An 84-ft rope is cut into three pieces. The second piece is twice the length of the first. The third piece is 4 ft longer than the first. Find the length of each piece.

9. A man traveled to and from a city over the same route. His average speed to the city was 60 mph. His return trip took 1 hr longer at an average speed of 45 mph. What was the distance of the trip (one way)?

10. Three times a negative number added to the square of the number produces 10. Find the number.

11. Find two integers whose sum is 3, if the sum of their squares is 29.

12. The product of two consecutive even integers is 16 more than twice the smaller integer. Find both integers.

13. The product of two consecutive even integers is twice the larger integer. Find both integers.

14. The product of two consecutive odd integers is 27 more than 4 times the larger integer. Find both integers.

15. The sum of a certain number and 25 is equal to 6 times the number. What is the number?

16. A certain number is four times another. Their sum is 135. Find the numbers.

17. An 8-ft board is cut into two pieces. One piece is 2 ft longer than the other. Find the length of each piece.

18. A 16-ft board is cut into three pieces. The second piece is twice as long as the first. The third piece is 4 ft longer than the first. Find the length of each piece.

19. The length of a rectangle is 6 m greater than its width. The perimeter is 44 m. Find the length and width.

20. The length of a rectangle is 4 ft more than its width. If the perimeter is 56 ft, find the dimensions.

21. The width of a rectangle is 4 ft less than its length. The area is 45 ft². Find the length and width.

22. The length of a rectangle is twice its width. If the width were increased by 2 in. and the length decreased by 1 in., the area would be 42 in². Find the dimensions of the original rectangle.

23. When each side of a square is increased by 3 cm(centimeters), the area is increased by 51 cm². Find the length of each side of the new square.

24. Two test rockets are fired over a 5600-mi range. One rocket travels twice the speed of the other. The faster rocket covers the distance in 2 hr less time than the slower. Find the speed of each rocket.

25. Two planes leave a certain airport in opposite directions. The speed of one plane averages 60 mph faster than that of the other. At the end of 5 hr, the planes are 1550 mi apart. Find the rate of the slower plane.

26. Two planes start from the same airport at the same time, one flying north, the other south. The ground speed of the first plane is 200 mph less than the ground speed of the other plane. After 2 hr, they are 1680 mi apart. What is the speed of each plane?

27. A negative number subtracted from the square of the same number produces 20. Find the number.

28. Three times a positive number added to the square of the same number produces 10. Find the number.

29. The length of a rectangle is 2 ft more than the width. The area of the rectangle is 15 ft². Find the length and the width.

30. The formula $s = 16t^2$ gives the distance an object falls due to gravity. s is distance in feet and t is time in seconds. How long will it take for an object to fall 64 ft?

31. A contractor wants to add a uniform border around a 20-by-30-ft rectangular pool. How wide should the border be if he has enough concrete for 600 ft²?

32. A landscape engineer wishes to sod a uniform strip of grass around a rectangular garden measuring 30 ft by 50 ft. She has enough sod to cover 336 ft². How wide should she make the strip of grass?

Summary

SYMBOLS

$P(x)$ Polynomial
FOIL   Shortcut for multiplying two binomials

DEFINITION OF A POLYNOMIAL

A polynomial is an expression containing a finite number of terms where each term fits the pattern ax^n. a represents a real number, x represents a variable, and n represents a whole number.

Example: $5x^4 + 2x^3 - 4x^2 + x - 5$

DEFINITION OF A MONOMIAL

A monomial is a polynomial consisting of one term.

DEFINITION OF A BINOMIAL

A binomial is a polynomial consisting of two terms.

DEFINITION OF A TRINOMIAL

A trinomial is a polynomial consisting of three terms.

SPECIAL PRODUCTS OF BINOMIALS

(1) The FOIL shortcut: $(a + b)(c + d) = ac + ad + bc + bd$
(2) Sum times a difference: $(a + b)(a - b) = a^2 - b^2$
(3) Squaring a binomial: $(a + b)^2 = a^2 + 2ab + b^2$

TYPES OF FACTORING

(1) Common monomial factoring.

Example: $6a^3b^2 - 9a^2b^3 + 3a^2b^2 = 3a^2b^2(2a - 3b + 1)$

(2) Difference of two squares.

Example: $25x^2 - 9y^2 = (5x + 3y)(5x - 3y)$

(3) Reverse of FOIL.

Example: $2x^2 - 5x - 3 = (2x + 1)(x - 3)$

THE ZERO FACTOR PROPERTY

If a and b represent real numbers, and if $a \cdot b = 0$, then $a = 0$ or $b = 0$

SOLVING A QUADRATIC EQUATION BY FACTORING

Step (1) Obtain all terms on one side of the equation. (The polynomial must equal zero.)

Step (2) Factor the polynomial.

Step (3) Use the zero factor property to set each factor equal to zero.

Step (4) Solve the resulting first degree equations.

Which of the following expressions are polynomials? (Answer yes or no.)

1. $x^4 - 2x^3 + 4x^2 - 3x + 1$

2. $\dfrac{3}{x}$

3. $4x^3y^2 - 3x^2y^3$

4. $\dfrac{a+1}{a-1}$

Classify each of these polynomials as a monomial, binomial, or trinomial.

5. $x^2 + 5x + 6$

6. $4a^2 - 9b^2$

7. $4m^2n$

8. 5

If $P(x) = x^3 - 2x^2 + 3x - 1$, find the following values.

9. $P(0)$

10. $P(1)$

11. $P(-1)$

12. $P(3)$

Combine the following polynomials by addition or subtraction. You may use either the horizontal or vertical method.

13. $(2x^2 - 5x + 1) + (4x^2 + 2x - 3) + (x^2 - x + 2)$

14. $(3m^2 - 2m + 3) - (m^2 - 4m + 5)$

15. $(5y^4 - 2y^2 + 6) - (2y^4 + y^2 - 1) + (3y^4 + 3y^2 + 2)$

Multiply the following polynomials vertically.

16. $(x + 3)(2x^2 - 4x + 3)$

17. $(m^2 + m - 2)(4m^3 + m^2 - 3m + 1)$

Multiply, using any shortcut which applies.

18. $(-3a^2b)(2ab^3c)$

19. $2x^2(3xy^2 + 2x^2y - 3y)$

20. $(x + 2)(x + 1)$

21. $(x - 3)(x - 2)$

22. $(2y - 3)(y + 2)$

23. $(3a - 4)(2a - 3)$

24. $(2x + 3)(2x - 3)$

25. $(3m + 5n)(3m - 5n)$

26. $(x + 3)^2$

27. $(2a - 3b)^2$

Divide the following polynomials.

28. $\dfrac{12m^3 + 6m^2 - 3m}{-3m}$

29. $\dfrac{2x^4 - 3x^3 + 4x^2 - x + 2}{x}$

30. $(6y^3 + 7y^2 - y + 1) \div (3y + 2)$

31. $(27a^3 - 1) \div (3a - 1)$

Factor each of these polynomials completely.

32. $3a - 6$

33. $2x^2 - 4x$

34. $15m^2n - 10mn^2 + 5mn$

35. $x^2 - 9$

36. $4a^2 - 25$

37. $16m^4 - 1$

38. $a^2 + 3a - 10$

39. $4x^2 - 12x + 9$

40. $2m^2 - 5m + 3$

41. $6a^3 - 3a^2 - 63a$

Solve the following equations.

42. $x + 6 = 0$

43. $3m + 2 = 7$

44. $3a + 5 = 2(a - 1) - a$

45. $x^2 + 3x + 2 = 0$

46. $n^2 - 2n = 8$

47. $2y^2 - 3y = 0$

48. $6x^2 + 19x = 7$

49. $(t + 2)(2t - 3) = t(t - 4)$

Solve the following applied problems.

50. A 16-ft board is cut into two pieces. One piece is 4 ft shorter than the other. How long is each piece?

51. Water tank A has a capacity of 100 gal more than water tank B. Water tank C has a capacity 4 times greater than tank B. Together the tanks have a total capacity of 1600 gal. Find the capacity of each tank.

52. A person made a trip at an average speed of 40 mph. A second person made the same trip in 2 hr less time at an average speed of 60 mph. Find the distance of the trip.

53. The length of a rectangle is 6 cm more than the width. The area of the rectangle is 27 cm². Find the length and width.

54. The product of two consecutive odd integers is 1 less than twice their sum. Find the integers.

55. Find two integers whose sum is 10, if the sum of their squares is 58.

Solutions to Self-Checking Exercise

1. Yes.

2. No.

3. Yes.

4. No.

5. Trinomial.

6. Binomial.

7. Monomial.

8. Monomial.

9. $P(0) = 0^3 - 2 \cdot 0^2 + 3 \cdot 0 - 1$
$= 0 - 0 + 0 - 1$
$= -1$

10. $P(1) = 1^3 - 2 \cdot 1^2 + 3 \cdot 1 - 1$
$= 1 - 2 + 3 - 1$
$= 1$

11. $P(-1) = (-1)^3 - 2(-1)^2 + 3(-1) - 1$
$= -1 - 2 - 3 - 1$
$= -7$

12. $P(3) = 3^3 - 2 \cdot 3^2 + 3 \cdot 3 - 1$
$= 27 - 18 + 9 - 1$
$= 17$

13. $7x^2 - 4x$

14. $2m^2 + 2m - 2$

15. $6y^4 + 9$

16. $2x^2 - 4x + 3$

17.
$$
\begin{array}{r}
4m^3 + m^2 - 3m + 1 \\
m^2 + m - 2 \\
\hline
4m^5 + m^4 - 3m^3 + m^2 \\
4m^4 + m^3 - 3m^2 + m \\
- 8m^3 - 2m^2 + 6m - 2 \\
\hline
4m^5 + 5m^4 - 10m^3 - 4m^2 + 7m - 2
\end{array}
$$

$$
\begin{array}{r}
x + 3 \\
\hline
2x^3 - 4x^2 + 3x \\
6x^2 - 12x + 9 \\
\hline
2x^3 + 2x^2 - 9x + 9
\end{array}
$$

18. $-6a^3b^4c$

19. $6x^3y^2 + 4x^4y - 6x^2y$

20. $x^2 + 3x + 2$

21. $x^2 - 5x + 6$

22. $2y^2 + y - 6$

23. $6a^2 - 17a + 12$

24. $4x^2 - 9$

25. $9m^2 - 25n^2$

26. $x^2 + 6x + 9$

27. $4a^2 - 12ab + 9b^2$

28. $-4m^2 - 2m + 1$

29. $2x^3 - 3x^2 + 4x - 1 + \dfrac{2}{x}$

30.
$$
\begin{array}{r}
2y^2 + y - 1 + \dfrac{3}{3y+2} \\
3y+2\overline{)6y^3 + 7y^2 - y + 1} \\
6y^3 + 4y^2 \\
\hline
3y^2 - y \\
3y^2 + 2y \\
\hline
-3y + 1 \\
-3y - 2 \\
\hline
3
\end{array}
$$

31.
$$
\begin{array}{r}
9a^2 + 3a + 1 \\
3a-1\overline{)27a^3 + 0a^2 + 0a - 1} \\
27a^3 - 9a^2 \\
\hline
9a^2 + 0a \\
9a^2 - 3a \\
\hline
3a - 1 \\
3a - 1 \\
\hline
0
\end{array}
$$

32. $3(a - 2)$

33. $2x(x - 2)$

34. $5mn(3m - 2n + 1)$

35. $(x + 3)(x - 3)$

36. $(2a + 5)(2a - 5)$

37. $(4m^2 + 1)(2m + 1)(2m - 1)$

38. $(a + 5)(a - 2)$

39. $(2x - 3)^2$

40. $(2m - 3)(m - 1)$

41. $3a(2a - 7)(a + 3)$

42. $x + 6 = 0$
$x = -6$

43. $3m + 2 = 7$
$3m = 5$
$m = \dfrac{5}{3}$

44. $3a + 5 = 2(a - 1) - a$
$3a + 5 = 2a - 2 - a$
$3a + 5 = a - 2$
$2a = -7$
$a = -\dfrac{7}{2}$

45. $x^2 + 3x + 2 = 0$
$(x + 2)(x + 1) = 0$
$x + 2 = 0 \quad$ or $\quad x + 1 = 0$
$x = -2 \qquad\qquad x = -1$
Solutions are -2 and -1

46. $n^2 - 2n = 8$
$n^2 - 2n - 8 = 0$
$(n - 4)(n + 2) = 0$
$n - 4 = 0 \quad$ or $\quad n + 2 = 0$
$n = 4 \qquad\qquad n = -2$
Solutions are 4 and -2

47. $2y^2 - 3y = 0$
$y(2y - 3) = 0$
$y = 0 \quad$ or $\quad 2y - 3 = 0$
$y = 0 \qquad\qquad 2y = 3$
$y = \dfrac{3}{2}$

Solutions are 0 and $\dfrac{3}{2}$

48.
$$6x^2 + 19x = 7$$
$$6x^2 + 19x - 7 = 0$$
$$(2x + 7)(3x - 1) = 0$$
$$2x + 7 = 0 \quad \text{or} \quad 3x - 1 = 0$$
$$2x = -7 \qquad\qquad 3x = 1$$
$$x = -\frac{7}{2} \qquad\qquad x = \frac{1}{3}$$

Solutions are $-\frac{7}{2}$ and $\frac{1}{3}$

49. $(t + 2)(2t - 3) = t(t - 4)$
$$2t^2 + t - 6 = t^2 - 4t$$
$$t^2 + 5t - 6 = 0$$
$$(t + 6)(t - 1) = 0$$
$$t + 6 = 0 \quad \text{or} \quad t - 1 = 0$$
$$t = -6 \qquad\qquad t = 1$$
Solutions are -6 and 1

50. $x + (x - 4) = 16$
$$2x - 4 = 16$$
$$2x = 20$$
$$x = 10$$
One piece is 10 ft and
the other is 6 ft.

51. $(b + 100) + b + 4b = 1600$
$$6b + 100 = 1600$$
$$6b = 1500$$
$$b = 250$$
Tank B holds 250 gal.
Tank A holds 350 gal.
Tank C holds 1000 gal.

52. $40t = 60(t - 2)$
$$40t = 60t - 120$$
$$-20t = -120$$
$$t = 6$$
Distance equals $40t$.
Thus, distance is 240 mi.

53.
$$(w + 6)w = 27$$
$$w^2 + 6w = 27$$
$$w^2 + 6w - 27 = 0$$
$$(w + 9)(w - 3) = 0$$
$$w + 9 = 0 \quad \text{or} \quad w - 3 = 0$$
$$\cancel{w = -9} \qquad\qquad w = 3$$
Width is 3 cm.
Length is 9 cm.

54.
$$n(n + 2) = 2[n + (n + 2)] - 1$$
$$n^2 + 2n = 2[2n + 2] - 1$$
$$n^2 + 2n = 4n + 4 - 1$$
$$n^2 - 2n - 3 = 0$$
$$(n - 3)(n + 1) = 0$$
$$n - 3 = 0 \quad \text{or} \quad n + 1 = 0$$
$$n = 3 \qquad\qquad n = -1$$
The two pairs of solutions are 3, 5
and -1, 1

55.
$$x^2 + (10 - x)^2 = 58$$
$$x^2 + 100 - 20x + x^2 = 58$$
$$2x^2 - 20x + 42 = 0$$
$$2(x^2 - 10x + 21) = 0$$
$$2(x - 7)(x - 3) = 0$$
$$x - 7 = 0 \quad \text{or} \quad x - 3 = 0$$
$$x = 7 \qquad\qquad x = 3$$
One integer is 7 and the other is 3

OPERATIONS WITH RATIONAL EXPRESSIONS

5

EXPANSION AND REDUCTION
OF RATIONAL EXPRESSIONS

5.1

rational expression

You learned in section 1.5 that a quotient of integers is called a *rational number*. Similarly, a quotient of two polynomials is called a **rational expression**. Since division by zero is impossible, the denominator of a rational expression can never represent zero. The domain of a rational expression is the set of all real numbers for which the expression is defined.

Example 1

$\dfrac{x + 5}{x - 2}$ is a rational expression. Its domain consists of all real numbers except 2

Example 2

$\dfrac{y^2 + 3y + 1}{y^2 - 9}$ is a rational expression. Its domain consists of all real numbers except 3 and −3

In section 3.1 you studied some basic properties of fractions. Rational expressions are fractions where the numerator and denominator are polynomials. Thus, the basic properties of fractions also hold true for rational expressions.

Review of the Basic Properties of Fractions

(1) The denominator of a fraction can never be zero.
   Example: $\dfrac{3x + 5}{0}$ does not represent a number.

(2) If the denominator of a fraction is 1, then the fraction is equal to the numerator.
   Example: $\dfrac{2y^2 + 3y + 7}{1} = 2y^2 + 3y + 7$

(3) If the numerator of a fraction is zero and the denominator is not zero, then the fraction is equal to zero.
   Example: $\dfrac{0}{m - 5} = 0$, where $m \neq 5$

(4) If the numerator and denominator of a fraction are equal but not zero, then the fraction is equal to 1.
   Example: $\dfrac{x - 2}{x - 2} = 1$, where $x \neq 2$

(5) The numerator and denominator of a fraction may be multiplied or divided by the same nonzero number (Fundamental Principle of Fractions).
   Examples: $\dfrac{a}{b} = \dfrac{a \cdot n}{b \cdot n}$ or $\dfrac{a}{b} = \dfrac{a \div n}{b \div n}$, where $b, n \neq 0$

The fundamental principle of fractions allows us to expand a rational expression to higher terms by multiplying the numerator and denominator by the same nonzero expression.

Expand $\dfrac{3a}{4b^2c}$ so that its denominator becomes $12b^3c^4$ **Example 3**

Solution: Since $3bc^3$ times $4b^2c$ is $12b^3c^4$, multiply numerator and denominator by $3bc^3$

$$\frac{3a}{4b^2c} = \frac{3a \cdot (3bc^3)}{4b^2c \cdot (3bc^3)} = \frac{9abc^3}{12b^3c^4}$$

Expand $\dfrac{2x}{5}$ so that its denominator becomes $5x + 5y$ **Example 4**

Solution: Factor $5x + 5y$ into $5(x + y)$. This shows that 5 times $(x + y)$ gives $5x + 5y$. Thus, the numerator and denominator must be multiplied by $(x + y)$

$$\frac{2x}{5} = \frac{2x\,(x + y)}{5\,(x + y)} = \frac{2x^2 + 2xy}{5x + 5y}$$

Expand $\dfrac{m - 3}{m + 2}$ so that its denominator becomes $m^2 - 4$ **Example 5**

Solution: Factor $m^2 - 4$ into $(m + 2)(m - 2)$. This shows that $(m + 2)$ times $(m - 2)$ gives $m^2 - 4$. Thus, the numerator and denominator must be multiplied by $(m - 2)$

$$\frac{m - 3}{m + 2} = \frac{(m - 3)(m - 2)}{(m + 2)(m - 2)} = \frac{m^2 - 5m + 6}{m^2 - 4}$$

Expand $\dfrac{y + 1}{y - 5}$ so that its denominator becomes $2y^2 - 7y - 15$ **Example 6**

Solution: Factor $2y^2 - 7y - 15$ into $(y - 5)(2y + 3)$. This shows that the numerator and denominator must be multiplied by $(2y + 3)$

$$\frac{y + 1}{y - 5} = \frac{(y + 1)(2y + 3)}{(y - 5)(2y + 3)} = \frac{2y^2 + 5y + 3}{2y^2 - 7y - 15}$$

The fundamental principle of fractions allows us to reduce a rational expression to lowest terms by dividing out common factors shared by the numerator and denominator. You should recall from section 3.3 that this procedure can be used only when the following two conditions are satisfied:

(1)   Both the numerator and denominator must be basic products.
(2)   The expressions divided out must be factors.

Therefore, to reduce a rational expression to lowest terms, we factor the numerator and denominator and then divide out the common factors.

Example 7

Reduce $\dfrac{6m}{9m^2n}$ to lowest terms.

Solution: Factor the numerator and denominator; then divide out the common factors.

$$\frac{6m}{9m^2n} = \frac{2 \cdot \cancel{3} \cdot \cancel{m}}{3 \cdot \cancel{3} \cdot \cancel{m} \cdot m \cdot n} = \frac{2}{3mn}$$

Example 8

Reduce $\dfrac{8a^2 + 4ab}{14a^2 + 7ab}$ to lowest terms.

Solution: Factor the numerator and denominator; then divide out the common factors.

$$\frac{8a^2 + 4ab}{14a^2 + 7ab} = \frac{4\cancel{a}\,(\cancel{2a + b})}{7\cancel{a}\,(\cancel{2a + b})} = \frac{4}{7}$$

Example 9

Reduce $\dfrac{4a^2 - 9}{(2a + 3)^2}$ to lowest terms.

Solution: $\dfrac{4a^2 - 9}{(2a + 3)^2} = \dfrac{(\cancel{2a + 3})(2a - 3)}{(\cancel{2a + 3})(2a + 3)} = \dfrac{2a - 3}{2a + 3}$

Example 10

Reduce $\dfrac{2x^2 - 13x - 7}{x^2 - 10x + 21}$ to lowest terms.

Solution: $\dfrac{2x^2 - 13x - 7}{x^2 - 10x + 21} = \dfrac{(2x + 1)(\cancel{x - 7})}{(x - 3)(\cancel{x - 7})} = \dfrac{2x + 1}{x - 3}$

Sometimes when reducing rational expressions, you will encounter binomial factors of the form $a - b$ and $b - a$. These binomials have their terms interchanged around a minus sign. Expressions of this type produce opposites (additive inverses). Consider the expressions $a - b$ and $b - a$ where $a = 5$ and $b = 2$:

$$
\begin{array}{cc}
a - b & b - a \\
\downarrow\ \ \downarrow & \downarrow\ \ \downarrow \\
5 - 2 & 2 - 5 \\
\downarrow & \downarrow \\
3 & -3
\end{array}
$$

Also consider the expression $2m - 5$ and $5 - 2m$ where $m = 3$:

$$2m - 5 \qquad 5 - 2m$$
$$2(3) - 5 \qquad 5 - 2(3)$$
$$6 - 5 \qquad 5 - 6$$
$$1 \qquad -1$$

Binomial expressions whose terms have been interchanged around a minus sign are opposites (additive inverses).

Each of the following pairs represents opposites.
(a) $x - y$ and $y - x$
(b) $m - 2$ and $2 - m$
(c) $2a - 3b$ and $3b - 2a$
(d) $r^2 - t^2$ and $t^2 - r^2$

Example 11

As illustrated by the next example, a non-zero expression when divided by its opposite produces -1.

Here are some illustrations of dividing opposites.

Example 12

(a) $\dfrac{-5}{5} = -1$

(b) $\dfrac{-x}{x} = -1$,   where $x \neq 0$

(c) $\dfrac{m - 5}{5 - m} = -1$,   where $m \neq 5$

Therefore, when reducing rational expressions, factors that represent opposites may be divided out provided a factor of -1 is left behind. It can appear in either the numerator or denominator, but not in both.

Reduce $\dfrac{ax - ay}{by - bx}$ to lowest terms.

Example 13

Solution: Factor the numerator and denominator; then divide opposites, producing a factor of -1.

$$\frac{ax - ay}{by - bx} = \frac{a(x - y)}{b(y - x)} = \frac{a \cdot (-1)}{b} = \frac{-a}{b}$$

Reduce $\dfrac{9 - 4x^2}{2x^2 - x - 3}$ to lowest terms.

Example 14

Solution (a): $\dfrac{9 - 4x^2}{2x^2 - x - 3} = \dfrac{(3 + 2x)(3 - 2x)}{(x + 1)(2x - 3)}$ divided opposites

$\qquad\qquad\quad = \dfrac{(3 + 2x)(-1)}{x + 1}$

$\qquad\qquad\quad = \dfrac{-3 - 2x}{x + 1}$

The factor of -1 could have been placed in the denominator rather than the numerator. This would give an alternate form of the answer, as shown in solution (b). Either form is correct.

Solution (b): $\dfrac{9 - 4x^2}{2x^2 - x - 3} = \dfrac{(3 + 2x)(3 - 2x)}{(x + 1)(2x - 3)}$ divided opposites

$\qquad\qquad\quad = \dfrac{3 + 2x}{(x + 1)(-1)}$

$\qquad\qquad\quad = \dfrac{3 + 2x}{-x - 1}$

You can see that these two answers are equivalent by using the equality test for fractions to show that their cross products are equal:

$$\frac{-3 - 2x}{x + 1} = \frac{3 + 2x}{-x - 1}$$

$$(-3 - 2x)(-x - 1) = (x + 1)(3 + 2x)$$

$$3x + 3 + 2x^2 + 2x = 3x + 2x^2 + 3 + 2x$$

$$2x^2 + 5x + 3 = 2x^2 + 5x + 3$$

Exercise 5.1

Change the following rational expressions as indicated. Assume that denominators do not represent zero.

1. Expand $\dfrac{3r}{7st}$ so that its denominator becomes $21s^4t^4$

2. Expand $\dfrac{3m}{7n}$ so that its denominator becomes $14mn^3$

3. Expand $\dfrac{2x}{3}$ so that its denominator becomes $3x + 3y$

4. Expand $\dfrac{5a}{3b}$ so that its denominator becomes $6b^2 - 3b$

5. Expand $\dfrac{x - 4}{x + 3}$ so that its denominator becomes $x^2 - 9$

6. Expand $\dfrac{2m - 1}{m - 2}$ so that its denominator becomes $3m^2 - 7m + 2$

7. Expand $\dfrac{2x + 1}{x + 3}$ so that its denominator becomes $x^2 + 5x + 6$.

8. Change $\dfrac{a - 5}{a - 6}$ so that its denominator becomes $6 - a$. (Hint: Multiply numerator and denominator by -1.)

9. Change $\dfrac{3a - 2}{2a - 1}$ so that its denominator becomes $1 - 2a$.

10. Expand $\dfrac{y + 3}{3y + 1}$ so that its denominator becomes $6y^2 - 10y - 4$

Reduce each of the following rational expressions to lowest terms. Assume that denominators do not represent zero.

11. $\dfrac{5m^2n}{15m}$

12. $\dfrac{8ab}{12ab^2}$

13. $\dfrac{12x^2y^3}{24x^3y}$

14. $\dfrac{-12rst^2}{21rst^3}$

15. $\dfrac{ab - ac}{ab + ac}$

16. $\dfrac{3a - 6b}{4a - 8b}$

17. $\dfrac{3a - 6b}{8b - 4a}$

18. $\dfrac{3x - 3y}{4x - 4y}$

19. $\dfrac{3m - 6n}{5m - 10n}$

20. $\dfrac{2a + 2b}{2a - 2b}$

21. $\dfrac{9a^2 - 4}{(3a - 2)^2}$

22. $\dfrac{x - 2}{x^2 - 4}$

23. $\dfrac{2 - x}{x^2 - 4}$

24. $\dfrac{10m^2 + 5mn}{10m^2 - 30mn}$

25. $\dfrac{5x - 10y}{14y - 7x}$

26. $\dfrac{3b - 2a}{4a^2 - 9b^2}$

27. $\dfrac{x^2 - 1}{x^2 - 2x + 1}$

28. $\dfrac{m^2 - m - 20}{m^2 + 7m + 12}$

29. $\dfrac{4 - y^2}{3y^2 - y - 10}$

30. $\dfrac{6x^2 - 11x - 10}{9x^2 - 4}$

31. $\dfrac{x^2 + 4x + 3}{x^2 + 5x + 6}$

32. $\dfrac{a^2 - 3a}{2a^2 - 11a + 15}$

33. $\dfrac{5y - y^2}{2y^2 - 9y - 5}$

34. $\dfrac{3 - m}{m^2 - 9}$

35. $\dfrac{a^2 - 9b^2}{3ab - 9b^2}$

36. $\dfrac{2x^2 + 9x + 4}{2x^2 - 5x - 3}$

37. $\dfrac{x^2 + 3x - 4}{x^2 + 5x + 4}$

38. $\dfrac{a^2 - 4a + 4}{a^2 - 5a + 6}$

39. $\dfrac{1 - a^2}{a^2 - 3a + 2}$

40. $\dfrac{x^2 - 8x + 7}{x^2 - 10x + 21}$

41. $\dfrac{m^2 - 3m}{m^2 - 2m - 3}$

42. $\dfrac{x^2 - 4xy + 4y^2}{x^2 - 4y^2}$

43. $\dfrac{4y^2 + 2y - 30}{4y^2 - 36}$

44. $\dfrac{4y^2 + 2y - 30}{50 - 8y^2}$

45. $\dfrac{(3 - a)(2a - 1)(a - 4)}{(a - 2)(4 - a)(a - 3)}$

46. $\dfrac{(3x - 1)(2x + 3)(2 - x)}{(x - 2)(1 - 3x)(3 + 2x)}$

47. $\dfrac{r^4 - 1}{1 - r^2}$

48. $\dfrac{2x^3 - 9x^2 - 18x}{x^3 - 36x}$

49. $\dfrac{(9m^2 - 1)(m^2 - 4)}{3m^2 - 5m - 2}$

50. $\dfrac{18t^2 + 9t - 2}{9t^2 + 12t + 4}$

MULTIPLICATION AND DIVISION OF RATIONAL EXPRESSIONS

5.2

Multiplication of Rational Expressions

In section 3.4 you learned that fractions can be multiplied by following these steps:

Step (1) Write each fraction in factored form and apply the fundamental principle of fractions to divide out any factors common to both a numerator and denominator.

Step (2) Multiply the remaining numerators and multiply the remaining denominators.

Rational expressions are fractions where the numerator and denominator are polynomials. Therefore, the same steps are used to multiply rational expressions.

Example 1

Multiply $\dfrac{4a^3}{3b} \cdot \dfrac{5b^2}{2a^2}$

Solution:

Step (1) $\dfrac{4a^3}{3b} \cdot \dfrac{5b^2}{2a^2} = \dfrac{2 \cdot 2 \cdot a \cdot a^2}{3 \cdot b} \cdot \dfrac{5 \cdot b \cdot b}{2 \cdot a^2}$

Step (2) $= \dfrac{10ab}{3}$

Example 2

Multiply $\dfrac{3x + 3y}{10x^2} \cdot \dfrac{15x}{2x + 2y}$

Solution:

Step (1) $\dfrac{3x + 3y}{10x^2} \cdot \dfrac{15x}{2x + 2y} = \dfrac{3 \cdot (x + y)}{2 \cdot 5 \cdot x \cdot x} \cdot \dfrac{3 \cdot 5 \cdot x}{2 \cdot (x + y)}$

Step (2) $= \dfrac{9}{4x}$

Multiply $\dfrac{a^2 - 4}{a^2 - 2a - 15} \cdot \dfrac{a^2 - 9}{a^2 - 6a + 8}$ **Example 3**

Solution:

Step (1) $\dfrac{a^2 - 4}{a^2 - 2a - 15} \cdot \dfrac{a^2 - 9}{a^2 - 6a + 8} = \dfrac{(a + 2) \cdot (a - 2)}{(a - 5) \cdot (a + 3)} \cdot \dfrac{(a + 3) \cdot (a - 3)}{(a - 2) \cdot (a - 4)}$

Step (2) $= \dfrac{(a + 2)(a - 3)}{(a - 5)(a - 4)}$

When multiplying rational expressions, do not multiply out the answers. They are usually left in factored form.

Multiply $\dfrac{3m^2 - 3m}{2m^2 + m - 6} \cdot \dfrac{m + 2}{3 - 3m}$ **Example 4**

Solution:

divided opposites

Step (1) $\dfrac{3m^2 - 3m}{2m^2 + m - 6} \cdot \dfrac{m + 2}{3 - 3m} = \dfrac{3 \cdot m \cdot (m - 1)}{(2m - 3)(m + 2)} \cdot \dfrac{(m + 2)}{3 \cdot (1 - m)}$

Step (2) $= \dfrac{(-1)m}{2m - 3}$ or $\dfrac{-m}{2m - 3}$

Division of Rational Expressions

To divide rational expressions, we use the same property that was developed in section 3.4. That is,

$$\dfrac{a}{b} \div \dfrac{c}{d} = \dfrac{a}{b} \cdot \dfrac{d}{c} \qquad \text{where } b, c, d \neq 0$$

reciprocals

You should recall that this property shows us how to divide fractions. Multiply by the multiplicative inverse (reciprocal) of the divisor. Or, more simply, invert the divisor and multiply.

Example 5

Divide $\dfrac{2a + 4}{3b} \div \dfrac{a + 2}{5b^2}$

Solution: $\dfrac{2a + 4}{3b} \div \dfrac{a + 2}{5b^2} = \dfrac{2a + 4}{3b} \cdot \dfrac{5b^2}{a + 2}$

$$= \dfrac{2 \cdot (\cancel{a + 2})}{3 \cdot \cancel{b}} \cdot \dfrac{5 \cdot b \cdot \cancel{b}}{(\cancel{a + 2})}$$

$$= \dfrac{10b}{3}$$

Example 6

Divide $\dfrac{4m^2 - 9}{3m} \div (3 - 2m)$

Solution: $\dfrac{4m^2 - 9}{3m} \div (3 - 2m) = \dfrac{4m^2 - 9}{3m} \cdot \dfrac{1}{3 - 2m}$

divided opposites

$$= \dfrac{(2m + 3) \cdot (2m - 3)}{3 \cdot m} \cdot \dfrac{1}{(3 - 2m)}$$

$$= \dfrac{2m + 3}{-3m}$$

Example 7

Divide $\dfrac{2y^2 + 3y - 2}{2y^2 - 11y + 5} \div \dfrac{y^2 + 7y + 10}{y - 5}$

Solution: $\dfrac{2y^2 + 3y - 2}{2y^2 - 11y + 5} \div \dfrac{y^2 + 7y + 10}{y - 5} = \dfrac{2y^2 + 3y - 2}{2y^2 - 11y + 5} \cdot \dfrac{y - 5}{y^2 + 7y + 10}$

$$= \dfrac{(2y - 1) \cdot (y + 2)}{(2y - 1) \cdot (y - 5)} \cdot \dfrac{(y - 5)}{(y + 2) \cdot (y + 5)}$$

$$= \dfrac{1}{y + 5}$$

Exercise 5.2

Multiply or divide as indicated. Assume that denominators and divisors do not represent zero.

1. $\dfrac{6m^3}{5n^2} \cdot \dfrac{7n}{12m^2}$

2. $\dfrac{4}{9m} \cdot \dfrac{3m^2}{7}$

3. $\dfrac{8a}{15} \cdot \dfrac{5}{a^2}$

4. $\dfrac{5x^2}{3} \div \dfrac{x}{6}$

5. $\dfrac{5y^2}{27x^2} \div \dfrac{25y}{3x}$

6. $\dfrac{3a^2b}{4ac^2} \cdot \dfrac{5a^3b^2}{6ab}$

7. $\dfrac{10xy^3}{9x^2y^2} \cdot \dfrac{72x^3y^4}{40}$

8. $\dfrac{6r^2}{5s} \cdot \dfrac{10rs^2}{2t} \div \dfrac{r}{3t^3}$

multiply or divide as indicated, assume the denominators and divisors do not represent zero

9. $\dfrac{5a + 5b}{8a^2} \cdot \dfrac{4a}{3a + 3b}$

10. $\dfrac{3m - 6}{5n^2} \div \dfrac{m - 2}{10n}$

11. $\dfrac{9x^2 - 1}{4x} \div (3x + 1)$

12. $\dfrac{9x^2 - 1}{4x} \div (1 - 3x)$

13. $\dfrac{a - 3}{15} \cdot \dfrac{5a + 10}{9 - 3a}$

14. $\dfrac{x^2 - 5x + 4}{x^2 - x - 6} \cdot \dfrac{x^2 - 4}{x^2 + 2x - 3}$

15. $\dfrac{m^2 - 9}{m^2 + 8m + 15} \cdot \dfrac{m^2 - 25}{m^2 - m - 6}$

16. $\dfrac{3m^2 + m - 2}{m^2 + 2m - 8} \div \dfrac{3m^2 + 4m - 4}{m^2 + m - 12}$

17. $\dfrac{x^2 - 1}{14x} \div \dfrac{3x + 3}{35x^2}$

18. $\dfrac{5a^2 - 10a}{a^2 + 4a + 3} \cdot \dfrac{a + 3}{20 - 10a}$

19. $\dfrac{3x^2 + 10x - 8}{x^2 - 7x + 10} \cdot \dfrac{x^2 + x - 6}{3x^2 + x - 2}$

20. $\dfrac{9y^2 - 16}{3y^2 + 2y - 8} \div \dfrac{3y^2 - 5y - 12}{y^2 - y - 6}$

21. $\dfrac{3a^2 - 13a - 10}{4a^2 - 4} \div \dfrac{a^2 - 6a + 5}{4a + 4}$

22. $\dfrac{5x^2 - 10x}{x^2 - 3x - 4} \cdot \dfrac{x + 1}{6 - 3x}$

23. $\dfrac{9m^2 - 4}{2m} \div (6m - 4)$

24. $\dfrac{2y^2 - y - 3}{3y^2 + y - 2} \div \dfrac{2y^2 - 5y + 3}{3y^2 - 5y + 2}$

25. $\dfrac{m^2 - n^2}{27} \cdot \dfrac{3}{m + n}$

26. $\dfrac{4x^2 - 9y^2}{x^2 - 1} \cdot \dfrac{x - 1}{6x - 9y}$

27. $\dfrac{x^2y^2 + y^2}{x^2y - xy^2} \div \dfrac{4x^2 + 4}{2xy^2 - 2x^2y}$

28. $\dfrac{r - 3}{r + 4} \div \dfrac{r^2 - 9}{r^2 - 16}$

29. $\dfrac{x^2 + 4x + 3}{x^2 + x - 6} \cdot \dfrac{x^2 - 2x}{x^2 + 3x + 2}$

30. $\dfrac{2a - 4b}{8a + 24b} \cdot \dfrac{2a + 6b}{4a - 8b}$

31. $\dfrac{x^2 + 9x + 14}{x^2 + 4x - 21} \div \dfrac{x^2 + 2x - 35}{x^2 - 3x - 10}$

32. $\dfrac{n^2 - 6n + 5}{n^2 + 8n + 7} \cdot \dfrac{n^2 + 3n + 2}{n^2 - 3n - 10}$

33. $\dfrac{a^2 + 3a}{a^2 - 3a - 4} \div \dfrac{a^2 + 2a - 3}{a^2 - 5a + 4}$

34. $\dfrac{x^2 - 2x - 3}{x^2 + 3x + 2} \div \dfrac{x^2 - 4}{x^2 - 5x + 6}$

35. $\dfrac{a^2 - b^2}{2a^2 - 5ab + 2b^2} \cdot \dfrac{2a^2 - 7ab + 3b^2}{a^2 + 2ab + b^2}$

36. $\dfrac{36 - x^2}{3x^2 - 2x - 1} \cdot \dfrac{1 - 9x^2}{3x^2 - 19x + 6}$

37. $\dfrac{x^2 - x - 6}{x^2 - 9} \cdot \dfrac{x^2 - 1}{x + 2} \cdot \dfrac{x + 3}{x^2 - x - 2}$

38. $\dfrac{4m^2 + 12m + 9}{2 - 6m} \cdot \dfrac{9m^2 - 1}{4m^2 - 9} \cdot \dfrac{4m - 6}{6m^2 + 11m + 3}$

39. $\dfrac{2}{m + 3} \cdot \dfrac{m + 3}{m} \div \dfrac{4m^2}{m^2 + 3m}$

40. $\dfrac{x - 4}{9 - x^2} \div \dfrac{2x - 8}{x + 1} \cdot \dfrac{2x^2}{2x + 6}$

ADDITION AND SUBTRACTION OF RATIONAL EXPRESSIONS

5.3

Rational expressions, being fractions of polynomials, are added and subtracted by obtaining common denominators and combining their numerators. Your success with this procedure depends upon your ability to find the lowest common denominator (LCD). Therefore, we review the method first discussed in section 3.6.

Example 1

Find the LCD of $\dfrac{3}{8a^2b^2}$ and $\dfrac{5}{6a^3b}$

Solution:

Step (1) Write each denominator in prime factored exponential form.

$$8a^2b^2 = 2^3 \cdot a^2 \cdot b^2$$
$$6a^3b = 2 \cdot 3 \cdot a^3 \cdot b$$

Step (2) Write each prime factor used as a base, but write it only once.

$$2 \cdot 3 \cdot a \cdot b$$

Step (3) From Step (1) find the largest exponent used on each base.

$$\text{LCD} = 2^3 \cdot 3 \cdot a^3 \cdot b^2$$

Step (4) Evaluate the exponential expression.

$$\text{LCD} = 24a^3b^2$$

Example 2

Find the LCD of $\dfrac{5}{6x - 6}$ and $\dfrac{3}{4x - 4}$

Solution:

Step (1) Factor.

$$6x - 6 = 2 \cdot 3 \cdot (x - 1)$$
$$4x - 4 = 2^2 \cdot (x - 1)$$

Step (2) Write the bases.

$$2 \cdot 3 \cdot (x - 1)$$

Step (3)   Apply the largest exponents.

$$LCD = 2^2 \cdot 3 \cdot (x - 1)$$

Step (4)   Evaluate.

$$LCD = 12(x - 1)$$

Find the LCD of $\dfrac{5}{x^2 + 6x + 9}$ and $\dfrac{7}{2x^2 + 7x + 3}$ **Example 3**

Solution:

Step (1)   Factor.

$$x^2 + 6x + 9 = (x + 3) \cdot (x + 3) = (x + 3)^2$$
$$2x^2 + 7x + 3 = (x + 3) \cdot (2x + 1)$$

Step (2)   Write the bases.

$$(x + 3) \cdot (2x + 1)$$

Step (3)   Apply the largest exponents.

$$LCD = (x + 3)^2 \cdot (2x + 1)$$

Step (4)   Evaluate. Binomial factors, however, are usually left in factored form.

$$LCD = (x + 3)^2(2x + 1)$$

After the denominators have been factored you will often be able to find the LCD mentally. However, when dealing with more complicated problems, don't hesitate to write out the complete process.

Adding and Subtracting Rational Expressions

To add or subtract rational expressions, we follow a procedure similar to the one developed for combining fractions in section 3.7:

Step (1)   Factor all denominators.

Step (2)   Find the LCD and expand all fractions to that denominator.

Step (3)   Combine the numerators and place the result over the LCD. If possible, reduce this answer to lowest terms.

Example 4

Add $\dfrac{5}{6r^2t} + \dfrac{7}{9rt^3}$

Solution:

Step (1) Factor all denominators.

$$\frac{5}{6r^2t} + \frac{7}{9rt^3} = \frac{5}{2 \cdot 3 \cdot r^2 \cdot t} + \frac{7}{3^2 \cdot r \cdot t^3}$$

Step (2) The LCD is $18r^2t^3$. Expand the fractions.

$$= \frac{5 \cdot (3t^2)}{6r^2t \cdot (3t^2)} + \frac{7 \cdot (2r)}{9rt^3 \cdot (2r)}$$

$$= \frac{15t^2}{18r^2t^3} + \frac{14r}{18r^2t^3}$$

Step (3) Combine the numerators. The answer cannot be reduced.

$$= \frac{15t^2 + 14r}{18r^2t^3}$$

Example 5

$$\frac{3x}{x^2 + 5x + 6} + \frac{6}{x + 2}$$

Solution:

Step (1) Factor all denominators.

$$\frac{3x}{x^2 + 5x + 6} + \frac{6}{x + 2} = \frac{3x}{(x + 3)(x + 2)} + \frac{6}{(x + 2)}$$

Step (2) The LCD is $(x + 3)(x + 2)$. Expand the fractions.

$$= \frac{3x}{(x + 3)(x + 2)} + \frac{6(x + 3)}{(x + 2)(x + 3)}$$

Step (3) Combine the numerators. Reduce, when possible.

$$= \frac{3x + 6(x + 3)}{(x + 3)(x + 2)}$$

$$= \frac{3x + 6x + 18}{(x + 3)(x + 2)}$$

$$= \frac{9x + 18}{(x + 3)(x + 2)} \qquad \begin{bmatrix} \text{Note: Denominator was} \\ \text{left in factored form.} \end{bmatrix}$$

$$= \frac{9(x + 2)}{(x + 3)(x + 2)}$$

$$= \frac{9}{x + 3}$$

Subtract $\dfrac{2m^2}{m^2 - 9} - \dfrac{m - 1}{m + 3}$ **Example 6**

Solution:

Step (1)   Factor all denominators.

$$\frac{2m^2}{m^2 - 9} - \frac{m - 1}{m + 3} = \frac{2m^2}{(m + 3)(m - 3)} - \frac{m - 1}{(m + 3)}$$

Step (2)   The LCD is $(m + 3)(m - 3)$. Expand the fractions.

$$= \frac{2m^2}{(m + 3)(m - 3)} - \frac{(m - 1)(m - 3)}{(m + 3)(m - 3)}$$

Step (3)   Combine the numerators.

$$= \frac{2m^2 - [(m - 1)(m - 3)]}{(m + 3)(m - 3)}$$

$$= \frac{2m^2 - [m^2 - 4m + 3]}{(m + 3)(m - 3)}$$

$$= \frac{2m^2 - m^2 + 4m - 3}{(m + 3)(m - 3)}$$

$$= \frac{m^2 + 4m - 3}{(m + 3)(m - 3)}$$

Add $\dfrac{3}{2 - a} + \dfrac{a - 1}{a^2 - 4}$ **Example 7**

Solution:

Step (1)   Factor all denominators.

$$\frac{3}{2 - a} + \frac{a - 1}{a^2 - 4} = \frac{3}{(2 - a)} + \frac{a - 1}{(a + 2)(a - 2)}$$

opposites

Multiply numerator and denominator of the first fraction by
-1. This will change $(2 - a)$ into $(a - 2)$

$$= \frac{-1 \cdot 3}{-1 \cdot (2 - a)} + \frac{a - 1}{(a + 2)(a - 2)}$$

$$= \frac{-3}{a - 2} + \frac{a - 1}{(a + 2)(a - 2)}$$

Step (2)   The LCD is $(a + 2)(a - 2)$. Expand the fractions.

$$= \frac{-3(a + 2)}{(a - 2)(a + 2)} + \frac{a - 1}{(a + 2)(a - 2)}$$

Step (3)   Combine the numerators.

$$= \frac{-3(a + 2) + (a - 1)}{(a + 2)(a - 2)}$$

$$= \frac{-3a - 6 + a - 1}{(a + 2)(a - 2)}$$

$$= \frac{-2a - 7}{(a + 2)(a - 2)}$$

Example 8

Subtract $\dfrac{y + 3}{y^2 + 2y + 1} - \dfrac{y + 4}{y^2 - 2y - 3}$

Solution:

Step (1)   Factor all denominators.

$$\frac{y + 3}{y^2 + 2y + 1} - \frac{y + 4}{y^2 - 2y - 3} = \frac{y + 3}{(y + 1)(y + 1)} - \frac{y - 4}{(y + 1)(y - 3)}$$

$$= \frac{y + 3}{(y + 1)^2} - \frac{y - 4}{(y + 1)(y - 3)}$$

Step (2)   The LCD is $(y + 1)^2(y - 3)$. Expand each fraction.

$$= \frac{(y + 3)(y - 3)}{(y + 1)^2(y - 3)} - \frac{(y + 4)(y + 1)}{(y + 1)(y - 3)(y + 1)}$$

$$= \frac{(y + 3)(y - 3)}{(y + 1)^2(y - 3)} - \frac{(y + 4)(y + 1)}{(y + 1)^2(y - 3)}$$

Step (3)   Combine the numerators.

$$= \frac{(y + 3)(y - 3) - [(y + 4)(y + 1)]}{(y + 1)^2(y - 3)}$$

$$= \frac{y^2 - 9 - [y^2 + 5y + 4]}{(y + 1)^2(y - 3)}$$

$$= \frac{y^2 - 9 - y^2 - 5y - 4}{(y + 1)^2(y - 3)}$$

$$= \frac{-5y - 13}{(y + 1)^2(y - 3)}$$

Combine $\dfrac{2x - 3}{x^2 - 2x - 3} - \dfrac{x - 1}{x + 1} + \dfrac{x}{3 - x}$ **Example 9**

Solution:

   Step (1)   Factor all denominators.

$$\frac{2x - 3}{x^2 - 2x - 3} - \frac{x - 1}{x + 1} + \frac{x}{3 - x} = \frac{2x - 3}{(x + 1)(x - 3)} - \frac{x - 1}{(x + 1)} + \frac{x}{(3 - x)}$$

opposites

$$= \frac{2x - 3}{(x + 1)(x - 3)} - \frac{x - 1}{(x + 1)} + \frac{-1 \cdot x}{-1 \cdot (3 - x)}$$

$$= \frac{2x - 3}{(x + 1)(x - 3)} - \frac{x - 1}{(x + 1)} + \frac{-x}{(x - 3)}$$

   Step (2)   The LCD is $(x + 1)(x - 3)$. Expand the fractions.

$$= \frac{2x - 3}{(x + 1)(x - 3)} - \frac{(x - 1)(x - 3)}{(x + 1)(x - 3)} + \frac{-x(x + 1)}{(x - 3)(x + 1)}$$

   Step (3)   Combine the numerators.

$$= \frac{2x - 3 - [(x - 1)(x - 3)] - x(x + 1)}{(x + 1)(x - 3)}$$

$$= \frac{2x - 3 - [x^2 - 4x + 3] - x^2 - x}{(x + 1)(x - 3)}$$

$$= \frac{2x - 3 - x^2 + 4x - 3 - x^2 - x}{(x + 1)(x - 3)}$$

$$= \frac{-2x^2 + 5x - 6}{(x + 1)(x - 3)}$$

Combine the following fractions by addition or subtraction. Assume that   **Exercise 5.3**
denominators do not represent zero.

1. $\dfrac{7}{18a^2b} + \dfrac{5}{12ab^2}$ **2.** $\dfrac{1}{10xy^2} - \dfrac{3}{5x^2y}$

3. $\dfrac{5}{3m^2n} + \dfrac{1}{6mn}$ **4.** $\dfrac{1}{6mn^2} - \dfrac{5}{3m^2n}$

OPERATIONS WITH RATIONAL EXPRESSIONS

combine the following fractions by addition or subtraction

5. $\dfrac{2}{a^2b} - \dfrac{3}{ab} + \dfrac{1}{ab^2}$

6. $\dfrac{5}{2x} + \dfrac{1}{3x} - \dfrac{4}{3}$

7. $\dfrac{2}{x^2y} - \dfrac{3}{2xy} - \dfrac{1}{3x^2}$

8. $\dfrac{3}{a^2b} + \dfrac{5}{b} - \dfrac{2}{a}$

9. $\dfrac{5}{2m - 2} + \dfrac{1}{m - 1}$

10. $\dfrac{4}{3a - 12} - \dfrac{2}{a - 4}$

11. $\dfrac{7}{4x - 8} + \dfrac{1}{2x - 4}$

12. $\dfrac{2}{3x + 3} - \dfrac{1}{x + 1}$

13. $\dfrac{2}{3 - x} + \dfrac{x - 2}{x^2 - 9}$

14. $\dfrac{a}{a + 3} + \dfrac{18}{a^2 - 9}$

15. $\dfrac{a}{a + 3} - \dfrac{18}{a^2 - 9}$

16. $\dfrac{3m}{m^2 - m - 6} - \dfrac{4}{m + 2}$

17. $\dfrac{5}{4 - x} + \dfrac{x - 3}{x^2 - 16}$

18. $\dfrac{m}{m + 2} - \dfrac{8}{m^2 - 4}$

19. $\dfrac{7}{a^2 - 3a - 10} + \dfrac{5}{a^2 - a - 6}$

20. $\dfrac{5x}{x + y} + \dfrac{7}{2x + 2y}$

21. $\dfrac{1}{m^2 + 2m + 1} - \dfrac{1}{m^2 - 1}$

22. $\dfrac{5x}{x^2 - x - 6} - \dfrac{7x - 2}{x^2 - 3x - 10}$

23. $\dfrac{x + 3}{2x^2 + x - 6} + \dfrac{2x + 1}{2x^2 - x - 3}$

24. $\dfrac{2t}{t^2 + 2t - 15} + \dfrac{3}{t + 5}$

25. $\dfrac{2a + 1}{3a^2 + 5a + 2} - \dfrac{a - 3}{3a^2 - 4a - 4}$

26. $\dfrac{2a}{2a^2 - 3a - 5} - \dfrac{3a}{2a^2 - a - 10}$

27. $\dfrac{2x - 3}{3x^2 + 10x + 8} - \dfrac{x + 1}{3x^2 - 5x - 12}$

28. $\dfrac{1}{n^2 + 3n + 2} - \dfrac{1}{n^2 - 1}$

29. $\dfrac{t}{t^2 - 9} + \dfrac{1}{t^2 + 4t - 21}$

30. $\dfrac{2}{x^2 - 5x + 6} + \dfrac{1}{x^2 - 4}$

31. $\dfrac{2a}{a^2 - 2a - 15} + \dfrac{3a}{a^2 + 2a - 3}$

32. $\dfrac{1}{x^2 - 4x + 4} - \dfrac{1}{x^2 - 4}$

33. $\dfrac{t + 7}{t^2 + 5t + 6} + \dfrac{t - 1}{t^2 + 3t + 2}$

34. $\dfrac{y + 3}{y^2 - y - 2} - \dfrac{y - 1}{y^2 + 2y + 1}$

35. $\dfrac{m - 3}{2m^2 - 3m - 2} + \dfrac{m - 2}{2m^2 - 5m - 3}$

36. $\dfrac{a + 1}{a^2 + a - 2} + \dfrac{a + 3}{a^2 - 4a + 3}$

37. $\dfrac{y}{y^2 - 9} - \dfrac{1}{y^2 + 4y - 21}$

38. $\dfrac{3}{x^2 - 9} + \dfrac{2}{x^2 - x - 6}$

39. $\dfrac{m - n}{m^2 + 4mn + 3n^2} + \dfrac{m + 3n}{m^2 + 2mn + n^2}$

40. $\dfrac{2x + y}{x^2 + 4xy + 3y^2} - \dfrac{x - 2y}{x^2 + 2xy + y^2}$

41. $\dfrac{2}{a + 1} - \dfrac{3}{a^2 - 1} + \dfrac{6}{(a + 1)^2}$

42. $\dfrac{2}{1 - a} - \dfrac{3}{a^2 - 1} + \dfrac{6}{(a + 1)^2}$

43. $\dfrac{1 - y}{2 - y} - \dfrac{2}{y + 2} + \dfrac{1}{y - 2}$

44. $\dfrac{2x - 3}{x^2 + 3x + 2} - \dfrac{x - 1}{x + 1} + \dfrac{x}{x + 2}$

45. $\dfrac{2}{m+n} - \dfrac{1}{m-n} + \dfrac{m-1}{m^2 - n^2}$ **46.** $\dfrac{2}{y^3 - y} + \dfrac{1}{y^2 + y} - \dfrac{1}{y^2 - y}$

47. $\dfrac{a+1}{a^2 - 9} + \dfrac{1}{3a+2} + \dfrac{2a}{3a^2 + 11a + 6}$

48. $\dfrac{6t+18}{8t^2 + 6t - 5} + \dfrac{2}{1 - 2t} + \dfrac{3}{4t+5}$

49. $\dfrac{5}{3-2x} + \dfrac{10x+24}{6x^2 - 5x - 6} + \dfrac{4}{3x+2}$

50. $\dfrac{3}{y-1} - \dfrac{y-1}{y^2 + 2y + 1} - \dfrac{y+1}{y^2 - 1}$

COMPLEX FRACTIONS

A fraction containing at least one rational expression in its numerator or denominator is called a **complex fraction**.

complex fraction

5.4

Each of these is a complex fraction.

Example 1

(a) $\dfrac{7}{\dfrac{2}{3}}$

(b) $\dfrac{m + \dfrac{2}{3}}{5}$

(c) $\dfrac{\dfrac{3}{a}}{\dfrac{1}{2a} - \dfrac{3}{a^2}}$

To simplify a complex fraction means to write it as a common fraction reduced to lowest terms. This is accomplished by using the fundamental principle of fractions. The numerator and denominator are multiplied by the LCD of all fractions contained within the complex fraction.

Example 2

Simplify $\dfrac{\dfrac{2}{3}}{\dfrac{1}{5}}$

Solution: The LCD of $\dfrac{2}{3}$ and $\dfrac{1}{5}$ is 15. Multiply the numerator and denominator by 15.

$$\frac{\frac{2}{3}}{\frac{1}{5}} = \frac{15\cdot\frac{2}{3}}{15\cdot\frac{1}{5}} = \frac{{}^{5}\cancel{15}\cdot\frac{2}{\cancel{3}}}{{}^{3}\cancel{15}\cdot\frac{1}{\cancel{5}}} = \frac{10}{3}$$

Example 3

Simplify $\dfrac{\dfrac{1}{2}}{1+\dfrac{1}{6}}$

Solution: The LCD of $\frac{1}{2}$ and $\frac{1}{6}$ is 6. Multiply the numerator and denominator by 6.

$$\frac{\frac{1}{2}}{1+\frac{1}{6}} = \frac{6\cdot\frac{1}{2}}{6\cdot\left(1+\frac{1}{6}\right)}$$

$$= \frac{6\cdot\frac{1}{2}}{6\cdot1+6\cdot\frac{1}{6}}$$

$$= \frac{{}^{3}\cancel{6}\cdot\frac{1}{\cancel{2}}}{6\cdot1+\cancel{6}\cdot\frac{1}{\cancel{6}}}$$

$$= \frac{3}{6+1}$$

$$= \frac{3}{7}$$

Example 4

Simplify $\dfrac{\dfrac{2a^2}{3b^2}}{\dfrac{5a}{9b^3}}$

Solution: Multiply the numerator and denominator by the LCD, $9b^3$.

$$\frac{\frac{2a^2}{3b^2}}{\frac{5a}{9b^3}} = \frac{(9\cancel{b^3})^{3b}\cdot\frac{2a^2}{3\cancel{b^2}}}{(9\cancel{b^3})\cdot\frac{5a}{9\cancel{b^3}}}$$

$$= \frac{6a^2b}{5a} \quad \text{[reduce to lowest terms]}$$

$$= \frac{6\cdot a\cdot\cancel{a}\cdot b}{5\cdot\cancel{a}}$$

$$= \frac{6ab}{5}$$

Example 5

Simplify $\dfrac{\dfrac{2}{m}}{\dfrac{1}{m} - \dfrac{1}{m^2}}$

Solution: Multiply the numerator and denominator by the LCD, m^2.

$$\frac{\dfrac{2}{m}}{\dfrac{1}{m} - \dfrac{1}{m^2}} = \frac{m^2 \cdot \dfrac{2}{m}}{m^2 \cdot \left(\dfrac{1}{m} - \dfrac{1}{m^2}\right)}$$

$$= \frac{m^2 \cdot \dfrac{2}{m}}{m^2 \cdot \dfrac{1}{m} - m^2 \cdot \dfrac{1}{m^2}}$$

$$= \frac{2m}{m - 1}$$

Example 6

Simplify $\dfrac{a - \dfrac{4}{a}}{1 + \dfrac{2}{a}}$

Solution: Multiply numerator and denominator by a

$$\frac{a - \dfrac{4}{a}}{1 + \dfrac{2}{a}} = \frac{a \cdot \left(a - \dfrac{4}{a}\right)}{a \cdot \left(1 + \dfrac{2}{a}\right)}$$

$$= \frac{a^2 - \cancel{a} \cdot \dfrac{4}{\cancel{a}}}{a + \cancel{a} \cdot \dfrac{2}{\cancel{a}}}$$

$$= \frac{a^2 - 4}{a + 2} \qquad \text{[reduce to lowest terms]}$$

$$= \frac{\cancel{(a + 2)}(a - 2)}{\cancel{(a + 2)}}$$

$$= a - 2$$

Simplify the following complex fractions. Assume that no denominator represents zero.   **Exercise 5.4**

1. $\dfrac{\dfrac{3}{4}}{\dfrac{5}{7}}$ **2.** $\dfrac{\dfrac{1}{3}}{5}$

Simplify the following complex fractions

3. $\dfrac{\dfrac{5}{2}}{\dfrac{2}{7}}$

4. $\dfrac{1 + \dfrac{2}{3}}{2}$

5. $\dfrac{1}{\dfrac{3}{5} + 3}$

6. $\dfrac{\dfrac{3}{4} + \dfrac{1}{2}}{\dfrac{3}{8} + \dfrac{1}{4}}$

7. $\dfrac{\dfrac{1}{2} - \dfrac{1}{3}}{\dfrac{1}{3} - \dfrac{1}{6}}$

8. $\dfrac{2 + \dfrac{2}{3}}{1 - \dfrac{1}{3}}$

9. $\dfrac{\dfrac{2}{5} + \dfrac{1}{10}}{\dfrac{1}{2} - \dfrac{3}{4}}$

10. $\dfrac{\dfrac{6x^2}{5y}}{\dfrac{2x}{y^2}}$

11. $\dfrac{\dfrac{5m^3}{2n}}{\dfrac{10m}{3n^2}}$

12. $\dfrac{\dfrac{9t^2}{2r}}{\dfrac{3t}{4r^2}}$

13. $\dfrac{m + \dfrac{1}{2}}{m}$

14. $\dfrac{3x + \dfrac{1}{3}}{3}$

15. $\dfrac{\dfrac{2}{x} + \dfrac{1}{2x}}{\dfrac{2}{x}}$

16. $\dfrac{2 - \dfrac{1}{n}}{4 - \dfrac{1}{n^2}}$

17. $\dfrac{a - \dfrac{1}{a}}{a + \dfrac{1}{a}}$

18. $\dfrac{m + \dfrac{m}{n}}{1 + \dfrac{1}{n}}$

19. $\dfrac{3 - \dfrac{r}{t}}{3 - \dfrac{t}{r}}$

20. $\dfrac{\dfrac{a - b}{b}}{\dfrac{1}{a} - \dfrac{1}{b}}$

21. $\dfrac{x + \dfrac{2}{x}}{\dfrac{x^2 + 2}{2}}$

22. $\dfrac{b - \dfrac{1}{b}}{\dfrac{1}{b} + 1}$

23. $\dfrac{m + \dfrac{1}{n}}{\dfrac{1}{m} + n}$

24. $\dfrac{a - \dfrac{1}{3}}{\dfrac{1}{a} - 3}$

25. $\dfrac{\dfrac{a}{2} + \dfrac{b}{3}}{\dfrac{a}{b} + 5}$

26. $\dfrac{3x + \dfrac{1}{3}}{\dfrac{1}{5}}$

27. $\dfrac{x + \dfrac{1}{2}}{\dfrac{1}{x} - 2}$

28. $\dfrac{\dfrac{a}{b} + 1}{\dfrac{a}{b} - 1}$

29. $\dfrac{\dfrac{1}{3} - \dfrac{1}{x}}{\dfrac{1}{3} + \dfrac{1}{x}}$

30. $\dfrac{\dfrac{2}{a} + \dfrac{3}{2a}}{5 + \dfrac{1}{a}}$

31. $\dfrac{\dfrac{1}{2} - \dfrac{a}{3b}}{\dfrac{2}{a} - \dfrac{4}{3b}}$

32. $\dfrac{\dfrac{1}{x} - \dfrac{1}{xy}}{\dfrac{1}{xy} - \dfrac{1}{y}}$

33. $\dfrac{m + 2}{1 - \dfrac{4}{m^2}}$

34. $\dfrac{2x - \dfrac{1}{3}}{3x + \dfrac{1}{6}}$

35. $\dfrac{x - 2 + \dfrac{3}{x - 6}}{x - 1 + \dfrac{6}{x - 6}}$

36. $\dfrac{\dfrac{4}{a^2 - b^2}}{\dfrac{1}{a - b}}$

37. $\dfrac{\dfrac{1}{x + 2}}{\dfrac{1}{x - 2}}$

38. $\dfrac{\dfrac{1}{a + 3}}{\dfrac{1}{2a + 1}}$

39. $\dfrac{\dfrac{4}{y^2 - 9}}{\dfrac{1}{y + 3} - \dfrac{1}{y - 3}}$

SOLVING FRACTIONAL EQUATIONS

5.5

Equations which contain at least one rational expression are called *fractional equations*.

Each of these is a fractional equation.   **Example 1**

(a) $\dfrac{x}{2} = \dfrac{3}{4}$

(b) $\dfrac{y}{2} + 7 = \dfrac{5y}{3}$

(c) $m - 2 = \dfrac{6}{m - 3}$

To solve a fractional equation, we use the multiplication property of equivalent equations to multiply each side by the LCD. This will transform the fractional equation into a simpler equation having no fractions. In our work, the resulting equation will either be linear or quadratic. Therefore, to solve a fractional equation, follow these steps:

Step (1) Find the lowest common denominator (LCD).

Step (2) Multiply both sides of the equation by the LCD.

Step (3) Classify the resulting equation as linear or quadratic. Then, solve accordingly.

Example 2

Solve $\dfrac{x}{5} = \dfrac{2}{3}$

Solution:

Step (1) The LCD of $\dfrac{x}{5}$ and $\dfrac{2}{3}$ is 15

Step (2) Multiply both sides of the equation by 15. This will clear the equation of all fractions.

$$\frac{x}{5} = \frac{2}{3}$$

$$15 \cdot \frac{x}{5} = 15 \cdot \frac{2}{3}$$

$$^{3}\cancel{15} \cdot \frac{x}{\cancel{5}} = {}^{5}\cancel{15} \cdot \frac{2}{\cancel{3}}$$

$$3x = 10$$

Step (3) The resulting equation is linear. Solve accordingly.

$$3x = 10$$

$$x = \frac{10}{3}$$

This solution can be checked by substituting it for x in the original equation, and seeing that the result is true.

$$\frac{x}{5} = \frac{2}{3}$$

$$\frac{\frac{10}{3}}{5} = \frac{2}{3}$$

$$\frac{\frac{10}{3} \cdot 3}{5 \cdot 3} = \frac{2}{3}$$

$$\frac{10}{15} = \frac{2}{3}$$

$$\frac{2}{3} = \frac{2}{3} \quad \text{[true]}$$

Solve $\dfrac{5m}{3} = \dfrac{m}{2} - 7$

Example 3

Solution:

Step (1) The LCD is 6

Step (2) Multiply both sides by 6. Be sure to multiply it times every term. This will clear the equation of all fractions.

$$\frac{5m}{3} = \frac{m}{2} - 7$$

$$6 \cdot \frac{5m}{3} = 6\left[\frac{m}{2} - 7\right]$$

$$6 \cdot \frac{5m}{3} = 6 \cdot \frac{m}{2} - 6 \cdot 7$$

$$10m = 3m - 42$$

Step (3) The resulting equation is linear.

$$10m = 3m - 42$$

$$10m - 3m = -42$$

$$7m = -42$$

$$m = -6$$

Check this solution by substituting it for *m* in the original equation.

In Examples 2 and 3, the variables were present only in the numerators. Variables can also be present in the denominators. However, as shown in Example 4, we must be careful to remember that denominators cannot represent zero.

Example 4   Solve $\dfrac{x + 2}{x - 3} + 3 = \dfrac{5}{x - 3}$

Solution:

Step (1)   The LCD is $x - 3$

Step (2)   Multiply both sides of the equation by $x - 3$. Be sure to multiply it times each term (many students forget to multiply the LCD times the 3).

$$\dfrac{x + 2}{x - 3} + 3 = \dfrac{5}{x - 3}$$

$$(x - 3) \cdot \left[\dfrac{x + 2}{x - 3} + 3\right] = (x - 3) \cdot \dfrac{5}{x - 3}$$

$$(x - 3) \cdot \dfrac{x + 2}{(x - 3)} + (x - 3) \cdot 3 = (x - 3) \cdot \dfrac{5}{(x - 3)}$$

$$x + 2 + (x - 3) \cdot 3 = 5$$

Step (3)   The resulting equation is linear.

$$x + 2 + (x - 3) \cdot 3 = 5$$
$$x + 2 + 3x - 9 = 5$$
$$4x - 7 = 5$$
$$4x = 12$$
$$x = 3$$

3 fails to check in the original equation because it turns both denominators of $x - 3$ into zero. Thus, the original equation has no solution.

A tentative solution that fails to check in the original equation because it turns a denominator into zero is said to be an **extraneous solution**. Whenever the denominator of a fractional equation contains a variable, there is danger of producing an extraneous solution.

extraneous solution

Example 5   Solve $\dfrac{1}{n - 2} = \dfrac{4}{n^2 - 4}$

Solution:

Step (1)   Find the LCD by factoring the denominators.

$$\frac{1}{(n-2)} = \frac{4}{(n+2)(n-2)}$$

The LCD is $(n+2)(n-2)$

Step (2)   Multiply both sides of the equation by $(n+2)(n-2)$

$$\frac{1}{(n-2)} = \frac{4}{(n+2)(n-2)}$$

$$(n+2)(n-2) \cdot \frac{1}{(n-2)} = (n+2)(n-2) \cdot \frac{4}{(n+2)(n-2)}$$

$$n + 2 = 4$$

Step (3)   The resulting equation is linear.

$$n + 2 = 4$$
$$n = 2$$

2 is an extraneous solution. It turns the denominators of $n - 2$ and $n^2 - 4$ into zero. The original equation has no solution.

Solve $m - 7 = \dfrac{21}{m-3}$

Example 6

Solution:

Step (1)   The LCD is $m - 3$

Step (2)   Multiply both sides of the equation by $m - 3$

$$m - 7 = \frac{21}{m-3}$$

$$(m-3) \cdot (m-7) = (m-3) \cdot \frac{21}{(m-3)}$$

$$m^2 - 10m + 21 = 21$$

Step (3)   The resulting equation is quadratic. Solve by factoring.

$$m^2 - 10m + 21 = 21$$
$$m^2 - 10m = 0$$
$$m(m - 10) = 0$$
$$m = 0 \quad \text{or} \quad m - 10 = 0$$
$$m = 10$$

The solutions to the original equation are 0 and 10. Neither of these will turn the denominator of $m - 3$ into zero. There are no extraneous solutions.

Example 7

Solve $\dfrac{2x}{x + 2} - \dfrac{x - 1}{x + 3} = \dfrac{7x + 17}{x^2 + 5x + 6}$

Solution:

Step (1)   Find the LCD by factoring the denominators.

$$\frac{2x}{(x + 2)} - \frac{x - 1}{(x + 3)} = \frac{7x + 17}{(x + 2)(x + 3)}$$

The LCD is $(x + 2)(x + 3)$

Step (2)   Multiply both sides of the equation by $(x + 2)(x + 3)$

$$(x + 2)(x + 3) \cdot \left[\frac{2x}{(x + 2)} - \frac{x - 1}{(x + 3)} \right] = (x + 2)(x + 3) \cdot \frac{7x + 17}{(x + 2)(x + 3)}$$

$$\cancel{(x + 2)}(x + 3) \cdot \frac{2x}{\cancel{(x + 2)}} - (x + 2)\cancel{(x + 3)} \cdot \frac{x - 1}{\cancel{(x + 3)}} = \cancel{(x + 2)}\cancel{(x + 3)} \cdot \frac{7x + 17}{\cancel{(x + 2)}\cancel{(x + 3)}}$$

$$(x + 3) \cdot 2x - (x + 2) \cdot (x - 1) = 7x + 17$$

$$2x^2 + 6x - [x^2 + x - 2] = 7x + 17$$

$$2x^2 + 6x - x^2 - x + 2 = 7x + 17$$

$$x^2 + 5x + 2 = 7x + 17$$

Step (3)   The resulting equation is quadratic. Solve by factoring.

$$x^2 + 5x + 2 = 7x + 17$$

$$x^2 - 2x - 15 = 0$$

$$(x + 3)(x - 5) = 0$$

$$x + 3 = 0 \quad \text{or} \quad x - 5 = 0$$

$$x = -3 \qquad\qquad x = 5$$

-3 is an extraneous solution since it turns the denominators of $x + 3$ and $x^2 + 5x + 6$ into zero. The only solution to the original equation is 5.

Solve each of these equations. Indicate any <u>extraneous solutions.</u>   **Exercise 5.5**

the solution turns the denominators to zero

1. $\dfrac{x}{10} = \dfrac{5}{6}$ **2.** $\dfrac{x}{7} = \dfrac{2}{3}$

3. $\dfrac{n}{3} = \dfrac{1}{2}$ **4.** $\dfrac{2m}{3} = \dfrac{1}{6}$

5. $\dfrac{2}{y} = \dfrac{3}{5}$ **6.** $\dfrac{9}{2t} = 5$

7. $\dfrac{4m}{5} = \dfrac{m}{2}$ **8.** $\dfrac{4r}{5} = \dfrac{r}{2} - 3$

9. $\dfrac{m}{2} - m = -3$ **10.** $\dfrac{y}{5} - \dfrac{y}{2} = 9$

11. $\dfrac{3x}{2} = \dfrac{x}{4} + 2$ **12.** $\dfrac{2}{5t} + \dfrac{3}{10} = \dfrac{1}{2}$

13. $\dfrac{7}{2} + \dfrac{2}{x} = 4$ **14.** $\dfrac{r}{5} - 7 = \dfrac{r}{3} - r$

15. $2m + \dfrac{1}{2} = \dfrac{1}{4}$ **16.** $\dfrac{7}{x} + 3 = \dfrac{10}{x^2}$

17. $\dfrac{2}{3y} + 4 = \dfrac{6}{y}$ **18.** $\dfrac{25}{3x} + \dfrac{1}{3} = \dfrac{10}{x}$

19. $\dfrac{n-1}{10} + \dfrac{19}{15} = \dfrac{n}{3}$ **20.** $\dfrac{y+6}{2} - 5 = 1$

21. $\dfrac{y-2}{y} + \dfrac{1}{3} = \dfrac{14}{3y}$ **22.** $\dfrac{2t-5}{t} + \dfrac{2}{3} = \dfrac{3}{t}$

23. $\dfrac{2}{r+4} = \dfrac{2}{3r}$ **24.** $\dfrac{6x+8}{8x-1} = \dfrac{4}{3}$

25. $\dfrac{y+2}{6y+9} = \dfrac{1}{3}$ **26.** $m + 5 = \dfrac{15}{m+3}$

27. $\dfrac{3}{x-3} - \dfrac{5}{x} = \dfrac{9}{x^2-3x}$ **28.** $m + 2 = \dfrac{6}{m+3}$

29. $\dfrac{5}{x-5} - \dfrac{2}{x} = \dfrac{25}{x^2-5x}$ **30.** $\dfrac{10}{y+5} = y + 2$

31. $\dfrac{2+x}{4} - \dfrac{5-6x}{3} = x$ **32.** $\dfrac{1}{2y-2} - \dfrac{1}{1-y} = \dfrac{3}{y+2}$

33. $\dfrac{2}{m+3} = \dfrac{5m}{m^2+4m+3}$ **34.** $\dfrac{1}{y-2} + \dfrac{1}{3} = \dfrac{4}{y^2-4}$

35. $\dfrac{3x}{4x^2-9} = \dfrac{3}{2x-3}$ **36.** $\dfrac{x+2}{x-2} - \dfrac{3x}{x-1} = \dfrac{6x-8}{x^2-3x+2}$

37. $\dfrac{x+3}{x-4} - \dfrac{4}{x+5} = \dfrac{2x^2+3x+19}{x^2+x-20}$ **38.** $\dfrac{5}{n-3} = \dfrac{n+2}{n-3} + 3$

39. $\dfrac{x^2 + 3}{x^2 + x - 12} = \dfrac{2x - 2}{x - 3} - \dfrac{x + 1}{x + 4}$ **40.** $\dfrac{1}{2y - 8} + \dfrac{2}{y - 4} = 5$

41. $\dfrac{m - 3}{m - 2} + \dfrac{m + 4}{m^2 + m - 6} = \dfrac{m + 2}{m + 3}$ **42.** $\dfrac{7}{2y - 1} - \dfrac{5y}{2y^2 - 3y + 1} = \dfrac{2}{4y - 2}$

43. $\dfrac{1}{x^2 - 3x + 2} + \dfrac{1}{x^2 - 5x + 6} = \dfrac{-1}{x^2 - 4x + 3}$

44. $\dfrac{x}{x^2 - 3x + 2} + \dfrac{x}{x^2 - 5x + 6} = \dfrac{-x}{x^2 - 4x + 3}$

45. $\dfrac{t + 3}{t + 2} + \dfrac{t + 1}{t - 2} = \dfrac{2}{t^2 - 4}$

SOLVING FORMULAS FOR A SPECIFIED SYMBOL

5.6
formulas

Equations having special applications or which express relationships between physical quantities are called **formulas**. Most fields of study use formulas. They help people find numbers to describe relationships in their field. Some common formulas are listed below.

(1) $A = lw$ — gives the area of a rectangle when the length and width are known.

(2) $A = \dfrac{1}{2}bh$ — gives the area of a triangle when the base and height are known.

(3) $d = rt$ — gives the distance traveled when the rate and time are known.

(4) $i = prt$ — gives the interest earned by a certain principal at a known interest rate for a specified time.

(5) $C = \dfrac{5}{9}(F - 32)$ — converts Fahrenheit temperature to Celsius.

(6) $S = 16t^2$ — gives the distance in feet that an object will fall in t seconds.

To use a formula, you substitute the known information into the equation and evaluate.

Example 1

Use the formula $A = lw$ to find the area of a rectangle having a length of 10 centimeters and a width of 6 centimeters.

Solution: $A = lw$
$= (10)(6)$
$= 60 \ cm^2$

Use the formula $C = \frac{5}{9}(F - 32)$ to convert a temperature of 68 °Fahren-   Example 2
heit to Celsius.

Solution: $C = \frac{5}{9}(F - 32)$

$ = \frac{5}{9}(68 - 32)$

$ = \frac{5}{9} \cdot 36$

$ = 20\ °C$

Use the formula $d = rt$ to find the average rate of an automobile which   Example 3
traveled 250 miles in 5 hours.

Solution: $d = rt$

$\ \ \ 250 = r \cdot (5)$

$\ \ \ \dfrac{250}{5} = r$

$\ \ \ \ 50 = r$

$\ \ \ \ \ \ r = 50\ \text{mph}$

Often it is necessary to solve a large number of problems involving the same formula. For instance, you might be called upon to solve several problems similar to Example 3. You could save time by rearranging the formula and solving for r. This means to use the properties of equations and get r, alone, on one side of the formula.

Solve the formula $d = rt$ for r. Then find the average rate of an au-   Example 4
tomobile which traveled 250 miles in 5 hours.

Solution: Divide both sides of the formula by t

$d = rt$

$\dfrac{d}{t} = \dfrac{r\cancel{t}}{\cancel{t}}$

$\dfrac{d}{t} = r \qquad \text{or} \qquad r = \dfrac{d}{t}, \quad \text{where } t \ne 0$

To find r, substitute 250 for d and 5 for t

$r = \dfrac{d}{t}$

$$= \frac{250}{5}$$
$$= 50 \text{ mph}$$

This procedure of rearranging a formula to isolate a symbol on one side is called *solving a formula for a specified symbol.*

Example 5   Solve the formula $A = \frac{1}{2}bh$ for h

Solution: The equation is fractional. Multiply both sides by the LCD, 2.

$$A = \frac{1}{2}bh$$

$$2A = 2 \cdot \frac{1}{2}bh$$

$$2A = bh$$

$$\frac{2A}{b} = \frac{bh}{b}$$

$$\frac{2A}{b} = h \qquad \text{or} \qquad h = \frac{2A}{b}, \quad \text{where } b \neq 0$$

Example 6   Solve the formula $C = \frac{5}{9}(F - 32)$ for F

Solution: The equation is fractional. Multiply both sides by the LCD, 9.

$$C = \frac{5}{9}(F - 32)$$

$$9C = 9 \cdot \frac{5}{9}(F - 32)$$

$$9C = 5(F - 32)$$

$$\frac{9}{5}C = \frac{5(F - 32)}{5} \qquad \text{[both sides divided by 5]}$$

$$\frac{9}{5}C = F - 32$$

$$\frac{9}{5}C + 32 = F \qquad \text{or} \qquad F = \frac{9}{5}C + 32$$

Example 7   Solve the formula $ax - ac = bx$ for x

Solution: All terms containing an x must be brought together on one side.

$$ax - ac = bx$$

$$ax - ac - bx = 0 \qquad \textit{[subtract bx from both sides]}$$

$$ax - bx = ac \qquad \textit{[add ac to both sides]}$$

$$x(a - b) = ac \qquad \textit{[factor out the x]}$$

$$\frac{x(a-b)}{(a-b)} = \frac{ac}{a-b} \qquad \textit{[divide by a - b]}$$

$$x = \frac{ac}{a-b}, \quad \text{where } a - b = 0$$

Solve the formula $\frac{1}{p} + \frac{1}{q} = \frac{1}{f}$ for f **Example 8**

Solution: The equation is fractional. Multiply both sides by the LCD of pqf.

$$\frac{1}{p} + \frac{1}{q} = \frac{1}{f}$$

$$pqf \cdot \left[\frac{1}{p} + \frac{1}{q}\right] = pqf \cdot \frac{1}{f}$$

$$pqf \cdot \frac{1}{p} + pqf \cdot \frac{1}{q} = pqf \cdot \frac{1}{f}$$

$$qf \quad + \quad pf \quad = \quad pq$$

$$f(q + p) \quad = \quad pq \qquad \textit{[factor out the f]}$$

$$\frac{f(q+p)}{(q+p)} \quad = \quad \frac{pq}{q+p} \qquad \textit{[divide by q + p]}$$

$$f = \frac{pq}{q+p}, \quad \text{where } q + p \neq 0$$

Evaluate the following formulas for the given information.   **Exercise 5.6**

1. The area of a triangle is given by the formula $A = \frac{1}{2}bh$. Find the area of a triangle whose base (b) is 6 cm and whose height (h) is 7 cm.
2. The formula for interest is $i = prt$. Find the interest earned by a principal (p) of \$5,000 at an interest rate (r) of 8% for a time (t) of 2 yr.

3. The formula $C = \frac{5}{9}(F - 32)$ converts Fahrenheit temperature to Celsius. Convert 50 °F to Celsius.

4. The formula $F = \frac{9}{5}C + 32$ converts Celsius temperature to Fahrenheit. Convert 100 °C to Fahrenheit.

5. The formula $S = 16t^2$ gives the distance in ft that an object will fall in t sec. How far has an object fallen 3 sec after it is dropped?

6. The formula $P = 2l + 2w$ gives the perimeter of a rectangle having a length of l and a width of w. Find the perimeter of a rectangle whose length is 5 ft and whose width is 3 ft.

7. The formula $V = \frac{4}{3}\pi r^3$ gives the formula for finding the volume of a sphere having a radius r. π is approximately 3.14, symbolized $\pi \approx 3.14$. Find the volume of a sphere having a radius of 3 cm.

8. The formula $V = \pi r^2 h$ gives volume of a right circular cylinder where r is the radius, h is the height, and $\pi \approx 3.14$. Find the volume of a tin can whose radius is 2 in. and whose height is 5 in.

9. The formula $C = 2\pi r$ gives the circumference of a circle. r is the radius and $\pi \approx 3.14$. Find the radius of a circle having a circumference of 18.84 in.

10. The formula $V = lwh$ gives the volume of a box having a length l, width w, and height h. Find the height of a box having a volume of 140 in³, a length of 7 in., and a width of 5 in.

Solve the following formulas for the specified symbol.

11. $f = ma$, for a **12.** $pr = c$, for p

13. $e = mc^2$, for m **14.** $b = \frac{1}{a}$, for a

15. $a = \frac{b}{x}$, for x **16.** $A = lw$, for w

17. $d = rt$, for t **18.** $i = prt$, for p

19. $i = prt$, for r **20.** $i = prt$, for t

21. $C = \pi d$, for d **22.** $C = 2\pi r$, for r

23. $V = lwh$, for l **24.** $V = lwh$, for w

25. $V = lwh$, for h **26.** $V = \pi r^2 h$, for h

27. $A = \frac{1}{2}bh$, for b **28.** $A = \frac{1}{2}bh$, for h

29. $P = 2l + 2w$, for l **30.** $P = 2l + 2w$, for w

31. $s = \frac{a}{1 - r}$, for r **32.** $s = \frac{a}{1 - r}$, for a

33. $A = \frac{h(b + c)}{2}$, for h **34.** $A = \frac{h(b + c)}{2}$, for b

35. $ab + ac = 1$, for a **36.** $xy + k = xw$, for x

37. $\dfrac{1}{a} + \dfrac{1}{b} = \dfrac{1}{c}$, for a **38.** $\dfrac{1}{R} = \dfrac{1}{R_1} + \dfrac{1}{R_2}$, for R

39. $\dfrac{1}{p} + \dfrac{1}{q} = \dfrac{1}{f}$, for p **40.** $R = \dfrac{gs}{g + s}$, for s

41. $F = \dfrac{mv^2}{gr}$, for m **42.** $F = \dfrac{mv^2}{gr}$, for r

43. $S = \dfrac{rl - a}{r - l}$, for r **44.** $T = \dfrac{E}{R + r}$, for R

45. $T = \dfrac{12(D - d)}{l}$, for D **46.** $s = at - \dfrac{1}{2}gt^2$, for a

INEQUALITIES AND THEIR PROPERTIES

5.7

Inequalities are mathematical statements whose expressions are related by any of these symbols:

   $<$ is less than
   $\leq$ is less than or equal to
   $>$ is greater than
   $\geq$ is greater than or equal to

Each of the following is an inequality.   **Example 1**
(a) $3x < 6$
(b) $5y + 2 \leq 17$
(c) $2(m + 3) > 0$
(d) $\dfrac{2t}{5} + \dfrac{3}{10} \geq \dfrac{t}{10} + \dfrac{1}{5}$

To solve an inequality means to find every real number which makes the inequality true. Often it is convenient to illustrate the solutions graphically on the standard number line. For example, $x \geq 3$ represents all real numbers that are greater than or equal to 3. This is graphed in figure 5.1 on the standard number line by placing a dot at 3, and then extending an arrow to the right.

$x \geq 3$

FIGURE 5.1

Graph $x < -2$ **Example 2**

Solution: This statement says that x is representing all real numbers which are less than negative 2. Since x cannot equal negative

2, we place an open circle at negative 2 and extend an arrow to the left (see figure 5.2).

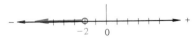

FIGURE 5.2

Example 3

Graph $-1 < x \le 4$.

Solution: To give meaning to statements of this type we begin with x, read to the left, and then read to the right, as follows: "x is greater than negative 1 and x is less than or equal to 4."

x is greater than -1, and

x is less than or equal to 4

This statement means that x represents all real numbers between -1 and 4. Also, x can equal 4, but not -1. Graphically, we have figure 5.3.

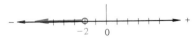

FIGURE 5.3

To solve inequalities we must discuss three important properties. These properties will enable us to change a given inequality into a simpler form.

The Addition–Subtraction Property of Inequalities

Given an inequality, the same real number may be added to both sides or subtracted from both sides without changing the solutions. Using symbols, this property states:

   If $a < b$ then $a + c < b + c$ and $a - c < b - c$

The addition-subtraction property holds true for all inequality symbols: $<$, $\le$, $>$, and $\ge$.

You can see that the property is true by considering this illustration: if $-4 < 1$, then adding 2 to both sides produces $-4 + 2 < 1 + 2$ or $-2 < 3$. Graphically, this is represented by a simple translation where both points are moved 2 units to the right. In figure 5.4 you should see that the relative positions are unchanged.

Examples 4 and 5 illustrate how the addition–subtraction property can be used to simplify the form of an inequality.

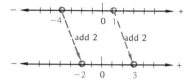

FIGURE 5.4

Simplify the inequality $x - 4 \geq 5$ by adding 4 to both sides.

Example 4

Solution:

$$x - 4 \geq 5$$
$$x - 4 + 4 \geq 5 + 4$$
$$x \geq 9$$

Simplify the inequality $3m + 2 < 1$ by subtracting 2 from both sides.

Example 5

Solution:

$$3m + 2 < 1$$
$$3m + 2 - 2 < 1 - 2$$
$$3m < -1$$

The next property of inequalities involves multiplication and division by *positive* numbers.

> **The Multiplication–Division Property of Inequalities (Positive Numbers)**
>
> Given an inequality, the same *positive* real number may be multiplied times both sides or divided into both sides without changing the solutions. In symbols, this property states:
>
> $$\text{If } a < b \text{ and } c \text{ is positive, then } ac < bc \text{ and } \frac{a}{c} < \frac{b}{c}$$
>
> This property holds true for all inequality symbols.

To see that this property is true, consider this illustration: if $-4 < 2$, then dividing 2 into both sides produces $\frac{-4}{2} < \frac{2}{2}$ or $-2 < 1$. Graphically you can see in figure 5.5 that the relative positions are unchanged.

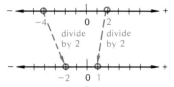

FIGURE 5.5

Examples 6 and 7 will show how this property is used to simplify the form of an inequality.

Example 6

Simplify the inequality $5x < 20$ by dividing 5 into both sides.

Solution: $5x < 20$

$$\frac{5x}{5} < \frac{20}{5}$$

$$x < 4$$

Example 7

Simplify the inequality $\dfrac{2x}{3} + \dfrac{1}{6} \geq \dfrac{1}{3}$ by multiplying 6 times each side.

Solution: $\dfrac{2x}{3} + \dfrac{1}{6} \geq \dfrac{1}{3}$

$$6\left[\frac{2x}{3} + \frac{1}{6}\right] \geq 6 \cdot \frac{1}{3}$$

$$6 \cdot \frac{2x}{3} + 6 \cdot \frac{1}{6} \geq 6 \cdot \frac{1}{3}$$

$$4x + 1 \geq 2$$

The final property of inequalities is especially important. It involves multiplication and division by *negative* numbers.

The Multiplication–Division Property of Inequalities (Negative Numbers)

Given an inequality, the same *negative* real number may be multiplied times both sides or divided into both sides without changing the solutions, provided the inequality symbol is reversed. Using variables, the property states:

If $a < b$ and c is negative, then $ac > bc$ and $\dfrac{a}{c} > \dfrac{b}{c}$

This property holds true for all inequality symbols.

To understand this property consider this illustration: if $-3 < 6$, then dividing negative 3 into both sides produces $\frac{-3}{-3} > \frac{6}{-3}$ or $1 > -2$. Notice that the inequality symbol was reversed because, graphically, the relative positions have changed (see figure 5.6). Observe in Examples 8 and 9 that this property can be used to simplify certain inequalities.

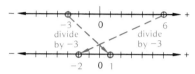

FIGURE 5.6

Simplify the inequality $-2x \leq 8$ by dividing -2 into both sides.

Example 8

Solution: $-2x \leq 8$ ⎤
                                        ⎣reverse inequality symbol

$$\frac{-2x}{-2} \geq \frac{8}{-2}$$

$$x \geq -4$$

Simplify the inequality $-m > -2$ by multiplying -1 times both sides.

Example 9

Solution: $-m > -2$ ⎤
                                        ⎣reverse inequality symbol

$$(-1)(-m) < (-1)(-2)$$
$$m < 2$$

Remember, when multiplying or dividing both sides of an inequality by a negative number, you must reverse the inequality symbol.

Graph each of these inequalities on the standard number line.

Exercise 5.7

1. $x > 1$ **2.** $x > -3$

3. $x < 2$ **4.** $x < 0$

5. $x < -4$ **6.** $x > 0$

7. $x \geq -3$ **8.** $x \leq -1$

9. $x \geq 0$ **10.** $x \geq 5$

11. $x \leq 0$ **12.** $x \leq 2$

13. $1 < x < 4$ **14.** $1 \leq x \leq 4$

15. $-2 \leq x \leq 3$ **16.** $-1 < x \leq 2$

17. $-3 \le x < 1$ **18.** $0 < x \le 5$

19. $-5 < x < -1$ **20.** $-3 \le x < 0$

21. $-2 \le x \le 2$

Write an inequality describing each of the following graphs.

22.

23. ⟵━━━━━━━━━━━●━━━━⟶
　　　　　　　　0　　4

24. ⟵━━━●━━━━━━━━━━●━━⟶
　　　　　−3　　0　　　　7

25. ⟵━━○━━━━━━○━━━━━⟶
　　　−6　　　0　2

26. ⟵━━━○━━━━━━●━━━⟶
　　　　−4　　0　　　5

Simplify the following inequalities as directed.

27. $x - 6 > 4$ [Add 6 to both sides.]

28. $y + 3 > -1$ [Subtract 3 from both sides.]

29. $4m + 2 > 6$ [Subtract 2 from both sides.]

30. $6t + 3 \le 12$ [Add -3 to both sides.]

31. $5x > 30$ [Divide 5 into both sides.]

32. $-4y < 12$ [Divide -4 into both sides.]

33. $\frac{5t}{3} \ge \frac{1}{6}$ [Multiply 6 times both sides.]

34. $\frac{2n}{5} + \frac{1}{10} < \frac{2}{5} + \frac{n}{10}$ [Multiply 10 times both sides.]

35. $-t \le 6$ [Multiply -1 times both sides.]

36. $-x > 0$ [Multiply -1 times both sides.]

37. $5x - 3 \ge 2x + 7$ [Subtract $2x$ from both sides.]

38. $9y - 5 < 3y - 7$ [Add 5 to both sides.]

39. $-2x \le 0$ [Divide -2 into both sides.]

40. $-7y > -42$ [Divide -7 into both sides.]

Write an inequality describing each of these word statements. Assume that the variables represent real numbers.

Example

My weekly salary (s) is less than $150.

Solution: $0 \le s < 150$

41. The amount of money (m) in my bank account is between $100 and $150.

42. The temperature (t) today will be more than 70 °F.

43. There are no less than 45 students (s) in my class.

44. A Boeing 727 airliner can hold more than 86 passengers (p).

45. My income tax (t) will be greater than $1,000 but no more than $1,500.

46. The gas mileage (g) for my car is between 21 and 28 mpg.

47. My weight (w) is always more than 160 lb but never more than 165 lb.

48. The width (w) of a room is between 12 ft and 13 ft.

49. The length of a rod (r) is measured as 6 cm. However, the measuring device has a possible error of no more than 0.01 cm.

50. The weight (w) of a Cadillac limousine is in excess of 5,000 lb.

SOLVING LINEAR INEQUALITIES

5.8

In section 5.7 we discussed three properties of inequalities. These properties will now be used to solve linear inequalities. A linear (or first degree) inequality is similar to a linear equation. The highest exponent on a variable is 1 and the inequality can be expressed in one of these forms:

$$ax + b < c \text{ or } ax + b \leq c$$
$$\text{where } a, b, \text{ and } c \text{ are constants and } a \neq 0$$

Each of these is a linear inequality.

Example 1

(a) $2x + 3 < 7$

(b) $5y - 4 > 6$

(c) $2(3m + 1) \leq 4(m - 2) + 3$

(d) $\dfrac{2x}{3} - \dfrac{5}{2} > \dfrac{5}{6}$

Linear inequalities are solved in the same manner as linear and fractional equations, except the inequality symbol must be reversed when each side is multiplied or divided by a negative number. We use the following three-step procedure:

Step (1)   Simplify the inequality. Perform indicated multiplications and combine like terms. If the inequality contains fractions, multiply both sides by the LCD.

Step (2)   Use the addition–subtraction property to obtain the terms containing a variable on one side and the numbers on the other side.

Step (3)   Use the multiplication–division property to divide both sides by the numerical coefficient of the variable. Remember, if you multiply or divide by a negative number, the inequality symbol must be reversed.

Example 2   Solve $2(x + 3) > 12$

Solution:

Step (1)   Simplify the inequality.

$$2(x + 3) > 12$$
$$2x + 6 > 12$$

Step (2)   Use the addition–subtraction property to obtain the variable on one side and the numbers on the other side.

$$2x + 6 > 12$$
$$2x + 6 - 6 > 12 - 6 \qquad \text{[subtract 6]}$$
$$2x > 6$$

Step (3)   Use the multiplication–division property to divide both sides by the numerical coefficient of the variable.

$$2x > 6$$
$$\frac{2x}{2} > \frac{6}{2} \qquad \text{[divide by 2]}$$
$$x > 3$$

Figure 5.7 shows the graph of this solution.

FIGURE 5.7

Example 3   Solve $\dfrac{2m}{3} - \dfrac{5}{2} \geq \dfrac{3m}{2} + \dfrac{5}{6}$

Solution:

Step (1)   Simplify the inequality. Multiply both sides by the LCD.

$$\frac{2m}{3} - \frac{5}{2} \geq \frac{3m}{2} + \frac{5}{6}$$
$$6\left[\frac{2m}{3} - \frac{5}{2}\right] \geq 6\left[\frac{3m}{2} + \frac{5}{6}\right]$$
$$6 \cdot \frac{2m}{3} - 6 \cdot \frac{5}{2} \geq 6 \cdot \frac{3m}{2} + 6 \cdot \frac{5}{6}$$
$$4m - 15 \geq 9m + 5$$

Step (2)   Use the addition–subtraction property to get the variable on one side and the numbers on the other side.

$$4m - 15 \geq 9m + 5$$

$$4m - 15 + 15 \geq 9m + 5 + 15 \qquad \text{[add 15]}$$

$$4m \geq 9m + 20$$

$$4m - 9m \geq 9m + 20 - 9m \qquad \text{[subtract 9m]}$$

$$-5m \geq 20$$

Step (3)   Use the multiplication–division property to divide both sides by the coefficient of the variable. In this case the number is negative. Be sure to reverse the inequality symbol.

$$-5m \geq 20$$

$$\frac{-5m}{-5} \leq \frac{20}{-5} \qquad \left[\begin{array}{l}\text{divide by } -5 \text{ and} \\ \text{reverse inequality symbol}\end{array}\right]$$

$$m \leq -4$$

Figure 5.8 shows the graph of this solution.

FIGURE 5.8

As you gain experience with this three-step procedure, it can be shortened. Study Examples 4, 5, and 6.

Solve $2y + 3 > 9$ **Example 4**

Solution: $2y + 3 > 9$

$$2y > 9 - 3$$

$$2y > 6$$

$$y > \frac{6}{2}$$

$$y > 3 \qquad \text{[See Figure 5.9]}$$

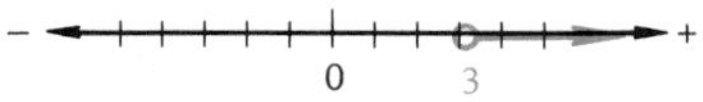

FIGURE 5.9

Solve $-4m - 5 > -9$ **Example 5**

Solution: $-4m - 5 > -9$

$$-4m > -9 + 5$$

$$-4m > -4$$

$$m < \frac{-4}{-4} \qquad \left[\begin{array}{l}\text{divide by } -4 \text{ and} \\ \text{reverse inequality symbol}\end{array}\right]$$

$$m < 1 \qquad \text{[see Figure 5.10]}$$

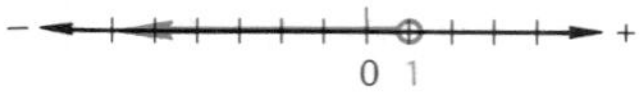

FIGURE 5.10

Example 6

Solve $\dfrac{3t}{2} - 3 \geq \dfrac{t}{5} + \dfrac{5t}{2} - \dfrac{1}{5}$

Solution: Multiply both sides by the LCD, which is 10.

$$\frac{3t}{2} - 3 \geq \frac{t}{5} + \frac{5t}{2} - \frac{1}{5}$$

$$10 \cdot \left[\frac{3t}{2} - 3\right] \geq 10 \cdot \left[\frac{t}{5} + \frac{5t}{2} - \frac{1}{5}\right]$$

$$10 \cdot \frac{3t}{2} - 10 \cdot 3 \geq 10 \cdot \frac{t}{5} + 10 \cdot \frac{5t}{2} - 10 \cdot \frac{1}{5}$$

$$15t - 30 \geq 2t + 25t - 2$$

$$15t - 30 \geq 27t - 2$$

$$15t - 27t \geq -2 + 30$$

$$-12t \geq 28 \qquad \left[\begin{array}{l}\text{divide by } -12 \text{ and} \\ \text{reverse inequality symbol}\end{array}\right]$$

$$t \leq \frac{28}{-12}$$

$$t \leq -\frac{7}{3} \quad \text{or} \quad t \leq -2\frac{1}{3} \qquad \text{[see Figure 5.11]}$$

FIGURE 5.11

When solving inequalities, you may encounter two special cases. It is easy to recognize these cases because during the solution process the variable disappears, resulting in either a false inequality or a true inequality.

Case 1

Solve $2(3x - 1) > 6x + 3$

Solution: $2(3x - 1) > 6x + 3$

$$6x - 2 > 6x + 3$$

$$6x - 6x > 3 + 2$$

$$0 > 5 \qquad \text{[false]}$$

Note that the variable disappeared, resulting in a false inequality (0 is not greater than 5). When this occurs the solution process is indicating that the original inequality has no solution.

Solve $2(3x - 1) > 6x - 5$ **Case 2**

Solution: $2(3x - 1) > 6x - 5$

$$6x - 2 > 6x - 5$$

$$6x - 6x > -5 + 2$$

$$0 > -3 \qquad \text{[true]}$$

Once again the variable disappeared, except in this case the resulting inequality is true (0 is greater than -3). When this occurs the solution process is indicating that the original inequality is always true and every real number is a solution. Graphically the solutions would be represented by the entire number line.

Solve the following inequalities. Also, graph each solution. **Exercise 5.8**

1. $2x < 6$

2. $5y > -15$

3. $4m \leq 8$

4. $-3t < 9$

5. $-3t > -9$

6. $-8x \geq 16$

7. $2n < 5$

8. $-2n < 5$

9. $2(x + 3) > 9$

10. $-3(m - 1) \geq 9$

11. $2y + 3 > -5$

12. $-2x + 3 \geq 9$

13. $3n + 5 > -10$

14. $-3n + 5 > -10$

15. $5x - 6 \leq 2 - x$

16. $10t - 3 - t > t - 20 - 23$

17. $5m + 19 > 3m - 6 - (3 + 2m)$

18. $5y + 3 \geq 2y + 12$

19. $4(x - 6) + 3 \leq 2(x - 5) + 3x$

20. $8m + 12 \leq 8 + 6m$

21. $4x + 3 \leq 6x - 6$

22. $\dfrac{x}{3} > \dfrac{x}{4}$

23. $\dfrac{3x}{2} + \dfrac{1}{3} < \dfrac{5x}{6} - \dfrac{2}{3}$

24. $\dfrac{a}{3} + \dfrac{1}{6} < \dfrac{a}{6} + \dfrac{2}{3}$

25. $4(2y - 1) > 8y + 3$

26. $6x + 7 \leq 6x + 9$

27. $4x + 3 \leq 6x - 6(x - 2)$

28. $\dfrac{7y}{6} + \dfrac{1}{2} \leq \dfrac{5y}{3} + \dfrac{3}{4}$

29. $6(x - 2) < 6x - 12$

30. $2y + 3 > 5y - 6$

31. $3(2y + 4) \geq 2(y - 1)$

32. $4m - 3 < 2m + 7$

33. $-2(n - 1) \leq 2$

34. $\dfrac{3m}{2} \geq \dfrac{5}{3}$

35. $5(x + 2) \leq 5x + 10$

36. $2b - 7 \leq 8$

37. $2(x + 3) \geq x + 3(x - 1)$

38. $-x \geq 2$

39. $-x < 0$

40. $\dfrac{3m}{5} + \dfrac{1}{10} < \dfrac{m}{5} - \dfrac{3}{10}$

41. $-3(x + 1) \geq -x - 2(x - 2)$

42. $\dfrac{y}{3} + 1 < 2$

43. $\dfrac{t}{4} + 2 > 4$

44. $\dfrac{5x - 7}{3} \leq 1$

45. $\dfrac{3m - 2}{2} \geq \dfrac{2m + 1}{5}$

46. $\dfrac{3x + 1}{2} < \dfrac{2x + 1}{5}$

47. $6 - \dfrac{5y}{3} \leq \dfrac{y}{2} - 7$

48. $3(2x + 1) + 6x < 5(x + 3) + 4x$

49. $5t < 7t$

50. $\dfrac{t}{2} + \dfrac{3}{4} \geq \dfrac{1}{2} - t$

PROBLEMS INVOLVING FRACTIONAL EQUATIONS AND LINEAR INEQUALITIES

5.9

The familiar four-step procedure will be used in this section to solve word problems involving fractional equations and linear inequalities.

Step (1)   Let some letter represent one of the unknown numbers. Then translate the word phrases into variable expressions.

Step (2)   Form the equation or inequality, using information given in the problem.

Step (3)   Solve the equation or inequality.

Step (4)   Interpret the solution. Often the solution to the equation or inequality is not the answer to the written problem. Something else may have to be done. Also, be sure to check each answer, using the wording of the given problem.

Example 1

The difference between two numbers is 3. If 5 times the smaller number is divided by the larger number, the quotient is 4. Find the numbers.

Solution:

Step (1) Let x represent the smaller number. Then $x + 3$ represents the larger number and $5x$ represents 5 times the smaller number.

Step (2) Form the equation by observing that the quotient is equal to 4.

$$\frac{5x}{x + 3} = 4$$

Step (3) Solve the equation.

$$\frac{5x}{x + 3} = 4$$

$$(x + 3) \cdot \frac{5x}{(x + 3)} = 4 \cdot (x + 3)$$

$$5x = 4x + 12$$

$$x = 12$$

Step (4) Interpret the solution. The smaller number is 12. The larger number is $x + 3$, or 15.

Two electric resistances are 20 ohms and 30 ohms respectively. How much must each resistance be increased so that their ratio is $\frac{5}{6}$? Assume that each resistance is increased the same amount.

Example 2

Solution:

Step (1) Let x represent the amount of increase. Then $20 + x$ represents the numerator and $30 + x$ represents the denominator.

Step (2) Form the equation, remembering the ratio must equal $\frac{5}{6}$

$$\frac{20 + x}{30 + x} = \frac{5}{6}$$

Step (3) Solve the equation. Multiply both sides by the LCD of $6(30 + x)$.

$$\frac{20 + x}{30 + x} = \frac{5}{6}$$

$$6(\cancel{30 + x}) \cdot \frac{20 + x}{(\cancel{30 + x})} = \frac{5}{\cancel{6}} \cdot \cancel{6}(30 + x)$$

$$6(20 + x) = 5(30 + x)$$

$$120 + 6x = 150 + 5x$$

$$6x - 5x = 150 - 120$$

$$x = 30$$

Step (4)   Interpret the solution. Each resistance must be increased by 30 ohms.

Example 3

A painter can paint the inside of a small house in 20 hours working alone. His assistant, who is somewhat slower, can paint the inside of the same house in 30 hours. How long would it take the painter and his assistant working together to paint the inside of the house?

Solution:

Step (1)   Let x represent the number of hours it takes for both, working together, to paint the house. Then, as shown in the following table, it would take the painter 20 hours to paint the house, working alone. He could complete $\frac{1}{20}$ of the job in 1 hour and $\frac{x}{20}$ of the job in x hours. Since it would take the assistant 30 hours to paint the house, he could complete $\frac{1}{30}$ of the job in 1 hour and $\frac{x}{30}$ of the job in x hours.

	Time Needed Alone	Fractional Part of Job Done in 1 Hour	Fractional Part of Job Done in X Hours
Painter	20 hr	$\frac{1}{20}$	$\frac{x}{20}$
Assistant	30 hr	$\frac{1}{30}$	$\frac{x}{30}$

Step (2)   We form the equation by noting that the portion of the job done by the painter in x hours plus the portion of the job done by his assistant in x hours must equal one whole job. Thus,

$$\frac{x}{20} + \frac{x}{30} = 1$$

Step (3)   Solve the equation.

$$\frac{x}{20} + \frac{x}{30} = 1$$

$$60 \cdot \frac{x}{20} + 60 \cdot \frac{x}{30} = 1 \cdot 60$$

$$3x + 2x = 60$$

$$5x = 60$$

$$x = 12$$

Step (4)   Interpret the solution. It would take the painter and his assistant, working together, 12 hours to paint the inside of the house.

How many milliliters of a 25% salt solution should be added to 30 milliliters of a 50% solution in order to obtain a 40% solution?

Example 4

Solution: Problems of this type are called *mixture problems*. They are solved by concerning ourselves with only one portion of the mixture — in this case, the salt. The amount of salt in the 50% solution together with the amount of salt in the 25% solution must be equal to the amount of salt in the resulting 40% solution. This relationship can be pictured as follows:

| amount of salt in the 50% solution | + | amount of salt in the 25% solution | = | amount of salt in the 40% solution |

Step (1)   Let x represent the number of milliliters of the 25% solution. There are 30 milliliters of the 50% solution, x milliliters of the 25% solution, and $(30 + x)$ milliliters of the resulting 40% solution. The amount of salt in the 50% solution is $50\% \cdot (30)$. The amount of salt in the 25% solution is $25\% \cdot (x)$. And the amount of salt in the resulting 40% solution is $40\% \cdot (30 + x)$. This information is pictured in the accompanying diagram.

| 30 milliliters of solution | | x milliliters of solution | | $(30 + x)$ milliliters of solution |

| amount of salt in the 50% solution is: | + | amount of salt in the 25% solution is: | = | amount of salt in the 40% solution is: |

| $50\% \cdot (30)$ | | $25\% \cdot (x)$ | | $40\% \cdot (30 + x)$ |

Step (2)   Form the equation by referring to the diagram.

$$50\% \cdot (30) + 25\% \cdot (x) = 40\% \cdot (30 + x)$$

Step (3)   Solve the equation. We will convert each percentage to a fraction. Then, each side of the equation will be multiplied by 100. This will change the percents to whole numbers.

$$50\% \cdot (30) + 25\% \cdot (x) = 40\% \cdot (30 + x)$$

$$\frac{50}{100} \cdot 30 + \frac{25}{100} \cdot x = \frac{40}{100} \cdot (30 + x)$$

$$\cancel{100} \cdot \frac{50}{\cancel{100}} \cdot 30 + \cancel{100} \cdot \frac{25}{\cancel{100}} \cdot x = \cancel{100} \cdot \frac{40}{\cancel{100}} \cdot (30 + x) \quad \left[\begin{array}{l}\text{multiply both sides}\\\text{by 100 and reduce.}\end{array}\right]$$

$$(50)(30) + 25x = 40(30 + x)$$

$$1{,}500 + 25x = 1{,}200 + 40x$$

$$25x - 40x = 1{,}200 - 1{,}500$$

$$-15x = -300$$

$$x = 20$$

Step (4)   Interpret the solution: 20 milliliters of the 25% solution must be added to the 50% solution.

Example 5   A student in an algebra class needs at least 360 points to obtain a grade of A. Her scores on three tests have been 82, 94, and 88. What can her score be on the fourth test to assure her of an A?

Solution:

Step (1)   Let x represent her score on the fourth test.

Step (2)   Form the inequality by noting that the sum of the four test scores must be at least 360 points.

$$x + 82 + 94 + 88 \geq 360$$

Step (3)   Solve the inequality.

$$x + 82 + 94 + 88 \geq 360$$

$$x + 264 \geq 360$$

$$x \geq 360 - 264$$

$$x \geq 96$$

Step (4)   Interpret the solution. She must obtain a score of 96 or higher on the fourth test.

Example 6

A bank offers two types of checking accounts, plan A and plan B. The monthly charge for plan A is a flat 15¢ per check. The charge for plan B is $1.00 per month plus 10¢ per check. After how many checks will the monthly charge for plan A exceed that for plan B?

Solution:

Step (1) Let x represent the number of checks. Then the monthly charge for plan A is $15x$ cents, and the monthly charge for plan B is $100 + 10x$ cents. ($1.00 = 100$ cents)

Step (2) Form the inequality. We want to know when the monthly charge for plan A exceeds plan B.

$$15x > 100 + 10x$$

Step (3) Solve the inequality.

$$15x > 100 + 10x$$

$$15x - 10x > 100$$

$$5x > 100$$

$$x > 20$$

Step (4) Interpret the solution. If more than 20 checks are written per month, the charge for plan A would exceed the charge for plan B.

Exercise 5.9

Solve the following problems.

1. The sum of two numbers is 18. If the smaller number is divided by the larger, the quotients is $\frac{4}{5}$. Find the numbers.

2. The difference between two numbers is 6. If 3 times the smaller number is divided by the larger, the quotient is 4. Find the numbers.

3. If $\frac{2}{3}$ of a certain number is subtracted from twice the number, the result is 20. Find the number.

4. Find two consecutive integers such that the sum of $\frac{1}{2}$ the first and $\frac{2}{3}$ of the next is 22.

5. The denominator of a certain fraction is 6 more than the numerator, and the fraction is equivalent to $\frac{5}{8}$. Find the numerator.

6. The denominator of a certain fraction is 8 more than the numerator, and the fraction is equivalent to $\frac{2}{3}$. Find the denominator.

7. The numerator of a fraction is 5 less than the denominator. If the numerator is decreased by 2 and the denominator is increased by 3, the value of the new fraction is $\frac{1}{3}$. Find the original fraction.

8. Two electric resistances are 2 ohms (Ω) and 3 Ω, respectively. How much must each resistance be increased so that the ratio is $\frac{3}{4}$? Assume that each resistance is increased by the same amount.

9. The width of a rectangular plate is $\frac{2}{3}$ of its length. If the perimeter is 8 in., find its dimensions.

10. How many liters of a 30% acid solution should be added to 20 liters of a 60% solution in order to obtain a 50% acid solution?

11. A 20-qt solution of acid and water is 30% acid. How many quarts of pure acid must be added so that the solution will be 65% acid?

12. How many quarts of an 80% salt solution must be added to 15 qt of a 12% solution to make a 30% salt solution?

13. How many liters of a 20% salt solution should be added to 30 liters of a 50% solution in order to obtain a 40% salt solution?

14. How much pure copper must be added to 20 lb of an alloy testing 15% copper to produce an alloy testing 35%?

15. How many grams of an alloy containing 40% aluminum must be melted with an alloy containing 70% aluminum in order to obtain 20 g (grams) of an alloy containing 45% aluminum?

16. How many pounds of an alloy containing 10% titanium must be melted with an alloy of 25% titanium in order to obtain 40 lb of an alloy containing 15% titanium?

17. How many pounds of walnuts at 49¢ per pound should a grocer mix with 20 lb of pecans at 58¢ a pound to give a mixture worth 54¢ a pound?

18. A man has an amount of money in savings at 5% and another investment at 6% simple interest. The amount invested at 6% is $2,000 more than that invested at 5%. The total annual interest is $395.00. How much money is invested at each rate?

19. Two investments produce an annual income of $910. $5,600 more is invested at 8% than at 6%. How much is invested at each interest rate?

20. Two investments produce an annual income of $3,090; $13,000 more is invested at 9% than at 7%. How much is invested at each interest rate?

21. Bob and Suzanne working together can paint a house in 10 days. When working alone, Bob can paint the house in 30 days. How long will it take Suzanne working alone to paint the house?

22. A man can build a table in 6 hr. It takes his son 12 hr to build the same table. How long would it take the man and his son working together to build the table?

23. Julie can mow a lawn in 5 hr. Jim can mow the same lawn in 10 hr. How long will it take the two together to mow the lawn?

24. Working alone an apprentice carpenter can complete a certain job in 6 days, while an experienced carpenter can complete the same job in 4 days. How long would it take the two working together to complete the job?

25. José can mow a lawn in 3 hr. His smaller brother needs 6 hr to do the same work. How long will it take them if they work together?

26. Maria can sew a dress in 8 hr, but she and Tom working together can sew the dress in 6 hr. How long would it take Tom working alone?

27. Two pumps when working together can fill a swimming pool in 24 hr. When operated alone, one pump requires 20 more hours than the other to fill the pool. How long will it take each pump to fill the pool?

28. A tank can be filled by an inlet pipe in 4 hr and emptied in 6 hr by a drain pipe. How long will it take to fill the tank if both pipes are open?

29. A tank can be filled by an inlet pipe in 8 hr when the drain pipe is open and in 5 hr if the drain pipe is closed. How long will it take to empty a full tank when the inlet pipe is closed?

30. The sum of two consecutive odd numbers is less than 80. What are the largest such numbers?

31. A student in a mathematics class needs at least 265 points to obtain a grade of A. What could she score on her third test so that she can make an A for the course? Her scores on the other two tests were 80 points and 96 points.

32. A student in a history class needs at least 360 points to obtain a grade of A. His scores on three tests have been 86, 90, and 94. What can his score be on the fourth test to assure him of an A?

33. A student received grades of 86, 75, and 80 on three tests. What must be his grade on a fourth test if his average on the four tests is to be at least 82?

34. A car rental agency has two rental plans. Plan A charges $25 plus 15¢ a mile. Plan B charges 25¢ a mile. After how many miles would the charges for plan B exceed the charges on plan A?

35. A man on retirement found that he needed more than $1,000 per year return on his investments. He invested $8,000 at 8%. How much did he have to invest at 9% to meet his goal?

Summary

A rational expression is a quotient of two polynomials. The divisor cannot represent zero.

DEFINITION OF A RATIONAL EXPRESSION

(1) The denominator of a fraction can never be zero.
(2) If the denominator of a fraction is 1, then the fraction is equal to the numerator.

BASIC PROPERTIES OF FRACTIONS

(3) If the numerator of a fraction is zero and the denominator is not zero, then the fraction is equal to zero.

(4) If the numerator and denominator of a fraction are equal, but not zero, then the fraction is equal to 1.

THE FUNDAMENTAL PRINCIPLE OF FRACTIONS

$$\frac{a}{b} = \frac{a \cdot n}{b \cdot n} \quad \text{and} \quad \frac{a}{b} = \frac{a \div n}{b \div n}, \text{ where } b, n \neq 0$$

MULTIPLICATION OF RATIONAL EXPRESSIONS

Step (1) Write each fraction in factored form and divide out all factors common to a numerator and denominator.

Step (2) Multiply the remaining numerators and multiply the remaining denominators.

DIVISION OF RATIONAL EXPRESSIONS

To divide rational expressions, multiply by the multiplicative inverse (reciprocal) of the divisor, or, more simply, invert the divisor and multiply.

ADDITION AND SUBTRACTION OF RATIONAL EXPRESSIONS

Step (1) Factor all denominators.

Step (2) Find the LCD and expand all fractions to that denominator.

Step (3) Combine the numerators and place the result over the LCD. If possible, reduce the answer to lowest terms.

DEFINITION OF A COMPLEX FRACTION

A fraction containing at least one rational expression in its numerator or denominator is called a *complex fraction*.

SIMPLIFYING A COMPLEX FRACTION

A complex fraction is simplified by multiplying its numerator and denominator by the LCD of all fractions contained within the complex fraction.

SOLVING FRACTIONAL EQUATIONS

Step (1) Find the LCD.

Step (2) Multiply both sides of the equation by the LCD.

Step (3) Classify the resulting equation as first degree or quadratic. Then solve accordingly.

EXTRANEOUS SOLUTIONS

A tentative solution that fails to check in the original equation is called an *extraneous solution*. These may occur whenever a denominator of a fractional equation contains a variable.

THE ADDITION-SUBTRACTION PROPERTY OF INEQUALITIES

Given an inequality, the same real number may be added to both sides or subtracted from both sides without changing the solutions.

$$\text{If } a < b, \text{ then } a + c < b + c \text{ and } a - c < b - c$$

Given an inequality, the same *positive* real number may be multiplied times both sides or divided into both sides without changing the solutions.

$$\text{If } a < b \text{ and } c > 0 \text{ then } ac < bc \text{ and } \frac{a}{c} < \frac{b}{c}$$

Given an inequality, the same *negative* real number may be multiplied times both sides or divided into both sides without changing the solutions, provided the inequality symbol is reversed.

$$\text{If } a < b \text{ and } c < 0 \text{ then } ac > bc \text{ and } \frac{a}{c} > \frac{b}{c}$$

Step (1)   Simplify the inequality. Perform multiplications and combine like terms. If the inequality contains fractions, multiply both sides by the LCD.

Step (2)   Use the addition–subtraction property to obtain the terms containing a variable on one side and the numbers on the other side.

Step (3)   Use the multiplication–division property to divide both sides by the numerical coefficient of the variable. Remember, if you multiply or divide by a negative number, the inequality symbol must be reversed.

Self-Checking Exercise

Expand the following rational expressions as indicated. Assume that denominators do not represent zero.

1. Expand $\dfrac{5a^2}{6b^3c}$ so that its denominator becomes $18b^5c^2$

2. Expand $\dfrac{x+2}{x-3}$ so that its denominator becomes $x^2 - 9$

3. Expand $\dfrac{2m-1}{3m+2}$ so that its denominator becomes $6m^2 - 5m - 6$

Reduce each of the following rational expressions to lowest terms. Assume that denominators do not represent zero.

4. $\dfrac{8m^3n^2}{20m^4n^2}$

5. $\dfrac{2m-8n}{20n-5m}$

6. $\dfrac{x^2+3x-4}{2x^2-5x+3}$

Multiply or divide as indicated. Assume that denominators and divisors do not represent zero.

7. $\dfrac{10m}{9n^2} \cdot \dfrac{3n^3}{5m^2}$

8. $\dfrac{4x^2-1}{2x} \div (2x+1)$

9. $\dfrac{5a^2-5a}{2a^2+5a-3} \cdot \dfrac{a+3}{5-5a}$

10. $\dfrac{4x^2-9}{x^2+x-2} \cdot \dfrac{x^2-2x+1}{2x^2-x-3}$

11. $\dfrac{a^2+a-6}{a^2-a-12} \div \dfrac{a^2-9}{a^2-16}$

12. $\left(\dfrac{t-3}{2t+1}\right)^2$

Add or subtract as indicated. Assume that denominators do not represent zero.

13. $\dfrac{5}{4m-12} + \dfrac{1}{2m-6}$

14. $\dfrac{5}{3-x} + \dfrac{x+2}{x^2-9}$

15. $\dfrac{n}{n+2} - \dfrac{8}{n^2-4}$

16. $\dfrac{3x-1}{x^2+5x+6} - \dfrac{x+1}{x+2} + \dfrac{x}{x+3}$

Simplify the following complex fractions. Assume that denominators do not represent zero.

17. $\dfrac{\dfrac{3m}{5n^2}}{\dfrac{7m^2}{10n^4}}$

18. $\dfrac{b - \dfrac{9}{b}}{1 + \dfrac{3}{b}}$

Solve the following equations.

19. $\dfrac{x}{6} = \dfrac{3}{5}$

20. $\dfrac{4x}{5} + 3 = \dfrac{x}{2}$

21. $y + 2 = \dfrac{6}{y + 3}$

22. $\dfrac{3}{4m} + \dfrac{1}{8} = \dfrac{1}{2}$

23. $\dfrac{9}{m^2 - 3m} + \dfrac{5}{m} = \dfrac{3}{m - 3}$

24. $\dfrac{2x^2 + 3x + 19}{x^2 + x - 20} + \dfrac{4}{x + 5} = \dfrac{x + 3}{x - 4}$

Solve the following formulas for the specified symbol.

25. $i = prt$, for r

26. $ab + ac = d$, for a

27. $\dfrac{1}{p} + \dfrac{1}{q} = \dfrac{1}{f}$, for p

28. $A = \dfrac{h(b + c)}{2}$, for c

Solve the following inequalities. If possible, graph the solution.

29. $3x - 5 < 10$

30. $2x + 3 > 4x - 1$

31. $\dfrac{3m}{2} + \dfrac{1}{3} \geq \dfrac{5}{6}$

32. $4a + 1 < 4(a + 2)$

33. $4a - 3 > 4(a + 1)$

34. $\dfrac{5t + 1}{3} \leq \dfrac{2t - 3}{2}$

Solve the following problems. (If you need help, review Examples 1 through 6 in section 5.9.)

35. The difference between two numbers is 5. If 6 times the smaller number is divided by the larger number, the quotient is 4. Find the numbers.

36. Two electric resistances are 40 Ω and 50 Ω, respectively. How much must each resistance be increased so that their ratio is $\frac{7}{8}$? Assume that each resistance is increased the same amount.

37. A painter can paint the outside of a house in 40 hr working alone. Her assistant can paint the same house in 50 hr. How long would it take both, working together, to paint the outside of the house?

38. How many milliliters of a 30% acid solution should be added to 50 ml (milliliters) of a 60% solution in order to obtain a 50% acid solution?

39. A student in an algebra class needs at least 450 points to obtain a grade of A. His scores on four tests have been 84, 86, 90, and 94. What can his score be on the fifth test to assure him of an A?

40. A bank offers two types of checking accounts, plan A and plan B. The monthly charge for plan A is a flat 17¢ per check. The charge for plan B is $1.50 per month plus 12¢ per check. After how many checks will the monthly charge for plan A exceed plan B?

---

1. $\dfrac{5a^2}{6b^3c} = \dfrac{5a^2 \cdot (3b^2c)}{6b^3c \cdot (3b^2c)} = \dfrac{15a^2b^2c}{18b^5c^2}$

2. $\dfrac{x + 2}{x - 3} = \dfrac{(x + 2)(x + 3)}{(x - 3)(x + 3)} = \dfrac{x^2 + 5x + 6}{x^2 - 9}$

3. $\dfrac{2m - 1}{3m + 2} = \dfrac{(2m - 1)(2m - 3)}{(3m + 2)(2m - 3)} = \dfrac{4m^2 - 8m + 3}{6m^2 - 5m - 6}$

Solutions to Self-Checking Exercise

4. $\dfrac{8m^3n^2}{20m^4n^2} = \dfrac{\cancel{4}\cdot 2\cdot \cancel{m^3}\cdot \cancel{n^2}}{\cancel{4}\cdot 5\cdot \cancel{m^3}\cdot m\cdot \cancel{n^2}} = \dfrac{2}{5m}$

5. $\dfrac{2m - 8n}{20n - 5m} = \dfrac{2\overset{-1}{\cancel{(m - 4n)}}}{5\cancel{(4n - m)}} = \dfrac{2(-1)}{5} = \dfrac{-2}{5}$

6. $\dfrac{x^2 + 3x - 4}{2x^2 - 5x + 3} = \dfrac{(x + 4)\cancel{(x - 1)}}{(2x - 3)\cancel{(x - 1)}} = \dfrac{x + 4}{2x - 3}$

7. $\dfrac{10m}{9n^2}\cdot \dfrac{3n^3}{5m^2} = \dfrac{2\cdot \cancel{5}\cdot \cancel{m}}{\cancel{3}\cdot 3\cdot \cancel{n^2}}\cdot \dfrac{\cancel{3}\cdot \cancel{n^2}\cdot n}{\cancel{5}\cdot \cancel{m}\cdot m} = \dfrac{2n}{3m}$

8. $\dfrac{4x^2 - 1}{2x} \div (2x + 1) = \dfrac{\cancel{(2x + 1)}(2x - 1)}{2x}\cdot \dfrac{1}{\cancel{(2x + 1)}} = \dfrac{2x - 1}{2x}$

9. $\dfrac{5a^2 - 5a}{2a^2 + 5a - 3}\cdot \dfrac{a + 3}{5 - 5a} = \dfrac{\overset{-1}{\cancel{5}a\cancel{(a - 1)}}}{(2a - 1)\cancel{(a + 3)}}\cdot \dfrac{\cancel{(a + 3)}}{\cancel{5}\cancel{(1 - a)}} = \dfrac{-a}{2a - 1}$

10. $\dfrac{4x^2 - 9}{x^2 + x - 2}\cdot \dfrac{x^2 - 2x + 1}{2x^2 - x - 3} = \dfrac{(2x + 3)\cancel{(2x - 3)}}{(x + 2)\cancel{(x - 1)}}\cdot \dfrac{(x - 1)\cancel{(x - 1)}}{\cancel{(2x - 3)}(x + 1)}$

$$= \dfrac{(2x + 3)(x - 1)}{(x + 2)(x + 1)} \quad \text{or} \quad \dfrac{2x^2 + x - 3}{x^2 + 3x + 2}$$

11. $\dfrac{a^2 + a - 6}{a^2 - a - 12} \div \dfrac{a^2 - 9}{a^2 - 16} = \dfrac{\cancel{(a + 3)}(a - 2)}{\cancel{(a + 3)}\cancel{(a - 4)}}\cdot \dfrac{(a + 4)\cancel{(a - 4)}}{(a + 3)(a - 3)}$

$$= \dfrac{(a - 2)(a + 4)}{(a + 3)(a - 3)} \quad \text{or} \quad \dfrac{a^2 + 2a - 8}{a^2 - 9}$$

12. $\left(\dfrac{t - 3}{2t + 1}\right)^2 = \dfrac{(t - 3)}{(2t + 1)}\cdot \dfrac{(t - 3)}{(2t + 1)} = \dfrac{t^2 - 6t + 9}{4t^2 + 4t + 1}$

13. $\dfrac{5}{4m - 12} + \dfrac{1}{2m - 6} = \dfrac{5}{4(m - 3)} + \dfrac{1}{2(m - 3)}$

$$= \dfrac{5}{4(m - 3)} + \dfrac{2\cdot 1}{2\cdot 2(m - 3)}$$

$$= \dfrac{5 + 2}{4(m - 3)}$$

$$= \dfrac{7}{4(m - 3)}$$

14. $\dfrac{5}{3 - x} + \dfrac{x + 2}{x^2 - 9} = \dfrac{5}{3 - x} + \dfrac{x + 2}{(x + 3)(x - 3)}$

$$= \dfrac{-5}{(x - 3)} + \dfrac{x + 2}{(x + 3)(x - 3)}$$

$$= \dfrac{-5(x + 3)}{(x - 3)(x + 3)} + \dfrac{x + 2}{(x + 3)(x - 3)}$$

$$= \dfrac{-5(x + 3) + x + 2}{(x + 3)(x - 3)}$$

$$= \dfrac{-5x - 15 + x + 2}{(x + 3)(x - 3)}$$

$$= \dfrac{-4x - 13}{(x + 3)(x - 3)}$$

15. $\dfrac{n}{n+2} - \dfrac{8}{n^2-4} = \dfrac{n}{n+2} - \dfrac{8}{(n+2)(n-2)}$

$\qquad\qquad = \dfrac{n(n-2)}{(n+2)(n-2)} - \dfrac{8}{(n+2)(n-2)}$

$\qquad\qquad = \dfrac{n(n-2)-8}{(n+2)(n-2)}$

$\qquad\qquad = \dfrac{n^2-2n-8}{(n+2)(n-2)}$

$\qquad\qquad = \dfrac{(n-4)\cancel{(n+2)}}{\cancel{(n+2)}(n-2)}$

$\qquad\qquad = \dfrac{n-4}{n-2}$

16. $\dfrac{3x-1}{x^2+5x+6} - \dfrac{x+1}{x+2} + \dfrac{x}{x+3} = \dfrac{3x-1}{(x+2)(x+3)} - \dfrac{x+1}{(x+2)} + \dfrac{x}{(x+3)}$

$\qquad\qquad\qquad\qquad = \dfrac{3x-1}{(x+2)(x+3)} - \dfrac{(x+1)(x+3)}{(x+2)(x+3)} + \dfrac{x(x+2)}{(x+3)(x+2)}$

$\qquad\qquad\qquad\qquad = \dfrac{3x-1-(x+1)(x+3)+x(x+2)}{(x+2)(x+3)}$

$\qquad\qquad\qquad\qquad = \dfrac{3x-1-x^2-4x-3+x^2+2x}{(x+2)(x+3)}$

$\qquad\qquad\qquad\qquad = \dfrac{x-4}{(x+2)(x+3)}$

17. $\dfrac{\dfrac{3m}{5n^2}}{\dfrac{7m^2}{10n^4}} = \dfrac{\dfrac{3m}{5n^2}\cdot(10n^4)}{\dfrac{7m^2}{10n^4}\cdot(10n^4)} = \dfrac{6mn^2}{7m^2} = \dfrac{6n^2}{7m}$

18. $\dfrac{b-\dfrac{9}{b}}{1+\dfrac{3}{b}} = \dfrac{b\left(b-\dfrac{9}{b}\right)}{b\left(1+\dfrac{3}{b}\right)} = \dfrac{b^2-9}{b+3} = \dfrac{\cancel{(b+3)}(b-3)}{\cancel{(b+3)}} = b-3$

19. $\dfrac{x}{6} = \dfrac{3}{5}$

$30\cdot\dfrac{x}{6} = 30\cdot\dfrac{3}{5}$

$5x = 18$

$x = \dfrac{18}{5}$

20. $\dfrac{4x}{5} + 3 = \dfrac{x}{2}$

$10\left[\dfrac{4x}{5} + 3\right] = 10\cdot\dfrac{x}{2}$

$8x + 30 = 5x$

$3x = -30$

$x = -10$

21. $y + 2 = \dfrac{6}{y+3}$

$(y+3)(y+2) = \dfrac{6}{\cancel{(y+3)}}\cdot\cancel{(y+3)}$

$y^2 + 5y + 6 = 6$

$y^2 + 5y = 0$

$y(y+5) = 0$

$y = 0 \quad$ or $\quad y+5 = 0$

$y = 0 \quad$ and $\quad y = -5$

22. $\dfrac{3}{4m} + \dfrac{1}{8} = \dfrac{1}{2}$

$8m\left[\dfrac{3}{4m} + \dfrac{1}{8}\right] = \dfrac{1}{2}\cdot 8m$

$6 + m = 4m$

$-3m = -6$

$m = 2$

23.
$$\frac{9}{m^2 - 3m} + \frac{5}{m} = \frac{3}{m - 3}$$

$$\frac{9}{m(m - 3)} + \frac{5}{m} = \frac{3}{(m - 3)}$$

$$m(m - 3)\left[\frac{9}{m(m - 3)} + \frac{5}{m}\right] = \frac{3}{(m - 3)} \cdot m(m - 3)$$

$$9 + 5(m - 3) = 3m$$
$$9 + 5m - 15 = 3m$$
$$2m = 6$$
$$m = 3$$

3 is an extraneous solution. The equation has no solutions.

24.
$$\frac{2x^2 + 3x + 19}{x^2 + x - 20} + \frac{4}{x + 5} = \frac{x + 3}{x - 4}$$

$$\frac{2x^2 + 3x + 19}{(x + 5)(x - 4)} + \frac{4}{(x + 5)} = \frac{x + 3}{(x - 4)}$$

$$(x + 5)(x - 4)\left[\frac{2x^2 + 3x + 19}{(x + 5)(x - 4)} + \frac{4}{(x + 5)}\right] = \frac{x + 3}{(x - 4)} \cdot (x + 5)(x - 4)$$

$$2x^2 + 3x + 19 + 4(x - 4) = (x + 3)(x + 5)$$
$$2x^2 + 3x + 19 + 4x - 16 = x^2 + 8x + 15$$
$$x^2 - x - 12 = 0$$
$$(x - 4)(x + 3) = 0$$

$$x - 4 = 0 \quad \text{or} \quad x + 3 = 0$$
$$x = 4 \quad \text{and} \quad x = -3$$

4 is an extraneous solution. −3 is the only solution to the equation.

25. $i = prt$

$$\frac{i}{pt} = \frac{prt}{pt}$$

$$\frac{i}{pt} = r$$

$$r = \frac{i}{pt}$$

26. $ab + ac = d$

$$a(b + c) = d$$

$$\frac{a(b + c)}{(b + c)} = \frac{d}{(b + c)}$$

$$a = \frac{d}{b + c}$$

27.
$$\frac{1}{p} + \frac{1}{q} = \frac{1}{f}$$

$$pqf\left[\frac{1}{p} + \frac{1}{q}\right] = \frac{1}{f} \cdot pqf$$

$$qf + pf = pq$$
$$pf - pq = -qf$$
$$p(f - q) = -qf$$

$$\frac{p(f - q)}{(f - q)} = \frac{-qf}{(f - q)}$$

$$p = \frac{-qf}{f - q} \quad \text{or}$$

$$p = \frac{qf}{q - f}$$

28.
$$A = \frac{h(b + c)}{2}$$

$$2A = 2 \cdot \frac{h(b + c)}{2}$$

$$2A = h(b + c)$$
$$2A = hb + hc$$
$$2A - hb = hc$$

$$\frac{2A - hb}{h} = c \quad \text{or}$$

$$c = \frac{2A - hb}{h}$$

29. $3x - 5 < 10$

$\qquad 3x < 15$

$\qquad x < 5$

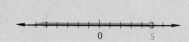

30. $2x + 3 > 4x - 1$

$\qquad 2x - 4x > -1 - 3$

$\qquad -2x > -4$

$\qquad x < \dfrac{-4}{-2}$

$\qquad x < 2$

31. $\dfrac{3m}{2} + \dfrac{1}{3} \geq \dfrac{5}{6}$

$6\left[\dfrac{3m}{2} + \dfrac{1}{3} \right] \geq \dfrac{5}{\cancel{6}} \cdot \cancel{6}$

$\qquad 9m + 2 \geq 5$

$\qquad 9m \geq 3$

$\qquad m \geq \dfrac{1}{3}$

32. $4a + 1 < 4(a + 2)$

$\quad 4a + 1 < 4a + 8$

$4a - 4a < 8 - 1$

$\qquad 0 < 7 \quad$ [true]

Every real number is a solution.
The graph is the entire
number line.

33. $4a - 3 > 4(a + 1)$

$\quad 4a - 3 > 4a + 4$

$4a - 4a > 4 + 3$

$\qquad 0 > 7 \quad$ [false]

The inequality has no solutions.

34. $\dfrac{5t + 1}{3} \leq \dfrac{2t - 3}{2}$

$6 \cdot \dfrac{5t + 1}{3} \leq 6 \cdot \dfrac{2t - 3}{2}$

$\quad 2(5t + 1) \leq 3(2t - 3)$

$\quad 10t + 2 \leq 6t - 9$

$\quad 10t - 6t \geq -9 - 2$

$\qquad 4t \leq -11$

$\qquad t \leq \dfrac{-11}{4} \quad$ or $\quad t \leq -2\dfrac{3}{4}$

35. $\dfrac{6x}{x + 5} = 4$

$(\cancel{x + 5}) \cdot \dfrac{6x}{(\cancel{x + 5})} = 4 \cdot (x + 5)$

$\qquad 6x = 4x + 20$

$\qquad 2x = 20$

$\qquad x = 10$

The smaller number is 10
The larger number is 15

36. $\dfrac{40 + x}{50 + x} = \dfrac{7}{8}$

$8(\cancel{50 + x}) \cdot \dfrac{40 + x}{(\cancel{50 + x})} = \dfrac{7}{\cancel{8}} \cdot \cancel{8}(50 + x)$

$\qquad 8(40 + x) = 7(50 + x)$

$\qquad 320 + 8x = 350 + 7x$

$\qquad 8x - 7x = 350 - 320$

$\qquad x = 30$

Each resistance must be
increased by 30 Ω

37. $\dfrac{t}{40} + \dfrac{t}{50} = 1$

$200\left[\dfrac{t}{40} + \dfrac{t}{50}\right] = 1 \cdot 200$

$5t + 4t = 200$

$9t = 200$

$t = \dfrac{200}{9}$

$t = 22\dfrac{2}{9}$

It would take the painter and her assistant $22\dfrac{2}{9}$ hr together.

38. $30\%(x) + 60\%(50) = 50\%(x + 50)$

$\dfrac{30}{100} \cdot x + \dfrac{60}{100} \cdot 50 = \dfrac{50}{100} \cdot (x + 50)$

$30 \cdot x + (60)(50) = 50(x + 50)$ [multiply both sides by 100]

$30x + 3000 = 50x + 2500$

$30x - 50x = 2500 - 3000$

$-20x = -500$

$x = \dfrac{-500}{-20}$

$x = 25$

25 ml of the 30% solution must be added.

39. $84 + 86 + 90 + 94 + x \geq 450$

$354 + x \geq 450$

$x \geq 450 - 354$

$x \geq 96$

The student must get a score of at least 96.

40. $17x > 150 + 12x$

$17x - 12x > 150$

$5x > 150$

$x > 30$

After 30 checks the charge for plan A exceeds the charge for plan B.

EXPONENTS AND RADICALS

6

NEGATIVE EXPONENTS

6.1

In section 2.1 we studied whole number exponents. We now extend our knowledge; we will define negative exponents so that they obey all of the properties of whole number exponents. To define negative exponents, we examine the product of $x^n \cdot x^{-n}$ where $x \neq 0$. Assuming that the addition property of exponents holds true, we have the following:

$$x^n \cdot x^{-n} = x^{n+(-n)}$$
$$= x^0$$
$$= 1$$

Since $x^n \cdot x^{-n} = 1$, we divide both sides by x^n and solve for x^{-n}:

$$x^n \cdot x^{-n} = 1$$
$$\frac{\cancel{x^n} \cdot x^{-n}}{\cancel{x^n}} = \frac{1}{x^n} \qquad \text{[where } x \neq 0 \text{]}$$
$$x^{-n} = \frac{1}{x^n}$$

This tells us that x^{-n} is the multiplicative inverse (reciprocal) of x^n and allows us to make the following definition.

Definition of Negative Exponents
$$x^{-n} = \frac{1}{x^n}, \qquad \text{where } x \neq 0$$

This definition allows us to convert a negative exponent to a positive exponent.

Example 1

Each of the following negative exponents has been converted to a positive exponent by applying the definition.

(a) $x^{-2} = \frac{1}{x^2}$,   where $x \neq 0$

(b) $3^{-2} = \frac{1}{3^2}$

(c) $(2m)^{-3} = \frac{1}{(2m)^3}$ or $\frac{1}{8m^3}$,   where $m \neq 0$

(d) $2m^{-3} = 2 \cdot \frac{1}{m^3}$ or $\frac{2}{m^3}$,   where $m \neq 0$

A number raised to a negative power can be evaluated by converting it to a positive exponent.

Example 2

Each of the following has been evaluated by using the definition to convert negative exponents to positive exponents.

(a) $5^{-2} = \dfrac{1}{5^2} = \dfrac{1}{25}$

(b) $3^{-1} = \dfrac{1}{3^1} = \dfrac{1}{3}$

(c) $10^{-1} = \dfrac{1}{10^1} = \dfrac{1}{10}$ or 0.1

(d) $10^{-3} = \dfrac{1}{10^3} = \dfrac{1}{1000}$ or 0.001

(e) $\left(\dfrac{2}{3}\right)^{-2} = \dfrac{1}{\left(\dfrac{2}{3}\right)^2} = \dfrac{1}{\dfrac{4}{9}} = \dfrac{9}{4}$

An exponential expression is said to be *simplified* when all negative exponents have been converted to positive exponents and when the resulting fraction, if any, is in lowest terms. We will learn to simplify the following two types of exponential expressions:

(1) Exponential expressions where the numerator and the denominator are basic products or single terms.

Example: $\dfrac{5a^{-4}}{2}$

(2) Exponential expressions where the numerator or denominator is a basic sum or difference.

Example: $x^{-2} + y^{-2}$

To simplify the first type of expression, where the numerator and denominator are basic products or single terms, we multiply numerator and denominator by the corresponding exponential raised to a positive power.

Example 3

Simplify $\dfrac{5a^{-4}}{2}$ where $a \neq 0$

Solution: Multiply numerator and denominator by a^4

$$\frac{5a^{-4}}{2} = \frac{5a^{-4} \cdot a^4}{2 \cdot a^4}$$

$$= \frac{5a^0}{2a^4}$$

$$= \frac{5 \cdot 1}{2a^4}$$

$$= \frac{5}{2a^4}$$

$$\text{Therefore, } \frac{5a^{-4}}{2} = \frac{5}{2a^4}$$

In effect, this procedure simply transfers the factor of a^{-4} across the fraction bar and changes the sign of its exponent.

Example 4

Simplify $\dfrac{n^{-2}}{3m}$ where $m, n \neq 0$

Solution: Multiply numerator and denominator by n^2

$$\frac{n^{-2}}{3m} = \frac{n^{-2} \cdot n^2}{3m \cdot n^2}$$

$$= \frac{n^0}{3mn^2}$$

$$= \frac{1}{3mn^2}$$

$$\text{Therefore, } \frac{n^{-2}}{3m} = \frac{1}{3mn^2}.$$

Again you should observe that this procedure transfers the factor of n^{-2} across the fraction bar and changes the sign of its exponent.

Examples 3 and 4 allow us to state a useful shortcut: If the numerator and denominator of an exponential expression are basic products or single terms, the factors having negative exponents can be transferred across the fraction bar, provided the negative exponent is changed to a positive exponent.

Example 5

Simplify $\dfrac{2x^{-3}}{5}$ where $x \neq 0$

Solution: Transfer x^{-3} across the fraction bar and change the sign of the exponent.

$$\frac{2x^{-3}}{5} = \frac{2}{5x^3}$$

Example 6

Simplify $\dfrac{b^{-5}}{4a}$ where $a, b \neq 0$

Solution: Transfer b^{-5} across the fraction bar and change the sign of the exponent. When a single term is transferred, a factor of 1 is left behind.

$$\frac{b^{-5}}{4a} = \frac{1}{4a \, b^5}$$

Simplify $\frac{3a^{-2}b^{-3}}{2c^{-4}d^2}$ where $a, b, c, d \neq 0$

Example 7

Solution: Transfer a^{-2}, b^{-3}, and c^{-4} across the fraction bar and change the sign of each exponent.

$$\frac{3a^{-2}b^{-3}}{2c^{-4}d^2} = \frac{3c^4}{2a^2b^3d^2}$$

Next we learn to simplify the second type of exponential expression, where the numerator or denominator is a basic sum or a basic difference. Unfortunately, there is no shortcut. We will convert negative exponents to positive exponents, and then simplify the resulting fraction or fractions.

Simplify $x^{-2} + y^{-2}$ where $x, y = 0$

Example 8

Solution: Convert x^{-2} to $\frac{1}{x^2}$ and y^{-2} to $\frac{1}{y^2}$. Then add the resulting fractions.

$$\begin{aligned}
x^{-2} + y^{-2} &= \frac{1}{x^2} + \frac{1}{y^2} \\
&= \frac{1 \cdot y^2}{x^2 \cdot y^2} + \frac{1 \cdot x^2}{y^2 \cdot x^2} \\
&= \frac{y^2}{x^2y^2} + \frac{x^2}{x^2y^2} \\
&= \frac{y^2 + x^2}{x^2y^2}
\end{aligned}$$

Simplify $\frac{a^{-3} - b^{-3}}{2}$ where $a, b \neq 0$

Example 9

Solution: Convert a^{-3} to $\frac{1}{a^3}$ and b^{-3} to $\frac{1}{b^3}$. Then simplify the resulting complex fraction.

$$\begin{aligned}
\frac{a^{-3} - b^{-3}}{2} &= \frac{\dfrac{1}{a^3} - \dfrac{1}{b^3}}{2} \\
&= \frac{a^3b^3 \cdot \left[\dfrac{1}{a^3} - \dfrac{1}{b^3}\right]}{a^3b^3 \cdot 2}
\end{aligned}$$

$$= \frac{a^{3}b^{3} \cdot \frac{1}{a^{3}} - a^{3}b^{3} \cdot \frac{1}{b^{3}}}{2a^{3}b^{3}}$$

$$= \frac{b^{3} - a^{3}}{2a^{3}b^{3}}$$

As stated earlier, negative exponents are defined in such a way that they obey the properties of whole number exponents. Therefore, the following properties hold true for both types of exponents:

Properties of Exponents

1. The Zero Power Property: $a^{0} = 1$ where $a \neq 0$
   Example: $(-5)^{0} = 1$

2. The Addition Property of Exponents: $a^{m} \cdot a^{n} = a^{m+n}$
   Example: $x^{6} \cdot x^{-4} = x^{2}$

3. The Subtraction Property of Exponents: $\dfrac{a^{m}}{a^{n}} = a^{m-n}$

   where $a \neq 0$
   Example: $\dfrac{x^{-2}}{x^{-5}} \times x^{(-2)-(-5)} = x^{-2+5} = x^{3}$

4. The Power to a Power Property of Exponents: $(a^{m})^{n} = a^{mn}$
   Example: $(x^{-2})^{-4} = x^{8}$

5. The Distributive Property of Exponents: $(a^{m} \cdot b^{n})^{x} = a^{mx} \cdot b^{nx}$

   or $\left(\dfrac{a^{m}}{b^{n}}\right)^{x} = \dfrac{a^{mx}}{b^{nx}}$ where $b \neq 0$

   Example: $(x^{-2}y^{3}z)^{-4} = x^{8}y^{-12}z^{-4}$

Using these properties, we can operate with negative exponents in the same manner as we operate with whole number exponents.

Example 10

Simplify $(3x^{-4})(2x^{6})$ where $x \neq 0$

Solution: Use the Addition Property of Exponents.

$$(3x^{-4})(2x^{6}) = 6x^{-4+6}$$
$$= 6x^{2}$$

Example 11

Simplify $\dfrac{8x^{-2}}{2x^{-5}}$ where $x \neq 0$

Solution: Use the Subtraction Property of Exponents.

$$\frac{8x^{-2}}{2x^{-5}} = 4x^{(-2)-(-5)}$$
$$= 4x^{-2+5}$$
$$= 4x^3$$

Simplify $(2a^{-2}b^4)^3$ where $a, b \neq 0$. Leave no negative exponents in the answer.

Example 12

Solution: Use the Distributive Property of Exponents.

$$(2a^{-2}b^4)^3 = 2^3a^{-6}b^{12}$$
$$= 8a^{-6}b^{12}$$
$$= \frac{8b^{12}}{a^6}$$

Exercise 6.1

Using the definition $x^{-n} = \frac{1}{x^n}$ where $x \neq 0$, convert the following negative exponents to positive exponents.

1. n^{-2} 2. y^{-1}
3. a^{-4} 4. $(2m)^{-3}$
5. $2m^{-3}$ 6. $(a + b)^{-1}$

Evaluate each of the these expressions.

7. 2^{-3} 8. 5^{-1}
9. 2^{-4} 10. 3^{-1}
11. 2^{-5} 12. 10^0
13. 10^{-2} 14. 10^{-4}
15. $\left(\frac{1}{10}\right)^{-1}$ 16. $\left(\frac{3}{5}\right)^{-2}$
17. $\left(\frac{3}{4}\right)^{-2}$ 18. $\left(\frac{1}{2}\right)^{-1}$

Simplify the following exponential expressions. Assume that no variables represent zero.

19. $\frac{x^{-2}}{3y}$ 20. $\frac{x^{-1}}{y^{-1}}$
21. $\frac{3m^{-4}}{2}$ 22. $\frac{y^{-2}}{5x}$
23. $\frac{2a}{b^{-3}}$ 24. $\frac{3x^{-1}}{2y^{-1}}$

25. $\dfrac{n^2}{3m^{-5}}$

26. $\dfrac{3a^{-4}}{5b^{-3}}$

27. $\dfrac{a^{-2}b^{-3}}{c^{-4}}$

28. $\dfrac{2m^{-2}n^3}{3p^{-4}q}$

29. $7x^{-3}y^{-5}$

30. $\dfrac{1}{a^{-3}b^{-5}}$

31. $\dfrac{6x^{-3}y^{-4}}{3z^{-2}w^3}$

32. $m^{-1} + n^{-1}$

33. $a^{-1} - b^{-1}$

34. $\dfrac{1}{x^{-1} + y^{-1}}$

35. $\dfrac{m^{-2} + n^{-2}}{5}$

36. $\dfrac{3}{x^{-2} + y^{-2}}$

37. $\dfrac{2}{x^{-2} + 1}$

38. $\dfrac{m^{-3} - n^{-3}}{4}$

39. $\dfrac{x^{-1} + 1}{y^{-1} + 1}$

Simplify these expressions using the properties of exponents. Final answers should not contain negative exponents. Assume the variables do not represent zero.

40. $m^{-3} \cdot m^{-4}$

41. $\dfrac{a^{-5}}{a^{-2}}$

42. $(2a^{-5})(7a^8)$

43. $(x^{-2})^3$

44. $\dfrac{15m^{-7}}{3m^{-2}}$

45. $(2a^{-3})(3a^{-1})(4a^2)$

46. $\dfrac{15x^2}{3x^{-5}}$

47. $\dfrac{24m^{-2}n^3}{8m^{-3}n^{-5}}$

48. $(3x^2y^{-1})(-2x^{-3}y^2)$

49. $\left(\dfrac{2a^3b^{-2}}{3c^{-4}}\right)^{-3}$

50. $\left(\dfrac{2m^{-1}}{3n^{-2}}\right)^2$

51. $(2a^{-1}b^2)^3(3ab^{-3})^2$

52. $(4x^2y)^{-1}(2xy^{-2})^3$

53. $\dfrac{(2a^{-2}b^{-1}c^3)^2(-3a^2b^{-3})^3}{(3a^3b^{-2}c)^2}$

54. $\dfrac{(3a^{-2}b^2c)^3(2ab^3c^{-4})^2}{(-a^{-2}b^3c^{-1})^3}$

55. $\dfrac{(-x^2yz^{-3})(-3xy^{-2})^3}{(x^{-3}y^{-2})(3x^{-1}y^2z^{-2})^2}$

Calculator Problems. With your instructor's approval, evaluate these problems with a calculator.

56. $(2.1)^{-2}$

57. $(0.604)^{-1}$

58. $(10.2)^{-3} + (8.9)^{-2}$

59. $(3.01)(2.63)^{-2}$

60. $\left(\dfrac{8.7}{6.2}\right)^{-2}$

61. $(0.03)^{-1} \cdot (0.05)^{-3}$

SCIENTIFIC NOTATION

6.2

An important application of exponentials is found in science. Some numbers occurring in science are extremely large or small and involve too many zeros to work with easily. For example, the speed of light is about 30,000,000,000 centimeters per second, and the wavelength of red light is about 0.000076 centimeters. Numbers of this type can be conveniently written using exponentials involving powers of ten. This method is called scientific notation. A number is written in **scientific notation** when it is expressed as the product of a number between 1 and 10, and a power of ten.

scientific notation

In scientific notation, 30,000,000,000 is written as 3×10^{10}. The number between 1 and 10 is 3. The power of ten is 10^{10}.

Example 1

In scientific notation, 0.000076 is written as 7.6×10^{-5}. The number between 1 and 10 is 7.6. The power of ten is 10^{-5}.

Example 2

To understand scientific notation, it is important that you be familiar with the powers of ten shown in Table 6.1.

TABLE 6.1 Powers of Ten

Positive Powers of 10	Negative Powers of 10
$10^1 = 10$	$10^{-1} = \dfrac{1}{10} = .1$
$10^2 = 100$	$10^{-2} = \dfrac{1}{100} = .01$
$10^3 = 1,000$	$10^{-3} = \dfrac{1}{1,000} = .001$
$10^4 = 10,000$	$10^{-4} = \dfrac{1}{10,000} = .0001$
$10^5 = 100,000$	$10^{-5} = \dfrac{1}{100,000} = .00001$
$10^6 = 1,000,000$	$10^{-6} = \dfrac{1}{1,000,000} = .000001$

Also remember that $10^0 = 1$.

From arithmetic you should recall these two rules:

(1) Each time a number is multiplied by 10, the decimal point is moved one place to the right.

   Examples: $2.6 \times 10 = 26$
   $2.6 \times 100 = 260$ (multiplied by 10 two times)
   $2.6 \times 10^3 = 2,600$ (multiplied by 10 three times)

(2)   Each time a number is divided by 10, the decimal point is moved one place to the left.

Examples: $4.3 \times 10^{-1} = \dfrac{4.3}{10} = 0.43$

$4.3 \times 10^{-2} = \dfrac{4.3}{100} = 0.043$ (divided by 10 two times)

$4.3 \times 10^{-3} = \dfrac{4.3}{1000} = 0.0043$ (divided by 10 three times)

Remembering the powers of ten and Rules 1 and 2 above, we can formulate a quick method of converting scientific notation to ordinary notation: If the exponent is positive, move the decimal point to the right. If the exponent is negative, move the decimal point to the left.

Example 3

Convert 2.39×10^3 to ordinary notation.

Solution: The exponent of 3 tells us to move the decimal point 3 places to the right.

$2.39 \times 10^3 = 2390.$

3 places

Example 4

Convert 5.91×10^{-2} to ordinary notation.

Solution: The exponent of -2 tells us to move the decimal point 2 places to the left.

$5.91 \times 10^{-2} = .0591$

2 places

Example 5

Convert 9.034×10^6 to ordinary notation.

Solution: Move the decimal point 6 places to the right.

$9.034 \times 10^6 = 9,034,000$

Example 6

Convert 1.6×10^{-5} to ordinary notation.

Solution: Move the decimal point 5 places to the left.

$1.6 \times 10^{-5} = .000016$

Example 7

Convert 8.05×10^0 to ordinary notation.

Solution: Since $10^0 = 1$, do not move the decimal point.

$$8.05 \times 10^0 = 8.05$$

Referring to Examples 4 and 6, you should see that a negative exponent in scientific notation indicates that the number is less than 1.

To convert ordinary notation to scientific notation, we reverse the procedure. That is, move the decimal point to create a number between 1 and 10. Then apply the corresponding exponent to form the correct power of 10. Remember, the exponent will be negative if the original number is less than 1.

Convert 0.000316 to scientific notation.   Example 8

Solution: Move the decimal point four places to form 3.16, which is between 1 and 10. The original number is less than 1, so the corresponding exponent is -4

$$.000316 = 3.16 \times 10^{-4}$$

4 places

Convert 4,380 to scientific notation.   Example 9

Solution: Move the decimal point three places to form 4.38, which is between 1 and 10. The original number is more than 1, so the corresponding exponent is 3

$$4.380 = 4.38 \times 10^3$$

3 places

Convert 6.5 to scientific notation.   Example 10

Solution: 6.5 is already between 1 and 10. Thus, the exponent is zero.

$$6.5 = 6.5 \times 10^0$$

Convert 805,600 to scientific notation.   Example 11

Solution: $805,600 = 8.056 \times 10^5$

Convert 0.013 to scientific notation.   Example 12

Solution: $0.013 = 1.3 \times 10^{-2}$

Some expressions can be easily evaluated by converting the numbers to scientific notation and applying the properties of exponents.

Example 13

Evaluate $\dfrac{(200)(0.006)}{3000}$

Solution: Convert to scientific notation. Then apply the addition and subtraction properties of exponents.

$$\frac{(200)(0.006)}{3000} = \frac{(2 \times 10^2)(6 \times 10^{-3})}{3 \times 10^3}$$

$$= \frac{12 \times 10^{-1}}{3 \times 10^3}$$

$$= 4 \times 10^{-1-3}$$

$$= 4 \times 10^{-4} \quad \text{or} \quad .0004$$

Exercise 6.2

Convert the following expressions from scientific notation to ordinary notation.

1. 5.91×10^2
2. 6.04×10^{-2}
3. 1.1×10^6
4. 9.6×10^{-4}
5. 4.871×10^0
6. 6.03×10^4
7. 9.2×10^1
8. 7.9×10^{-5}
9. 6×10^{-7}
10. 4×10^{12}
11. 5.06×10^{-4}
12. 1.21×10^{-1}
13. 3.8×10^0
14. 4.653×10^3

Convert the following expressions from ordinary notation to scientific notation.

15. 364
16. 5,300,000
17. 0.000402
18. 0.56
19. 2.91
20. 0.00000103
21. 651,000,000,000
22. 0.002031
23. 0.158
24. 63.9
25. 800
26. 9,000,000
27. 0.0000000094
28. 8,621,000,000,000

For each statement, express the numbers in scientific notation.

29. The sun is 93,000,000 miles from the earth.
30. The human brain contains more than 10,000,000,000 nerve cells.

31. The age of the earth is about 100,000,000,000,000,000 seconds.

32. Light waves travel 1 foot in approximately 0.000000001 seconds.

33. Light waves visible to the human eye have a wavelength between 0.00004 and 0.00007 centimeters.

For each statement, express the numbers in ordinary notation.

34. The budget of the United States government is more than 3.5×10^{11} dollars.

35. The average human lifetime is 1×10^9 seconds.

36. The population of the earth is more than 3.5×10^9.

37. The circumference of the earth is approximately 2.4×10^4 miles.

38. An angstrom is equivalent to 4×10^{-9} inches.

39. The wavelength of yellow-green light is approximately 6×10^{-5} centimeters.

Use scientific notation and properties of exponents to evaluate the following expressions.

40. $\dfrac{(0.08)(2000)}{0.0004}$

41. $\dfrac{(40)(6000)}{800}$

42. $\dfrac{(0.006)(0.8)}{(0.04)(20)}$

43. $\dfrac{(0.006)(80)}{(0.02)((0.3)}$

44. $\dfrac{8000}{(20)(0.1)}$

45. $\dfrac{(0.006)(0.9)}{(200)(0.003)}$

46. $\dfrac{(0.05)(20)}{50}$

47. $\dfrac{(300)(40)(0.8)}{(200)(30)}$

ROOTS AND RADICALS

6.3

In this section we discuss even roots and odd roots. Even roots include square roots, fourth roots, sixth roots, and so on. Odd roots include cube roots, fifth roots, seventh roots, and so on. Even roots are best characterized by studying square roots. A **square root** of a number is one of its two equal factors. Thus, x is a square root of N if $x \cdot x = N$.

square root

Here are some illustrations of square root.

Example 1

(a) One square root of 16 is 4, because $4 \cdot 4 = 16$. A second square root of 16 is -4, because $(-4)(-4) = 16$

(b) One square root of 36 is 6, because $6 \cdot 6 = 36$. Likewise, a second square root of 36 is -6, because $(-6)(-6) = 36$

(c) 0 is the only square root of 0 because $0 \cdot 0 = 0$, and these are the only two equal factors whose product is zero.

(d)   Since the product of two equal factors is never negative (but is always positive or zero), a negative number such as -25 cannot have a real square root.

Looking at the illustrations in Example 1, we state the following three properties.

Properties of Square Roots

(1)   Each positive number has two real square roots.
     Example: The two square roots of 25 are 5 and -5.

(2)   Zero has only one square root.
     Example: The square root of 0 is 0.

(3)   Negative numbers do not have real square roots.
     Example: -9 has no real square roots.

---

principal square roots

Nonnegative square roots are said to be **principal square roots**. They are symbolized by the radical sign, $\sqrt{\ }$. For example, the nonnegative square root of 4 is 2. Thus, $\sqrt{4} = 2$. The number inside the radical sign is called the **radicand**. The entire expression consisting of the radical sign and radicand is called a **radical**. To symbolize a negative square root, a minus sign is placed in front of the radical. Hence, $-\sqrt{4} = -2$.

radicand
radical

Example 2

Here are some illustrations of the radical sign.

(a)   The principal square root of 49 is symbolized by $\sqrt{49}$, where $\sqrt{49} = 7$
(b)   The negative square root of 49 is symbolized by $-\sqrt{49}$, where $-\sqrt{49} = -7$
(c)   The principal square root of 0 is symbolized by $\sqrt{0}$, where $\sqrt{0} = 0$
(d)   The principal square root of -16 does not exist as a real number. Thus, $\sqrt{-16}$ is not a real number.

Other even roots are similar to square roots. Thus, a fourth root of a number is one of its four equal factors. And a sixth root of a number is one of its six equal factors. Nonnegative even roots are called principal roots and are symbolized by a radical sign with the proper **index**. Thus, the principal fourth root is symbolized by a radical sign with an index of 4, $\sqrt[4]{\ }$. And the principal sixth root is symbolized by a radical sign with an index of 6, $\sqrt[6]{\ }$. Other even roots follow the same format.

index

Example 3

Here are some illustrations of even roots.

(a)   The two fourth roots of 16 are 2 and -2, because $2 \cdot 2 \cdot 2 \cdot 2 = 16$ and $(-2)(-2)(-2)(-2) = 16$
(b)   The principal fourth root of 16 is symbolized by $\sqrt[4]{16}$, where $\sqrt[4]{16} = 2$

(c) The negative fourth root of 16 is symbolized by $-\sqrt[4]{16}$, where $-\sqrt[4]{16} = -2$

(d) The principal sixth root of 729 is symbolized by $\sqrt[6]{729}$ where $\sqrt[6]{729} = 3$, because $3 \cdot 3 \cdot 3 \cdot 3 \cdot 3 \cdot 3 = 729$

Even roots obey the same properties as square roots.

Properties of Even Roots

The following properties are true for each even root greater than or equal to 2

(1) Each positive number has two even roots.
   Example: The two fourth roots of 81 are 3 and -3

(2) Zero has only one even root.
   Example: The sixth root of 0 is 0

(3) Negative numbers have no real even roots.
   Example: -16 has no real fourth roots.

Next we discuss odd roots. Odd roots include cube roots, fifth roots, seventh roots, and, so on. They are best characterized by examining cube roots. A **cube root** of a number is one of its three equal factors.

cube root

Here are some illustrations of cube roots.

Example 4

(a) The only real cube root of 8 is 2, because $2 \cdot 2 \cdot 2 = 8$
(b) The only real cube root of 27 is 3, because $3 \cdot 3 \cdot 3 = 27$
(c) The only real cube root of -27 is -3, because $(-3)(-3)(-3) = -27$
(d) The only real cube root of 0 is 0, because $0 \cdot 0 \cdot 0 = 0$

Looking at the illustrations in Example 4, we state the following four properties.

Properties of Cube Roots

(1) Every real number has exactly one real cube root.
   Examples: The cube root of 8 is 2; the cube root of 0 is 0; the cube root of -27 is -3

(2) The cube root of a positive number is a positive number.
   Example: The cube root of 8 is 2

(3) The cube root of a negative number is a negative number.
   Example: The cube root of -27 is -3

(4) The cube root of zero is zero.

Compare these with the properties of square roots and note the following differences:

(1)   Positive numbers have two square roots but only one cube root.

(2)   Negative numbers have no real square roots, but each has one cube root.

Since each number has only one real cube root, there is no need to define a principal cube root. Thus, the symbol for cube root is the radical sign with an index of 3, $\sqrt[3]{\ }$.

Example 5

Here are some illustrations of the cube root symbol.

(a) $\sqrt[3]{64} = 4$

(b) $\sqrt[3]{-27} = -3$

(c) $-\sqrt[3]{8} = -2$

(d) $-\sqrt[3]{-1} = 1$

(e) $\sqrt[3]{0} = 0$

Other odd roots are similar to cube roots. For example, the fifth root of a number is one of its five equal factors and is symbolized by a radical sign with an index of 5, $\sqrt[5]{\ }$. All odd roots follow the same pattern. Other odd roots obey the same properties as cube roots. Compare these with the properties for even roots listed on page 297.

Properties of Odd Roots

The following properties are true for each odd root.

(1)   Every real number has exactly one real odd root.
   Examples: $\sqrt[5]{32} = 2$; $\sqrt[7]{0} = 0$; $\sqrt[3]{-8} = -2$

(2)   An odd root of a positive number is a positive number.
   Example: $\sqrt[3]{64} = 4$

(3)   An odd root of a negative number is a negative number.
   Example: $\sqrt[3]{-1} = -1$

(4)   An odd root of zero is zero.
   Example: $\sqrt[5]{0} = 0$

Most of our work with radicals will involve perfect squares and perfect cubes. Thus, Tables 6.2 and 6.3 are particularly important and should be memorized.

When the radicand is not a perfect square or a perfect cube you can refer to the table in the back of this book entitled "Powers and Roots" (see page A-44). The answers there have been rounded to the nearest thousandth.

TABLE 6.2 Partial Listing of Perfect Squares	
Perfect Squares	Principal Square Root
0	0
1	1
4	2
9	3
16	4
25	5
36	6
49	7
64	8
81	9
100	10
121	11
144	12

TABLE 6.3 Partial Listing of Perfect Cubes	
Perfect Cubes	Cube Root
−216	−6
−125	−5
−64	−4
−27	−3
−8	−2
−1	−1
0	0
1	1
8	2
27	3
64	4
125	5
216	6

Use the Powers and Roots table to approximate the principal square root of 10.

Example 6

Solution: $\sqrt{10} \approx 3.162$ ($\approx$ means "approximately equal to")

Define the following terms.

Exercise 6.3

1. Square root.

2. Principal square root.

3. Cube root.

4. Fourth root.

True or false.

5. 3 is a square root of 9

6. −3 is a square root of 9

7. −3 is the principal square root of 9

8. $\sqrt{9} = 3$

9. $-\sqrt{25} = -5$

10. $\sqrt{-25} = -5$

11. 4 is the cube root of 64

12. $\sqrt[3]{-27} = 3$

13. $-\sqrt[3]{-8} = 2$

14. $\sqrt[4]{81} = 3$

15. $\sqrt[5]{-32} = -2$

16. Given the radical $\sqrt[3]{125}$, the index is 125

17. Given the radical $\sqrt[4]{16}$, the radicand is 16

18. Each negative number has exactly one real square root.

19. Each positive number has two cube roots.

20. Principal roots do not apply when the root is odd.

Evaluate the following expressions. If the root does not exist, write "not a real number."

21. $\sqrt{81}$ **22.** $\sqrt[3]{-125}$

23. $\sqrt[4]{16}$ **24.** $\sqrt{-9}$

25. $\sqrt[5]{32}$ **26.** $-\sqrt{16}$

27. $-\sqrt[3]{8}$ **28.** $\sqrt{144}$

29. $\sqrt{1}$ **30.** $\sqrt[3]{-1}$

31. $\sqrt[4]{0}$ **32.** $\sqrt{121}$

33. $-\sqrt[4]{81}$ **34.** $\sqrt[6]{-1}$

35. $-\sqrt[3]{-216}$ **36.** $-\sqrt{100}$

37. $\sqrt[3]{64}$ **38.** $\sqrt{64}$

39. $-\sqrt{-16}$ **40.** $\sqrt[6]{64}$

Use the Powers and Roots table to evaluate the following expressions.

41. $\sqrt{2}$ **42.** $\sqrt[3]{2}$

43. $\sqrt{33}$ **44.** $\sqrt[3]{-5}$

45. $\sqrt[3]{21}$ **46.** $\sqrt{13}$

47. $\sqrt[3]{75}$ **48.** $\sqrt{92}$

49. $\sqrt[3]{-84}$ **50.** $\sqrt{23}$

Find the square roots of these decimals using trial and error.

Example $\sqrt{0.25} = 0.5$ because $(0.5)(0.5) = 0.25$

51. $\sqrt{0.36}$ **52.** $\sqrt{0.04}$

53. $\sqrt{0.0081}$ **54.** $\sqrt{1.44}$

55. $\sqrt{26.01}$ **56.** $\sqrt{0.5041}$

Calculator Problems. With your instructor's approval, evaluate the problems using a calculator.

57. $\sqrt{305}$ **58.** $\sqrt{0.621}$

59. $\sqrt[3]{12.6}$ **60.** $\sqrt[4]{2048}$

61. $3.2\sqrt{14.8}$ **62.** $6.5\sqrt[5]{9.8}$

63. $\sqrt[5]{17} + \sqrt[8]{84}$ **64.** $\dfrac{\sqrt[3]{6.21}}{\sqrt[5]{17.8}}$

RATIONAL EXPONENTS

In this section we define rational exponents. Remember, rational numbers have a form of $\frac{m}{n}$ where m and n are integers and $n \neq 0$. Thus, we will study exponentials such as $4^{1/2}$, $9^{-1/2}$, $8^{2/3}$, and $25^{-3/2}$. Assuming that the properties of exponents hold true for rational numbers, we will first define exponents of the type $1/n$ where n is an integer and $n \geq 2$. This will give meaning to exponentials such as $4^{1/2}$, $27^{1/3}$, and $16^{1/4}$.

In general, consider the exponential $x^{1/n}$. Multiplying n factors of $x^{1/n}$ we have the following:

$$\overbrace{x^{1/n} \cdot x^{1/n} \cdot x^{1/n} \cdots x^{1/n}}^{n \text{ factors}} = (x^{1/n})^n$$
$$= x^{n/n}$$
$$= x^1$$
$$= x$$

This shows that $x^{1/n}$ is one of the n equal factors of x. Therefore, it is reasonable to define $x^{1/n}$ as the nth root of x. If n is an even integer, then $x^{1/n}$ is defined to be the principal nth root of x. This gives the following definition.

Definition of an Exponent of the Form 1/n

$x^{1/n} = \sqrt[n]{x}$ where n is an integer and $n \geq 2$

If n is even, then $x \geq 0$

This definition allows us to convert exponential form to a radical form or radical form to exponential form.

These equations show how to convert exponentials to radical form.   **Example 1**
(a) $x^{1/2} = \sqrt{x}$ where $x \geq 0$
(b) $y^{1/3} = \sqrt[3]{y}$
(c) $(2m)^{1/5} = \sqrt[5]{2m}$
(d) $2m^{1/5} = 2\sqrt[5]{m}$
(e) $(x + y)^{1/2} = \sqrt{x + y}$ where $x + y \geq 0$

These equations show how to convert radical form to exponential form.   **Example 2**
(a) $\sqrt{m} = m^{1/2}$ where $m \geq 0$
(b) $\sqrt[3]{5y} = (5y)^{1/3}$
(c) $3\sqrt[4]{x} = 3x^{1/4}$ where $x \geq 0$
(d) $\sqrt[5]{m + n} = (m + n)^{1/5}$

The definition also enables us to evaluate certain expressions.

Example 3

These equations illustrate how to evaluate exponentials.
(a) $4^{1/2} = \sqrt{4} = 2$
(b) $27^{1/3} = \sqrt[3]{27} = 3$
(c) $(-8)^{1/3} = \sqrt[3]{-8} = -2$
(d) $16^{1/4} = \sqrt[4]{16} = 2$
(e) $(-25)^{1/2} = \sqrt{-25} = $ no real number
(f) $-25^{1/2} = -\sqrt{25} = -5$

Using the previous definition, we can now formulate a definition for rational exponents of the form m/n where m and n are integers and $n \geq 2$. Assuming that the properties of exponents hold true for rational exponents, we use the power to a power property to write the following:

$$x^{m/n} = (x^m)^{1/n} \quad \text{or} \quad (x^{1/n})^m$$

Converting to radicals, we have:

$$x^{m/n} = \sqrt[n]{x^m} \quad \text{or} \quad (\sqrt[n]{x})^m$$

To avoid even roots of negative numbers, the base x must be nonnegative when n is an even integer. Thus, we arrive at the following definition.

Definition of Rational Exponents

If m and n are integers where $n \geq 2$, then $x^{m/n} = \sqrt[n]{x^m}$ or $(\sqrt[n]{x})^m$
If n is even, then $x \geq 0$

The definition lets us convert back and forth between exponential and radical form. It is easy to convert if you remember that the numerator of a rational exponent indicates the power, while the denominator indicates the root (index).

$$\underset{\text{root}}{\overset{\text{power}}{x^{m/n}}} = \underset{\text{power}}{\overset{\text{root}}{\sqrt[n]{x^m}}} = \overset{\text{root}}{(\sqrt[n]{x})^m} \leftarrow \text{power}$$

Example 4

Here are some illustrations of converting exponentials to radical form.
(a) $x^{2/3} = \sqrt[3]{x^2}$ or $(\sqrt[3]{x})^2$
(b) $m^{3/4} = \sqrt[4]{m^3}$ or $(\sqrt[4]{m})^3$ where $m \geq 0$

Example 5

Here are some illustrations of converting radical form to exponential form.

(a) $\sqrt[3]{y^2} = y^{2/3}$

(b) $(\sqrt[3]{y})^2 = y^{2/3}$

(c) $\sqrt[6]{b^5} = b^{5/6}$ where $b \geq 0$

We can also use the definition of rational exponents to evaluate certain exponentials.

Evaluate $8^{2/3}$ **Example 6**

Solution (a): $8^{2/3} = \sqrt[3]{8^2}$

$$= \sqrt[3]{64}$$

$$= 4$$

Solution (b): $8^{2/3} = (\sqrt[3]{8})^2$

$$= (2)^2$$

$$= 4$$

Either solution is correct. However, it is usually easier to determine the root first and the power second, as shown in solution (b).

Evaluate $16^{3/4}$ **Example 7**

Solution: $16^{3/4} = (\sqrt[4]{16})^3$

$$= 2^3$$

$$= 8$$

Evaluate $(-32)^{3/5}$ **Example 8**

Solution: $(-32)^{3/5} = (\sqrt[5]{-32})^3$

$$= (-2)^3$$

$$= -8$$

Evaluate $(-16)^{3/4}$ **Example 9**

Solution: There is no answer. The definition does not hold when the root is even and the base is negative.

Exponentials involving a negative exponent are evaluated by changing the negative exponent to positive according to this definition from section 6.1:

$$x^{-n} = \frac{1}{x^n} \qquad \text{where } x \neq 0$$

Evaluate $9^{-1/2}$ **Example 10**

Solution: Change to a positive exponent; then evaluate.

$$9^{-1/2} = \frac{1}{9^{1/2}}$$

$$= \frac{1}{\sqrt{9}}$$

$$= \frac{1}{3}$$

Example 11   Evaluate $16^{-3/2}$

Solution: Change to a positive exponent; then evaluate.

$$16^{-3/2} = \frac{1}{16^{3/2}}$$

$$= \frac{1}{(\sqrt{16})^3}$$

$$= \frac{1}{4^3}$$

$$= \frac{1}{64}$$

Exercise 6.4   Convert the following exponentials to radical form. Assume that the base is greater than or equal to zero whenever the root is even.

1. $a^{1/2}$ **2.** $b^{1/3}$

3. $m^{1/5}$ **4.** $(3x)^{1/2}$

5. $3x^{1/2}$ **6.** $(m + n)^{1/3}$

7. $y^{2/3}$ **8.** $n^{4/5}$

9. $z^{3/2}$ **10.** $(5m)^{2/5}$

11. $5m^{2/5}$ **12.** $(x + y)^{3/4}$

Convert the following expressions to exponential form. Assume that each base is greater than or equal to zero.

13. $\sqrt{b}$ **14.** $\sqrt[3]{m}$

15. $\sqrt[5]{z}$ **16.** $7\sqrt{n}$

17. $\sqrt{7n}$ **18.** $\sqrt{x^3}$

19. $(\sqrt{x})^3$ **20.** $2\sqrt[3]{x^2}$

21. $5\sqrt[5]{a^4}$

Evaluate these exponentials.

22. $36^{1/2}$

23. $-36^{1/2}$

24. $(-36)^{1/2}$

25. $36^{-1/2}$

26. $8^{1/3}$

27. $(-8)^{1/3}$

28. $8^{-1/3}$

29. $0^{1/2}$

30. $1^{1/3}$

31. $\left(\dfrac{9}{16}\right)^{1/2}$

32. $\left(\dfrac{1}{8}\right)^{1/3}$

33. $\left(\dfrac{27}{64}\right)^{1/3}$

34. $\left(\dfrac{25}{36}\right)^{-1/2}$

35. $\left(\dfrac{1}{9}\right)^{-1/2}$

36. $\left(\dfrac{125}{8}\right)^{-1/3}$

37. $16^{3/2}$

38. $-16^{3/2}$

39. $(-16)^{3/2}$

40. $16^{-3/2}$

41. $0^{5/6}$

42. $1^{4/5}$

43. $(-64)^{2/3}$

44. $(-27)^{2/3}$

45. $(-27)^{-2/3}$

46. $(-1)^{3/5}$

47. $100^{1/2}$

48. $100^{3/2}$

49. $1000^{1/3}$

50. $1000^{2/3}$

51. $1000^{-2/3}$

52. $4^{-3/2}$

53. $32^{4/5}$

54. $(-32)^{4/5}$

55. $(-32)^{-4/5}$

56. $(-8)^{5/3}$

57. $81^{-3/4}$

58. $\left(\dfrac{4}{9}\right)^{3/2}$

59. $\left(-\dfrac{8}{27}\right)^{-2/3}$

60. $\left(\dfrac{16}{81}\right)^{3/4}$

Calculator Problems. With your instructor's approval evaluate these problems with a calculator.

61. $(2.62)^{2/3}$

62. $(0.512)^{3/2}$

63. $(1.08)^{-4/5}$

64. $(16.3)^{-6/5}$

65. $(1.23)^{1/3} + (2.64)^{1/4}$

66. $5.1(6.2)^{2/5}$

67. $(19.8)^{2.1}$

68. $(0.173)^{-3.6}$

SIMPLIFYING RADICALS

6.5

In this section we develop an important property called the Basic Property of Radicals. It will be used to simplify radical expressions and to multiply radical expressions. To derive this property we convert a radical expression containing a product to a product containing radicals. This is accomplished, as shown in the following, by first converting the radical expression to an equivalent expression having a rational exponent. Then the distributive property of exponents is used to share the exponent with each factor in the group. Thus,

$$\sqrt[n]{a \cdot b} = (a \cdot b)^{1/n} \quad \text{[converting to a rational exponent]}$$
$$= a^{1/n} \cdot b^{1/n} \quad \text{[distributive property of exponents]}$$
$$= \sqrt[n]{a} \cdot \sqrt[n]{b} \quad \text{[converting to radicals]}$$

We now state the Basic Property of Radicals with appropriate restrictions.

The Basic Property of Radicals
$$\sqrt[n]{a \cdot b} = \sqrt[n]{a} \cdot \sqrt[n]{b}$$

where (1) n is even and $a, b \geq 0$
(2) n is odd and a, b represent any real numbers.

This property states that the radical of a product is equal to the product of the radicals. The radicals, must, of course, have the same index.

Example 1

Change $\sqrt[3]{5x}$ to a product of radicals.

Solution: Apply the basic property of radicals.

$$\sqrt[3]{5x} = \sqrt[3]{5} \cdot \sqrt[3]{x}$$

Example 2

Simplify $\sqrt{4 \cdot 5}$ by changing to a product of radicals.

Solution: $\sqrt{4 \cdot 5} = \sqrt{4} \cdot \sqrt{5}$
$$= 2 \cdot \sqrt{5}$$

Example 3

Simplify $\sqrt[3]{27 \cdot 2}$ by changing to a product of radicals.

Solution: $\sqrt[3]{27 \cdot 2} = \sqrt[3]{27} \cdot \sqrt[3]{2}$
$$= 3 \cdot \sqrt[3]{2}$$

To *simplify* a radical means to write an equivalent form with the smallest possible radicand.

Example 4

Simplify $\sqrt{18}$

Solution: Factor the radicand 18 into two factors, one of which is the largest possible perfect square. Then use the basic property of radicals. Notice that 18 is evenly divisible by the perfect square 9

$$\sqrt{18} = \sqrt{9 \cdot 2}$$
$$= \sqrt{9} \cdot \sqrt{2}$$
$$= 3 \cdot \sqrt{2}$$

Simplify $\sqrt[3]{54}$ **Example 5**

Solution: Factor the radicand 54 into two factors, one of which is the largest possible perfect cube. Then use the basic property of radicals. Notice that 54 is evenly divisible by the perfect cube 27

$$\sqrt[3]{54} = \sqrt[3]{27 \cdot 2}$$
$$= \sqrt[3]{27} \cdot \sqrt[3]{2}$$
$$= 3 \cdot \sqrt[3]{2}$$

Simplify $2\sqrt{75}$ **Example 6**

Solution: 75 is evenly divisible by the perfect square 25

$$2\sqrt{75} = 2 \cdot \sqrt{25 \cdot 3}$$
$$= 2 \cdot \sqrt{25} \cdot \sqrt{3}$$
$$= 2 \cdot 5 \cdot \sqrt{3}$$
$$= 10 \cdot \sqrt{3}$$

Simplify $3\sqrt[3]{-40}$ **Example 7**

Solution: $3\sqrt[3]{-40} = 3 \cdot \sqrt[3]{-8 \cdot 5}$
$$= 3 \cdot \sqrt[3]{-8} \cdot \sqrt[3]{5}$$
$$= 3 \cdot (-2) \cdot \sqrt[3]{5}$$
$$= -6 \cdot \sqrt[3]{5}$$

Simplify $5\sqrt{13}$ **Example 8**

Solution: 13 is not even evenly divisible by a perfect square. Hence, $5\sqrt{13}$ cannot be simplified.

Radicals often involve variables. Examples 9 and 10 lead to an important property that will help us to simplify radicals containing variables.

Example 9

Simplify $\sqrt[3]{x^3}$

Solution: $\sqrt[3]{x^3} = x$,   because $x \cdot x \cdot x = x^3$

Example 10

Simplify $\sqrt{x^2}$ where $x \geq 0$

Solution: $\sqrt{x^2} = x$ where $x \geq 0$,   because $x \cdot x = x^2$

The restriction of $x \geq 0$ is important because the even index calls for the principal (nonnegative) root.

Generalizing, we state the following property.

$$\sqrt[n]{x^n} = x$$

where (1) n is odd and x represents any real number
      (2) n is even and $x \geq 0$

Combining this property with the basic property of radicals lets us simplify many radicals containing variables.

Example 11

Simplify $2\sqrt[3]{x^4}$

Solution: $2\sqrt[3]{x^4} = 2 \cdot \sqrt[3]{x^3 \cdot x}$
$\qquad\qquad\quad = 2 \cdot \sqrt[3]{x^3} \cdot \sqrt[3]{x}$
$\qquad\qquad\quad = 2 \cdot \;\; x \cdot \sqrt[3]{x}$ or $2x\sqrt[3]{x}$

Example 12

Simplify $2\sqrt{98a^3b}$ where $a, b \geq 0$

Solution: $2\sqrt{98a^3b} = 2 \cdot \sqrt{49 \cdot 2 \cdot a^2 \cdot a \cdot b}$
$\qquad\qquad\quad\;\; = 2 \cdot \sqrt{49 \cdot a^2 \cdot 2 \cdot a \cdot b}$
$\qquad\qquad\quad\;\; = 2 \cdot \sqrt{49} \cdot \sqrt{a^2} \cdot \sqrt{2ab}$
$\qquad\qquad\quad\;\; = 2 \cdot \;\; 7 \cdot \;\; a \cdot \sqrt{2ab}$
$\qquad\qquad\quad\;\; = 14a\sqrt{2ab}$

Use the basic property of radicals to simplify the following expressions.   **Exercise 6.5**

1. $\sqrt{16 \cdot 3}$

2. $\sqrt[3]{27 \cdot 2}$

3. $\sqrt{9 \cdot 5}$

4. $\sqrt[3]{-64 \cdot 3}$

5. $\sqrt[4]{16 \cdot 7}$

6. $\sqrt[5]{-32 \cdot 7}$

7. $\sqrt{8}$

8. $\sqrt[3]{16}$

9. $\sqrt[3]{-16}$

10. $\sqrt{12}$

11. $\sqrt{27}$

12. $\sqrt[3]{54}$

13. $\sqrt[3]{128}$

14. $\sqrt{200}$

15. $\sqrt{48}$

16. $3\sqrt{72}$

17. $2\sqrt[3]{-24}$

18. $3\sqrt[3]{128}$

19. $3\sqrt{48}$

20. $2\sqrt[3]{40}$

21. $3\sqrt{8}$

22. $4\sqrt{150}$

23. $3\sqrt[3]{-54}$

24. $3\sqrt{75}$

25. $5\sqrt{28}$

26. $2\sqrt{18}$

27. $4\sqrt[3]{125}$

28. $3\sqrt[4]{32}$

29. $3\sqrt[5]{64}$

30. $2\sqrt[4]{162}$

31. $-4\sqrt{45}$

32. $-3\sqrt{80}$

33. $-4\sqrt[3]{-80}$

34. $5\sqrt{288}$

35. $3\sqrt{200}$

36. $6\sqrt[3]{56}$

37. $3\sqrt{20}$

38. $2\sqrt{1100}$

39. $4\sqrt[3]{2000}$

Simplify the following expressions. If the root is even, assume that the variables represent nonnegative numbers.

40. $\sqrt[3]{x^3}$

41. $\sqrt[5]{m^5}$

42. $\sqrt[6]{y^6}$

43. $\sqrt[3]{x^5}$

44. $\sqrt[7]{n^{10}}$

45. $\sqrt{a^3}$

46. $\sqrt{y^4}$

47. $\sqrt[3]{x^6}$

48. $\sqrt[5]{m^{11}}$

49. $\sqrt{x^3y^2}$

50. $\sqrt[3]{m^4n}$

51. $\sqrt[4]{a^4b^5c^6}$

52. $\sqrt[3]{x^3y^5z^6}$

53. $\sqrt{4m^3n^2}$

54. $\sqrt{12a^3}$

55. $\sqrt[3]{16x^3y^2}$

56. $\sqrt{12a^3b^2}$

57. $\sqrt{36n^4}$

58. $\sqrt[3]{54a^5b^4}$

59. $3\sqrt{8x^2y}$

60. $2\sqrt[3]{80a^6b^4}$

61. $5x\sqrt{28x^3y}$

62. $3\sqrt{200x^4y^3}$

63. $5\sqrt[3]{-80m^9n^7}$

64. $3y\sqrt{72y^5z^3}$

ADDITION AND SUBTRACTION
OF RADICAL EXPRESSIONS

6.6

like radicals

You should recall that like terms can be added or subtracted by combining their numerical coefficients. Thus, $2x + 5x = 7x$ where x represents any real number. In particular, x could represent a radical such as $\sqrt{3}$; then $2\sqrt{3} + 5\sqrt{3} = 7\sqrt{3}$. We call $2\sqrt{3}$ and $5\sqrt{3}$ *like radicals*. Thus, radicals having the same radicand and the same index are called **like radicals**. Like radicals can be added or subtracted by combining their numerical coefficients.

Example 1

Combine these radicals, where possible.

(a) $3\sqrt{7} + 9\sqrt{7} = 12\sqrt{7}$

(b) $9\sqrt[3]{2} - 4\sqrt[3]{2} = 5\sqrt[3]{2}$

(c) $6\sqrt{7} + 3\sqrt{5}$ cannot be combined because the radicands are different.

(d) $2\sqrt[3]{6} - 8\sqrt[4]{6}$ cannot be combined because each radical has a different index.

Example 2

Combine $5\sqrt[3]{4} + 3\sqrt[3]{11} + 2\sqrt[3]{4} - 6\sqrt[3]{11}$

Solution: Combine only the like radicals.

$$5\sqrt[3]{4} + 3\sqrt[3]{11} + 2\sqrt[3]{4} - 6\sqrt[3]{11} = 7\sqrt[3]{4} - 3\sqrt[3]{11}$$

It is important to simplify radicals before attempting to add or subtract.

Example 3

Combine $2\sqrt{75} + 5\sqrt{27}$

Solution: Simplify each radical; then combine like radicals.

$$
\begin{aligned}
2\sqrt{75} + 5\sqrt{27} &= 2 \cdot \sqrt{25 \cdot 3} + 5 \cdot \sqrt{9 \cdot 3} \\
&= 2 \cdot \sqrt{25} \cdot \sqrt{3} + 5 \cdot \sqrt{9} \cdot \sqrt{3} \\
&= 2 \cdot 5 \cdot \sqrt{3} + 5 \cdot 3 \cdot \sqrt{3} \\
&= 10\sqrt{3} + 15\sqrt{3} \\
&= 25\sqrt{3}
\end{aligned}
$$

Example 4

Combine $\sqrt[3]{16} - 5\sqrt[3]{54}$

Solution: Simplify each radical; then combine like radicals.

$$
\begin{aligned}
\sqrt[3]{16} - 5\sqrt[3]{54} &= \sqrt[3]{8 \cdot 2} - 5 \cdot \sqrt[3]{27 \cdot 2} \\
&= \sqrt[3]{8} \cdot \sqrt[3]{2} - 5 \cdot \sqrt[3]{27} \cdot \sqrt[3]{2}
\end{aligned}
$$

$$= 2 \cdot \sqrt[3]{2} - 5 \cdot 3 \cdot \sqrt[3]{2}$$
$$= 2\sqrt[3]{2} - 15\sqrt[3]{2}$$
$$= -13\sqrt[3]{2}$$

Combine $3\sqrt{24} + \sqrt{6} - 2\sqrt{27} + 5\sqrt{12}$

Example 5

Solution: Simplify each radical; then combine only the like radicals.

$$3\sqrt{24} + \sqrt{6} - 2\sqrt{27} + 5\sqrt{12} = 3 \cdot \sqrt{4 \cdot 6} + \sqrt{6} - 2 \cdot \sqrt{9 \cdot 3} + 5 \cdot \sqrt{4 \cdot 3}$$
$$= 3 \cdot 2 \cdot \sqrt{6} + 1 \cdot \sqrt{6} - 2 \cdot 3 \cdot \sqrt{3} + 5 \cdot 2 \cdot \sqrt{3}$$
$$= 6\sqrt{6} + 1\sqrt{6} - 6\sqrt{3} + 10\sqrt{3}$$
$$= 7\sqrt{6} + 4\sqrt{3}$$

Simplify and combine like radicals.

Exercise 6.6

1. $2\sqrt{5} + 4\sqrt{5}$

2. $2\sqrt{3} + \sqrt{3}$

3. $5\sqrt[3]{6} - 9\sqrt[3]{6}$

4. $\sqrt[3]{7} - 5\sqrt[3]{7}$

5. $\sqrt{5} - 3\sqrt{5} + 4\sqrt{5}$

6. $2\sqrt{7} - 5\sqrt{7} + 2\sqrt{7}$

7. $\sqrt[3]{9} - 3\sqrt[3]{9} + 4\sqrt[3]{9}$

8. $\sqrt[5]{6} + 4\sqrt[5]{6} - \sqrt[5]{6}$

9. $3\sqrt{2} + 4\sqrt{3} + \sqrt{2}$

10. $5\sqrt{11} - 7\sqrt{3} + \sqrt{11} - 2\sqrt{3}$

11. $\sqrt[3]{5} + 2\sqrt{3} - 4\sqrt[3]{5} + \sqrt{3}$

12. $5\sqrt[4]{10} + 3\sqrt[3]{2} - 6\sqrt[4]{10} - \sqrt[3]{2}$

13. $3\sqrt{5} + 6\sqrt{7} - \sqrt{5} + 2\sqrt{7}$

14. $\sqrt{8} + \sqrt{18}$

15. $\sqrt[3]{16} - \sqrt[3]{54}$

16. $3\sqrt{8} + 7\sqrt{50}$

17. $4\sqrt{3} + 7\sqrt{27}$

18. $3\sqrt{27} - \sqrt{75}$

19. $5\sqrt[3]{54} + 2\sqrt[3]{128}$

20. $5\sqrt{18} + \sqrt{8}$

21. $\sqrt{3} - 2\sqrt{27}$

22. $3\sqrt{72} - 2\sqrt{50}$

23. $5\sqrt{18} + 3\sqrt{32}$

24. $2\sqrt{75} - 3\sqrt{12}$

25. $3\sqrt[3]{-24} - 5\sqrt[3]{81}$

26. $2\sqrt[3]{250} - 6\sqrt[3]{128}$

27. $3\sqrt{20} - 5\sqrt{45} + 3\sqrt{5}$

28. $\sqrt{12} - 5\sqrt{3} - 4\sqrt{27}$

29. $2\sqrt{54} - 3\sqrt{20} + 9\sqrt{24}$

30. $5\sqrt[3]{54} + 2\sqrt[3]{128} - \sqrt[3]{16}$

31. $5\sqrt{12} - \sqrt{27} + 2\sqrt{48}$

32. $2\sqrt{8} + 3\sqrt{18} - \sqrt{32}$

33. $3\sqrt{24} - \sqrt{128} + 3\sqrt{6} - 2\sqrt{54}$

34. $4\sqrt{8} - 2\sqrt{12} + \sqrt{50} + 3\sqrt{75}$

35. $3\sqrt[3]{16} - \sqrt[3]{54} + \sqrt{2} + \sqrt{8}$

36. $4\sqrt{18} - 7\sqrt{98} + 5\sqrt[3]{16} - \sqrt[3]{54}$

Simplify and combine like radicals. If the root is even, assume that each variable represents nonnegative numbers.

37. $\sqrt{25x} - \sqrt{36x} + \sqrt{49x}$

38. $\sqrt{9m} - \sqrt{25m} - \sqrt{81m}$

39. $\sqrt{8y^2} + 3y\sqrt{2}$

40. $2\sqrt{45x^2} + 3x\sqrt{20}$

41. $\sqrt[3]{8m} + \sqrt[3]{27m} - \sqrt[3]{64m}$

42. $6\sqrt[3]{2x} + 3\sqrt[3]{16x}$

43. $2\sqrt{8x} + 5\sqrt{27y} - 3\sqrt{32x} - \sqrt{48y}$
44. $\sqrt{8x} - \sqrt{32x^3} + \sqrt{2x}$
45. $5\sqrt{3m} + 7\sqrt{3m} - 8\sqrt{2n} + 10\sqrt{2n}$
46. $2\sqrt{50a} + 3\sqrt{32a} - 3\sqrt{2a}$
47. $2\sqrt{16x^3} - 5\sqrt{8x^2} + 7\sqrt{64x^3} + 12\sqrt{32x^2}$
48. $2\sqrt[3]{54x^4} + 3\sqrt[3]{128x^4} - 5x\sqrt[3]{2x}$
49. $\sqrt{75a^2} - 2\sqrt[3]{16a^4} + 2\sqrt{48a^2} - 3\sqrt[3]{54a^4}$

MULTIPLICATION AND DIVISION OF RADICALS

6.7

In section 6.5 we discussed the basic property of radicals, which states that:

$$\sqrt[n]{ab} = \sqrt[n]{a} \cdot \sqrt[n]{b}$$

where (1) n is even and a, $b \geq 0$
(2) n is odd and a, b represent any two real numbers.

Reversing the property, we have:

$$\sqrt[n]{a} \cdot \sqrt[n]{b} = \sqrt[n]{ab}$$

where (1) n is even and a, $b \geq 0$
(2) n is odd and a, b represent any two real numbers.

Written this way, the basic property of radicals gives us the following rule for multiplying radicals.

Multiplication of Radicals
Radicals having the same index can be multiplied by finding the product of the radicands. This product is then placed under the corresponding radical sign.

Example 1

Here are some examples of multiplying radicals.
(a) $\sqrt{6} \cdot \sqrt{5} = \sqrt{6 \cdot 5} = \sqrt{30}$
(b) $\sqrt[3]{5} \cdot \sqrt[3]{7} = \sqrt[3]{5 \cdot 7} = \sqrt[3]{35}$
(c) $\sqrt{3} \cdot \sqrt[3]{5}$ cannot be multiplied by this property because the radicals have a different index.

This property can be combined with the special multiplication techniques we studied in chapter 5 to let us multiply a variety of radical expressions.

Multiply $(2\sqrt{5})(3\sqrt{7})$ **Example 2**

Solution: Multiply the coefficients. Then multiply the radicals.

$$(2\sqrt{5})(3\sqrt{7}) = 2 \cdot 3 \cdot \sqrt{5 \cdot 7}$$
$$= 6\sqrt{35}$$

Multiply $(2\sqrt[3]{18x})(4\sqrt[3]{3x})$ **Example 3**

Solution: Multiply the coefficients and multiply the radicals. Always simplify the answer, if possible.

$$(2\sqrt[3]{18x})(4\sqrt[3]{3x}) = 2 \cdot 4 \cdot \sqrt[3]{(18x)(3x)}$$
$$= 8 \cdot \sqrt[3]{54x^2} \qquad \text{[simplify the radical]}$$
$$= 8 \cdot \sqrt[3]{27 \cdot 2x^2}$$
$$= 8 \cdot 3 \cdot \sqrt[3]{2x^2}$$
$$= 24\sqrt[3]{2x^2}$$

Multiply $\sqrt{2}(\sqrt{5} + 2\sqrt{7})$ **Example 4**

Solution: Use the distributive property of multiplication to share $\sqrt{2}$ with each term of the binomial.

$$\sqrt{2}(\sqrt{5} + 2\sqrt{7}) = \sqrt{2} \cdot \sqrt{5} + \sqrt{2} \cdot 2\sqrt{7}$$
$$= \sqrt{10} + 2\sqrt{14}$$

Multiply $(5 + \sqrt{3})(5 - \sqrt{3})$ **Example 5**

Solution: Use the "sum times a difference" shortcut.

$$(5 + \sqrt{3})(5 - \sqrt{3}) = 5 \cdot 5 - \sqrt{3} \cdot \sqrt{3}$$
$$= 25 - \sqrt{9}$$
$$= 25 - 3$$
$$= 22$$

Multiply $(5 - \sqrt{3})(3 + 2\sqrt{3})$.   **Example 6**

Solution: Use the FOIL shortcut.

$$(5 - \sqrt{3})(3 + 2\sqrt{3}) = \overset{F}{5 \cdot 3} + \overset{O}{5 \cdot 2\sqrt{3}} - \overset{I}{3 \cdot \sqrt{3}} - \overset{L}{\sqrt{3} \cdot 2\sqrt{3}}$$
$$= 15 + 10\sqrt{3} - 3\sqrt{3} - 2\sqrt{9}$$
$$= 15 + 10\sqrt{3} - 3\sqrt{3} - 2 \cdot 3$$
$$= 15 + 7\sqrt{3} - 6$$
$$= 9 + 7\sqrt{3}$$

Next we develop a property which will allow us to divide two radical expressions. We use our knowledge of rational exponents, as follows:

$$\frac{\sqrt[n]{a}}{\sqrt[n]{b}} = \frac{a^{1/n}}{b^{1/n}} \qquad \text{[converting to rational exponents]}$$

$$= \left(\frac{a}{b}\right)^{1/n} \qquad \text{[distributive property of exponents]}$$

$$= \sqrt[n]{\frac{a}{b}} \qquad \text{[converting to a radical]}$$

We now state the Division Property of Radicals with the appropriate restrictions.

The Division Property of Radicals

$$\frac{\sqrt[n]{a}}{\sqrt[n]{b}} = \sqrt[n]{\frac{a}{b}}$$

where (1) n is even, $a \geq 0$, and $b > 0$
        (2) n is odd and $b \neq 0$

In words, this property says the quotient of two radicals is the radical of the quotient. It provides us with the following rule for dividing radicals, which is similar to the rule for multiplying radicals.

Division of Radicals

Radicals having the same index can be divided by taking the quotient of the radicands. This quotient is then placed under the corresponding radical sign.

Example 7

Here are some illustrations of dividing radicals.

(a) $\dfrac{\sqrt{10}}{\sqrt{2}} = \sqrt{\dfrac{10}{2}} = \sqrt{5}$

(b) $\dfrac{\sqrt[3]{30}}{\sqrt[3]{5}} = \sqrt[3]{\dfrac{30}{5}} = \sqrt[3]{6}$

(c) $\dfrac{\sqrt[3]{10}}{\sqrt{2}}$ cannot be divided by this property because the radicals have a different index.

Example 8

Divide $\dfrac{12\sqrt{15}}{6\sqrt{3}}$

Solution: Divide the numerical coefficients. Then divide the radicals.

$$\frac{12\sqrt{15}}{6\sqrt{3}} = 2 \cdot \sqrt{\frac{15}{3}}$$

$$= 2\sqrt{5}$$

The division property of radicals can be reversed to simplify a radical involving a quotient. Thus, $\sqrt[n]{\dfrac{a}{b}} = \dfrac{\sqrt[n]{a}}{\sqrt[n]{b}}$ where $b \neq 0$.

Simplify $\sqrt{\dfrac{36}{25}}$

Example 9

Solution: $\sqrt{\dfrac{36}{25}} = \dfrac{\sqrt{36}}{\sqrt{25}} = \dfrac{6}{5}$

Simplify $\sqrt{\dfrac{5}{16}}$

Example 10

Solution: $\sqrt{\dfrac{5}{16}} = \dfrac{\sqrt{5}}{\sqrt{16}} = \dfrac{\sqrt{5}}{4}$

Multiply the following. If possible, simplify the answer. Assume that variables used with even roots represent nonnegative numbers.

Exercise 6.7

1. $\sqrt{5} \cdot \sqrt{6}$
3. $\sqrt{3} \cdot \sqrt{15}$
5. $\sqrt{2} \cdot \sqrt{5} \cdot \sqrt{3}$
7. $\sqrt{7} \cdot \sqrt{7}$
9. $\sqrt{x} \cdot \sqrt{y}$
11. $\sqrt{ab} \cdot \sqrt{a}$
13. $(2\sqrt{6})(3\sqrt{5})$
15. $(2\sqrt{5})(6\sqrt{3})$
17. $(2\sqrt[3]{a})(5\sqrt[3]{6a})$
19. $(5\sqrt[3]{4})(2\sqrt[3]{6})$
21. $(4\sqrt{3})(\sqrt{2})(\sqrt{6})$
23. $(5\sqrt{10})(3\sqrt{5})(\sqrt{4})$
25. $(3\sqrt[3]{4})(\sqrt[3]{2})(2\sqrt[3]{3})$
27. $\sqrt{3}(2 + \sqrt{5})$
29. $2\sqrt{3}(3\sqrt{2} - 4\sqrt{5})$

2. $\sqrt[3]{3} \cdot \sqrt[3]{7}$
4. $\sqrt[3]{6} \cdot \sqrt[3]{9}$
6. $\sqrt[4]{3} \cdot \sqrt[4]{5} \cdot \sqrt[4]{2}$
8. $\sqrt[3]{2} \cdot \sqrt[3]{2} \cdot \sqrt[3]{2}$
10. $\sqrt[3]{x^2 y} \cdot \sqrt[3]{xy^2}$
12. $\sqrt{3mn} \cdot \sqrt{2n}$
14. $(2\sqrt[3]{3})(4\sqrt[3]{2})$
16. $(3\sqrt{5x})(6\sqrt{7y})$
18. $(3\sqrt{6})(5\sqrt{2})$
20. $(2\sqrt{5})(3\sqrt{2})(4\sqrt{3})$
22. $(3\sqrt{a})(2\sqrt{a})(\sqrt{a})$
24. $(2\sqrt[3]{3})(3\sqrt[3]{2})(\sqrt[3]{5})$
26. $\sqrt{2}(2 + \sqrt{3})$
28. $\sqrt{3}(\sqrt{5} - \sqrt{6})$
30. $4\sqrt{5}(\sqrt{7} + 2\sqrt{3} - 3\sqrt{2})$

31. $\sqrt{3}(\sqrt{2} - 4\sqrt{5} + \sqrt{6})$ | **32.** $(3 + \sqrt{2})(3 - \sqrt{2})$
33. $(\sqrt{5} + 2)(\sqrt{5} - 2)$ **34.** $(2\sqrt{3} + 5)(2\sqrt{3} - 5)$
35. $(\sqrt{6} + \sqrt{3})(\sqrt{6} - \sqrt{3})$ **36.** $(3\sqrt{2} + 5\sqrt{3})(3\sqrt{2} - 5\sqrt{3})$
37. $(2 + \sqrt{5})(3 + 4\sqrt{5})$ **38.** $(2 + \sqrt{3})(1 - \sqrt{3})$
39. $(5 - \sqrt{2})(2 + \sqrt{2})$ **40.** $(\sqrt{3} + 1)(\sqrt{3} - 3)$
41. $(3 + 2\sqrt{5})(2 - 3\sqrt{5})$ **42.** $(2\sqrt{6} + 3)(\sqrt{6} - 1)$
43. $(\sqrt{5} + 2\sqrt{3})(2\sqrt{5} - \sqrt{3})$ **44.** $(\sqrt{2} - 3\sqrt{5})(4\sqrt{2} + \sqrt{5})$
45. $(\sqrt{2} + \sqrt{5})(\sqrt{2} + 3\sqrt{5})$ **46.** $(5\sqrt{7} - 2\sqrt{3})(2\sqrt{7} + 4\sqrt{3})$
47. $(\sqrt{3} + \sqrt{2})^2$ **48.** $(2\sqrt{5} - 3\sqrt{2})^2$

Divide the following expressions. If possible, simplify the answer.

49. $\dfrac{\sqrt{30}}{\sqrt{6}}$ **50.** $\dfrac{\sqrt[3]{50}}{\sqrt[3]{5}}$

51. $\dfrac{\sqrt{24}}{\sqrt{3}}$ **52.** $\dfrac{\sqrt[4]{20}}{\sqrt[4]{4}}$

53. $\dfrac{6\sqrt{15}}{2\sqrt{3}}$ **54.** $\dfrac{\sqrt{72}}{\sqrt{8}}$

55. $\dfrac{20\sqrt{10}}{5\sqrt{2}}$ **56.** $\dfrac{33\sqrt[3]{32}}{11\sqrt[3]{2}}$

Simplify these expressions using the reverse of the division property.

57. $\sqrt{\dfrac{49}{64}}$ **58.** $\sqrt[3]{\dfrac{8}{27}}$

59. $\sqrt{\dfrac{100}{36}}$ **60.** $\sqrt{\dfrac{1}{4}}$

61. $\sqrt{\dfrac{1}{16}}$ **62.** $\sqrt[3]{\dfrac{1}{8}}$

63. $\sqrt{\dfrac{5}{16}}$ **64.** $\sqrt{\dfrac{2}{9}}$

65. $\sqrt[3]{\dfrac{2}{27}}$ **66.** $\sqrt{\dfrac{11}{25}}$

67. $\sqrt[4]{\dfrac{7}{16}}$ **68.** $\sqrt{\dfrac{3}{100}}$

69. $\sqrt[3]{\dfrac{7}{1000}}$ **70.** $3\sqrt{\dfrac{2}{9}}$

71. $10\sqrt{\dfrac{7}{4}}$ **72.** $4\sqrt[3]{\dfrac{3}{8}}$

RATIONALIZING DENOMINATORS

6.8

We often find in mathematics that it is inconvenient to work with fractions whose denominators contain radicals. For example, suppose you were asked to evaluate the fraction $\dfrac{2}{\sqrt{3}}$. You would proceed to the Powers and Roots table in the back of the text and find that $\sqrt{3}$ is approximately 1.732. Then you would perform the following long division:

$$
\begin{array}{r}
1.154 \\
1.732_x\overline{)2.000_x000} \\
\underline{1\ 732} \\
268\ 0 \\
\underline{173\ 2} \\
94\ 80 \\
\underline{86\ 60} \\
8\ 200 \\
\underline{6\ 928}
\end{array}
$$

Thus, $\dfrac{2}{\sqrt{3}}$ is approximately 1.154.

The expression $\dfrac{2}{\sqrt{3}}$ can be evaluated more easily if the denominator is converted to a rational number. This is done by multiplying numerator and denominator by $\sqrt{3}$, as follows:

$$\frac{2}{\sqrt{3}} = \frac{2 \cdot \sqrt{3}}{\sqrt{3} \cdot \sqrt{3}} = \frac{2 \cdot \sqrt{3}}{\sqrt{9}} = \frac{2\sqrt{3}}{3}$$

Now you can evaluate the fraction in your head by performing a simple multiplication and a short division.

$$\frac{2}{\sqrt{3}} = \frac{2\sqrt{3}}{3} \approx \frac{2 \cdot (1.732)}{3}$$
$$\approx \frac{3.464}{3}$$
$$\approx 1.154$$

Notice that we obtained the same answer as with the first more complicated computation.

The process of converting a denominator from a radical (irrational number) to a rational number is called *rationalizing the denominator*. In this section you will learn how to rationalize these two types of denominators:

(1)   Monomial denominators containing a radical;

Examples: $\dfrac{3}{\sqrt{5}}$ and $\dfrac{\sqrt[3]{5}}{3\sqrt[3]{5}}$

(2)   Binomial denominators containing square roots.

Example: $\dfrac{5}{3 + \sqrt{2}}$

Rationalizing Monomial Denominators

To rationalize a monomial denominator consisting of an nth root, multiply the numerator and denominator by a radical which will produce a perfect n power in the denominator.

Example 1

Rationalize the denominator of $\dfrac{3}{\sqrt{5}}$

Solution: Multiplying numerator and denominator by $\sqrt{5}$ will produce a perfect square of 25 in the denominator.

$$\frac{3}{\sqrt{5}} = \frac{3 \cdot \sqrt{5}}{\sqrt{5} \cdot \sqrt{5}}$$
$$= \frac{3 \cdot \sqrt{5}}{\sqrt{25}}$$
$$= \frac{3\sqrt{5}}{5}$$

Example 2

Rationalize the denominator of $\dfrac{\sqrt[3]{5}}{3\sqrt[3]{2}}$

Solution: Multiplying numerator and denominator by $\sqrt[3]{4}$ will produce a perfect cube of 8 in the denominator.

$$\frac{\sqrt[3]{5}}{3\sqrt[3]{2}} = \frac{\sqrt[3]{5} \cdot \sqrt[3]{4}}{3\sqrt[3]{2} \cdot \sqrt[3]{4}}$$
$$= \frac{\sqrt[3]{20}}{3 \cdot \sqrt[3]{8}}$$
$$= \frac{\sqrt[3]{20}}{3 \cdot 2}$$
$$= \frac{\sqrt[3]{20}}{6}$$

Example 3

Rationalize the denominator of $\dfrac{2\sqrt{5}}{3\sqrt{6}}$

Solution: Multiplying numerator and denominator by $\sqrt{6}$ will produce a perfect square of 36 in the denominator.

$$\frac{2\sqrt{5}}{3\sqrt{6}} = \frac{2\sqrt{5} \cdot \sqrt{6}}{3\sqrt{6} \cdot \sqrt{6}}$$

$$= \frac{2 \cdot \sqrt{30}}{3 \cdot \sqrt{36}}$$

$$= \frac{\overset{1}{2} \cdot \sqrt{30}}{3 \cdot \underset{3}{6}} \qquad \text{[always reduce to lowest terms]}$$

$$= \frac{\sqrt{30}}{9}$$

Rationalize the denominator of $\sqrt{\dfrac{2}{3}}$

Example 4

Solution: Use the division property of radicals to write $\sqrt{\dfrac{2}{3}}$ as $\dfrac{\sqrt{2}}{\sqrt{3}}$. Then multiply numerator and denominator by $\sqrt{3}$

$$\sqrt{\frac{2}{3}} = \frac{\sqrt{2}}{\sqrt{3}}$$

$$= \frac{\sqrt{2} \cdot \sqrt{3}}{\sqrt{3} \cdot \sqrt{3}}$$

$$= \frac{\sqrt{6}}{\sqrt{9}}$$

$$= \frac{\sqrt{6}}{3}$$

Rationalize the denominator of $\sqrt[3]{\dfrac{7}{9}}$

Example 5

Solution: Use the division property of radicals to write $\sqrt[3]{\dfrac{7}{9}}$ as $\dfrac{\sqrt[3]{7}}{\sqrt[3]{9}}$. Then multiply numerator and denominator by $\sqrt[3]{3}$

$$\sqrt[3]{\frac{7}{9}} = \frac{\sqrt[3]{7}}{\sqrt[3]{9}}$$

$$= \frac{\sqrt[3]{7} \cdot \sqrt[3]{3}}{\sqrt[3]{9} \cdot \sqrt[3]{3}}$$

$$= \frac{\sqrt[3]{21}}{\sqrt[3]{27}}$$

$$= \frac{\sqrt[3]{21}}{3}$$

Rationalizing Binomial Denominators

Next we discuss how to rationalize a binomial denominator containing square roots. Remember, a sum times a difference produces a difference of two squares. Therefore, we will multiply the numerator and denominator by a binomial identical to the denominator, except for its middle sign.

Example 6

Rationalize the denominator of $\dfrac{5}{3 + \sqrt{2}}$

Solution: Multiply numerator and denominator by $3 - \sqrt{2}$ This will produce a sum times a difference in the denominator.

$$\frac{5}{3 + \sqrt{2}} = \frac{5 \cdot (3 - \sqrt{2})}{(3 + \sqrt{2}) \cdot (3 - \sqrt{2})}$$
$$= \frac{5 \cdot (3 - \sqrt{2})}{9 - \sqrt{4}}$$
$$= \frac{5 \cdot (3 - \sqrt{2})}{9 - 2}$$
$$= \frac{5(3 - \sqrt{2})}{7} \quad \text{or} \quad \frac{15 - 5\sqrt{2}}{7}$$

Example 7

Rationalize the denominator of $\dfrac{2}{\sqrt{5} - \sqrt{3}}$

Solution: Multiply numerator and denominator by $\sqrt{5} + \sqrt{3}$

$$\frac{2}{\sqrt{5} - \sqrt{3}} = \frac{2 \cdot (\sqrt{5} + \sqrt{3})}{(\sqrt{5} - \sqrt{3}) \cdot (\sqrt{5} + \sqrt{3})}$$
$$= \frac{2 \cdot (\sqrt{5} + \sqrt{3})}{\sqrt{25} - \sqrt{9}}$$
$$= \frac{2 \cdot (\sqrt{5} + \sqrt{3})}{5 - 3}$$
$$= \frac{\cancel{2} \cdot (\sqrt{5} + \sqrt{3})}{\cancel{2}} \qquad \text{[always reduce to lowest terms]}$$
$$= \sqrt{5} + \sqrt{3}$$

Exercise 6.8

Rationalize the denominators of the following fractions. Reduce your answers to lowest terms.

1. $\dfrac{2}{\sqrt{5}}$

2. $\dfrac{6}{\sqrt{5}}$

3. $\dfrac{3}{\sqrt[3]{2}}$

4. $\dfrac{5}{\sqrt[3]{4}}$

5. $\dfrac{3}{\sqrt{3}}$

6. $\sqrt{\dfrac{5}{3}}$

7. $\sqrt{\dfrac{1}{3}}$

8. $\sqrt[3]{\dfrac{7}{2}}$

9. $\sqrt[3]{\dfrac{5}{4}}$

10. $\sqrt[3]{\dfrac{1}{3}}$

11. $\sqrt{\dfrac{3}{2}}$

12. $\dfrac{2\sqrt{3}}{\sqrt{7}}$

13. $\dfrac{2\sqrt{3}}{3\sqrt{7}}$

14. $\dfrac{5}{2\sqrt[3]{2}}$

15. $\dfrac{-4}{\sqrt[3]{4}}$

16. $\dfrac{3}{\sqrt[3]{-9}}$

17. $\dfrac{2\sqrt[3]{2}}{3\sqrt[3]{25}}$

18. $\dfrac{2\sqrt{6}}{3\sqrt{5}}$

19. $\dfrac{4\sqrt{15}}{5\sqrt{2}}$

20. $\dfrac{2\sqrt{10}}{3\sqrt{3}}$

21. $\dfrac{4\sqrt[3]{3}}{5\sqrt[3]{4}}$

22. $\sqrt{\dfrac{7}{2}}$

23. $\sqrt[3]{\dfrac{6}{5}}$

24. $\sqrt[4]{\dfrac{1}{2}}$

25. $\dfrac{3}{2+\sqrt{6}}$

26. $\dfrac{5}{1-\sqrt{3}}$

27. $\dfrac{1}{4+\sqrt{5}}$

28. $\dfrac{5}{\sqrt{6}+\sqrt{2}}$

29. $\dfrac{3}{\sqrt{7}+\sqrt{5}}$

30. $\dfrac{4}{3-\sqrt{5}}$

31. $\dfrac{1}{\sqrt{3}+\sqrt{2}}$

32. $\dfrac{6}{3-\sqrt{7}}$

33. $\dfrac{3}{\sqrt{7}-\sqrt{3}}$

34. $\dfrac{2}{\sqrt{5}-\sqrt{7}}$

35. $\dfrac{1}{2\sqrt{3}+3\sqrt{2}}$

36. $\dfrac{3}{3\sqrt{5}-4\sqrt{2}}$

37. $\dfrac{2\sqrt{7}}{3\sqrt{2}-5\sqrt{7}}$

38. $\dfrac{\sqrt{2}+\sqrt{3}}{3\sqrt{2}-2\sqrt{3}}$

39. $\dfrac{2\sqrt{3}+\sqrt{5}}{\sqrt{3}-3\sqrt{5}}$

Rationalize each denominator; then evaluate using the Powers and Roots table. Round answers to the nearest thousandth.

40. $\dfrac{1}{\sqrt{2}}$

41. $\dfrac{1}{\sqrt{3}}$

42. $\dfrac{2}{\sqrt{5}}$ **43.** $\dfrac{1}{\sqrt[3]{4}}$

44. $\dfrac{1}{\sqrt[3]{2}}$ **45.** $\dfrac{2}{\sqrt[3]{4}}$

46. $\dfrac{\sqrt{5}}{\sqrt{6}}$ **47.** $\dfrac{3\sqrt{2}}{2\sqrt{3}}$

48. $\dfrac{2\sqrt[3]{3}}{3\sqrt[3]{2}}$ **49.** $\dfrac{3}{2 - \sqrt{6}}$

50. $\dfrac{1}{\sqrt{3} - \sqrt{2}}$ **51.** $\dfrac{3}{\sqrt{5} + \sqrt{6}}$

SOLVING EQUATIONS CONTAINING RADICALS

6.9

In this section you will learn how to solve equations containing square roots, such as $\sqrt{2x + 1} = 3$. This is done by squaring both sides of the equation to eliminate the radical. However, this procedure can lead to extraneous solutions. You should remember from section 5.5 that an extraneous solution is brought about by the solution process and will not check in the original equation. For example, if $x = 3$ we can square both sides to obtain $x^2 = 9$. The original equation, $x = 3$, has one solution—the number 3. The second equation, $x^2 = 9$, has two solutions—the numbers 3 and -3. Notice that -3 will not check in the original equation and is therefore extraneous. In summary, we state the squaring property of equations.

> **The Squaring Property of Equations**
> Squaring both sides of an equation will preserve every solution of the original equation, but may produce some extraneous solutions that will not check in the original equation.

To solve equations containing square roots we will follow this procedure:

Step (1) Isolate the term containing the radical on one side of the equation.

Step (2) Square both sides of the equation to eliminate the radical. Then solve the resulting equation.

Step (3) Check all answers in the original equation to find extraneous solutions.

Example 1   Solve $\sqrt{2x + 1} - 3 = 0$

Solution:

Step (1)   Isolate the radical on one side of the equation.

$$\sqrt{2x + 1} - 3 = 0$$
$$\sqrt{2x + 1} = 3$$

Step (2)   Square both sides of the equation; then solve.

$$(\sqrt{2x + 1})^2 = (3)^2$$
$$2x + 1 = 9$$
$$2x = 8$$
$$x = 4$$

Step (3)   Check for extraneous solutions.

$$\sqrt{2x + 1} = 3$$
$$\sqrt{2 \cdot 4 + 1} = 3$$
$$\sqrt{9} = 3$$
$$3 = 3 \quad \text{[true]}$$

It is true that 3 equals 3. Thus, the solution checks. There are no extraneous solutions. The solution is 4.

Solve $m = 5 + \sqrt{m - 3}$ **Example 2**

Solution:

Step (1)   Isolate the radical on one side of the equation.

$$m = 5 + \sqrt{m - 3}$$
$$m - 5 = \sqrt{m - 3}$$

Step (2)   Square both sides of the equation; then solve.

$$(m - 5)^2 = (\sqrt{m - 3})^2$$
$$m^2 - 10m + 25 = m - 3 \quad \text{[quadratic equation]}$$
$$m^2 - 11m + 28 = 0$$
$$(m - 7)(m - 4) = 0 \quad \text{[factor; then set each equal to 0]}$$
$$m = 7 \quad \text{or} \quad m = 4$$

Step (3) Check for extraneous solutions.

Check for 7	**Check for 4**
$m = 5 + \sqrt{m - 3}$	$m = 5 + \sqrt{m - 3}$
$7 = 5 + \sqrt{7 - 3}$	$4 = 5 + \sqrt{4 - 3}$
$7 = 5 + \sqrt{4}$	$4 = 5 + \sqrt{1}$
$7 = 5 + 2$	$4 = 5 + 1$
$7 = 7$ [true]	$4 = 6$ [false]

Since 4 does not check, it is an extraneous solution. The only solution is 7.

This procedure can be shortened as illustrated in the following examples.

Example 3

Solve $y - \sqrt{y - 1} = 3$

Solution:
$$y - \sqrt{y - 1} = 3$$
$$y - 3 = \sqrt{y - 1} \qquad \text{[isolate the radical]}$$
$$(y - 3)^2 = (\sqrt{y - 1})^2 \qquad \text{[square both sides]}$$
$$y^2 - 6y + 9 = y - 1 \qquad \text{[quadratic equation]}$$
$$y^2 - 7y + 10 = 0$$
$$(y - 5)(y - 2) = 0 \qquad \text{[solve by factoring]}$$
$$y = 5 \quad \text{or} \quad y = 2$$

Check for 5	**Check for 2**
$y - \sqrt{y - 1} = 3$	$y - \sqrt{y - 1} = 3$
$5 - \sqrt{5 - 1} = 3$	$2 - \sqrt{2 - 1} = 3$
$5 - \sqrt{4} = 3$	$2 - \sqrt{1} = 3$
$5 - 2 = 3$	$2 - 1 = 3$
$3 = 3$ [true]	$1 = 3$ [false]

Since 2 does not check, it is extraneous. The only solution is 5.

Example 4

Solve $\sqrt{x + 1} = -2$

Solution:
$$\sqrt{x + 1} = -2$$
$$(\sqrt{x + 1})^2 = (-2)^2 \qquad \text{[square both sides]}$$

$$x + 1 = 4$$

$$x = 3$$

Check for 3

$$\sqrt{x + 1} = -2$$

$$\sqrt{3 + 1} = -2$$

$$\sqrt{4} = -2$$

$$2 = -2 \quad \text{[false]}$$

Since 3 does not check, it is extraneous. The equation has no real number solutions whatsoever. (Note: you might have observed at the outset that no solution exists because $\sqrt{x + 1} \geq 0$ while $-2 < 0$.)

Solve each of these equations.   **Exercise 6.9**

1. $\sqrt{x} = 3$ **2.** $\sqrt{y} = 5$

3. $\sqrt{m} = 1$ **4.** $\sqrt{x} = -1$

5. $\sqrt{3y} = 6$ **6.** $\sqrt{2n} = 4$

7. $\sqrt{5x} = 10$ **8.** $\sqrt{5x} = -10$

9. $\sqrt{m} - 2 = 0$ **10.** $\sqrt{4y} - 2 = 0$

11. $2\sqrt{x} - 3 = 0$ **12.** $3\sqrt{n} - 1 = 0$

13. $\sqrt{x + 3} = 2$ **14.** $\sqrt{y + 3} = 0$

15. $\sqrt{m + 2} - 3 = 0$ **16.** $\sqrt{n - 3} - 2 = 0$

17. $\sqrt{x + 6} = 1$ **18.** $\sqrt{a - 3} = 1$

19. $\sqrt{x - 3} = 0$ **20.** $\sqrt{x + 3} = 0$

21. $\sqrt{2y - 5} = 3$ **22.** $\sqrt{2m - 3} - 1 = 0$

23. $\sqrt{m + 2} = 3$ **24.** $\sqrt{3n - 5} = 3$

25. $\sqrt{7x + 5} - 3 = 0$ **26.** $\sqrt{3m + 4} - 5 = 0$

27. $3 + \sqrt{4x + 1} = 8$ **28.** $8 + \sqrt{2t - 7} = 11$

29. $y - 3 = \sqrt{y - 1}$ **30.** $\sqrt{x - 3} = x - 5$

31. $t = 5 + \sqrt{t - 3}$ **32.** $\sqrt{4x - 11} = x - 2$

33. $m - 7 = \sqrt{2m + 1}$ **34.** $\sqrt{5y + 1} = y + 1$

35. $t + 1 = \sqrt{t + 1}$ **36.** $3 + \sqrt{2x - 3} = x$

37. $\sqrt{y + 6} = \sqrt{2y - 3}$ **38.** $\sqrt{x + 8} = \sqrt{3x - 2}$

39. $\sqrt{t + 10} - \sqrt{3t - 4} = 0$ **40.** $6 + \sqrt{4y + 5} = 2y + 1$

41. $\sqrt{x^2 + 21} = x + 3$ **42.** $t + \sqrt{t^2 + 3} = 3$

43. $4 = n + \sqrt{n^2 - 8}$ **44.** $\sqrt{x^2 + 5} = x + 1$

45. $\sqrt{x + 5} = \sqrt{x} + 1$
(*Hint:* square both
sides of the equation
twice.)

46. $\sqrt{x + 5} = 5 - \sqrt{x}$
(*Hint:* square both
sides of the equation
twice.)

Solve the following word problems.

47. The square root of the sum of a certain number and 2 is 3. Find the
number.

48. The square root of twice a certain number is 4. Find the number.

49. The square root of 4 times a certain number equals the sum of that
number and 1. Find the number.

50. $t = \sqrt{\dfrac{2s}{g}}$ is a formula giving the time (t) in seconds that it takes
an object starting at rest to fall a distance of s feet when acted upon
by a gravitational attraction of g feet per second per second.
  (a) How long would it take an object to fall 25 ft if the gravitational
attraction is 32 ft/sec²?
  (b) Solve the formula for s.
  (c) In 2 sec, how far would an object fall from rest if the gravitational attraction is 32 ft/sec²?

SOLVING QUADRATIC EQUATIONS
BY COMPLETING THE SQUARE

6.10

In section 4.8 you learned to solve quadratic equations by factoring.
However, many trinomials are not factorable. Thus, it is impossible to
solve all quadratic equations by factoring. In this section we develop a
procedure called *completing the square,* which can be used to solve
any quadratic equation. This procedure is important because it will be
used in the next section to derive the quadratic formula. The quadratic
formula will let us solve any quadratic equation by simply evaluating a
formula. To aid us in completing the square we must first discuss perfect
square trinomials and the square root property of equations. A **perfect
square trinomial** is a trinomial that is the square of a binomial.

perfect square trinomial

Example 1

$x^2 + 6x + 9$ is a perfect square trinomial because $x^2 + 6x + 9 = (x + 3)^2$

Example 2

$x^2 - 10x + 25$ is a perfect square trinomial because $x^2 - 10x + 25 = (x - 5)^2$

We will find it convenient to know how the constant term of a perfect
square trinomial is related to the coefficient of the middle term.

Example 3

$x^2 + 6x + 9$ is a perfect square trinomial. The constant term is 9. It is
equal to the square of one-half the coefficient of x.

$x^2 + 6x + 9$

$$\left(\frac{1}{2} \cdot 6\right)^2$$

Example 4

$x^2 - 10x + 25$ is a perfect square trinomial. The constant term is 25. It is equal to the square of one-half the coefficient of x.

$x^2 - 10x + 25$

$$\left(\frac{1}{2} \cdot 10\right)^2$$

In general, we can say that the constant term of a perfect square trinomial is equal to the square of one-half the coefficient of the middle term, provided the coefficient of the first term is 1.

Next we develop the square root property of equations. If $x^2 = 4$, we know by inspection that there are two solutions, $x = 2$ and $x = -2$. Or, since $2 = \sqrt{4}$, we can state it another way: if $x^2 = 4$, then $x = \sqrt{4}$ or $x = -\sqrt{4}$. Using this as a guide, we state the following property.

The Square Root Property of Equations
If $x^2 = k$, where x is the variable and k is a nonnegative constant, then $x = \sqrt{k}$ or $x = -\sqrt{k}$. (For convenience, the two equations can be written as $x = \pm\sqrt{k}$.)

As illustrated by Examples 5 through 9, the square root property can be used to solve quadratic equations where the variable is contained within a perfect square.

Solve $x^2 = 9$

Example 5

Solution: Use the square root property of equations. If $x^2 = 9$, then $x = \pm\sqrt{9}$. Thus, $x = \pm 3$. The solutions are 3 and -3

Solve $y^2 = 10$

Example 6

Solution: $y^2 = 10$

$y = \pm\sqrt{10}$

The solutions are $\sqrt{10}$ and $-\sqrt{10}$

Example 7

Solve $(x + 3)^2 = 25$

Solution: $(x + 3)^2 = 25$

$x + 3 = \pm 5$

$x = -3 \pm 5$

Using the plus sign: $x = -3 + 5$ or   2
Using the minus sign: $x = -3 - 5$ or -8
The two solutions are 2 and -8

Example 8

Solve $(y - 5)^2 = 7$

Solution: $(y - 5)^2 = 7$

$y - 5 = \pm\sqrt{7}$

$y = 5 \pm \sqrt{7}$

The solutions are $5 + \sqrt{7}$ and $5 - \sqrt{7}$

Example 9

Solve $x^2 - 10x + 25 = 3$

Solution: $x^2 - 10x + 25$ is a perfect square trinomial and is equal to $(x - 5)^2$

$x^2 - 10x + 25 = 3$

$(x - 5)^2 = 3$

$x - 5 = \pm\sqrt{3}$

$x = 5 \pm \sqrt{3}$

The solutions are $5 + \sqrt{3}$ and $5 - \sqrt{3}$

Using our knowledge of perfect square trinomials and the square root property of equations, we show how to solve quadratic equations by completing the square.

Example 10

Solve $x^2 + 6x - 7 = 0$ by completing the square.

Solution:

Step (1) Write the equation so that all terms containing a variable appear on one side and the constants appear on the other side.

$x^2 + 6x - 7 = 0$

$x^2 + 6x \qquad = 7$

Step (2) Determine what the constant term must be in order to turn the left side into a perfect square trinomial. Then add that number to both sides of the equation. Remember, the constant equals the square of one-half the coefficient of the middle term.

$$x^2 + 6x \quad\;\; = 7$$
$$x^2 + 6x + 9 = 7 + 9$$
$$\left(\frac{1}{2}\cdot 6\right)^2$$
$$x^2 + 6x + 9 = 16$$

Step (3) Write the perfect square trinomial as the square of a binomial. Then use the square root property of equations.

$$x^2 + 6x + 9 = 16$$
$$(x + 3)^2 = 16$$
$$x + 3 = \pm\sqrt{16}$$
$$x + 3 = -3 \pm 4$$
$$x = 1 \quad\text{or}\quad x = -7$$

The solutions are 1 and -7

Solve $2y^2 - 6y - 3 = 0$ by completing the square.

Example 11

Solution:

Step (1) Obtain all terms containing a variable on one side and all constant terms on the other.

$$2y^2 - 6y - 3 = 0$$
$$2y^2 - 6y \quad\;\; = 3$$

Step (2) Divide both sides of the equation by 2. Then turn the left side into a perfect square trinomial. Remember, the rule relating the constant to the coefficient of the middle term is valid only when the coefficient of the first term is 1.

$$2y^2 - 6y \quad\;\; = 3$$
$$y^2 - 3y \quad\;\; = \frac{3}{2} \qquad \text{[divide both sides by 2]}$$

$$y^2 - 3y + \frac{9}{4} = \frac{3}{2} + \frac{9}{4}$$

$$\left(\frac{1}{2} \cdot 3\right)^2$$

$$y^2 - 3y + \frac{9}{4} = \frac{6}{4} + \frac{9}{4}$$

$$y^2 - 3y + \frac{9}{4} = \frac{15}{4}$$

Step (3)   Write the perfect square trinomial as the square of a binomial. Then use the square root property of equations.

$$y^2 - 3y + \frac{9}{4} = \frac{15}{4}$$

$$\left(y - \frac{3}{2}\right)^2 = \frac{15}{4}$$

$$\left(y - \frac{3}{2}\right) = \pm\sqrt{\frac{15}{4}}$$

$$y - \frac{3}{2} = \pm\frac{\sqrt{15}}{2}$$

$$y = \frac{3}{2} \pm \frac{\sqrt{15}}{2}$$

$$y = \frac{3 \pm \sqrt{15}}{2}$$

The solutions are $\dfrac{3 + \sqrt{15}}{2}$ and $\dfrac{3 - \sqrt{15}}{2}$

Example 12   Solve $3m^2 + 6m + 4 = 0$ by completing the square.

Solution:

Step (1)   Obtain variables on one side and constants on the other.

$$3m^2 + 6m + 4 = 0$$
$$3m^2 + 6m = -4$$

Step (2)   Divide both sides by 3. Then turn the left side into a perfect square trinomial.

$$3m^2 + 6m = -4$$

$$m^2 + 2m = -\frac{4}{3} \qquad \text{[divide both sides by 3]}$$

$$m^2 + 2m + 1 = \frac{-4}{3} + 1$$

$$\left(\frac{1}{2} \cdot 2\right)^2$$

$$m^2 + 2m + 1 = \frac{-4}{3} + \frac{3}{3}$$

$$m^2 + 2m + 1 = -\frac{1}{3}$$

Step (3) Write as a binomial squared. Then use the square root property of equations.

$$m^2 + 2m + 1 = -\frac{1}{3}$$

$$(m + 1)^2 = -\frac{1}{3}$$

$$m + 1 = \pm\sqrt{-\frac{1}{3}}$$

The square root of a negative number is meaningless in the real number system. Therefore, this equation has no real number solutions.

Fill in the blanks so that each trinomial is a perfect square. **Exercise 6.10**

1. $x^2 + 12x +$ _____

2. $x^2 + 2x +$ _____

3. $x^2 - 6x +$ _____

4. $y^2 + 4y +$ _____

5. $m^2 + 20m +$ _____

6. $n^2 - 8n +$ _____

7. $x^2 + 14x +$ _____

8. $y^2 - 30y +$ _____

9. $y^2 + y +$ _____

10. $y^2 - y +$ _____

11. $x^2 - 3x +$ _____

12. $x^2 + 5x +$ _____

13. $r^2 + 9r +$ _____

14. $t^2 - 7t +$ _____

15. $x^2 + \frac{3}{4}x +$ _____

16. $m^2 + \frac{1}{2}m +$ _____

17. $x^2 + kx +$ _____

18. $x^2 + \frac{b}{a}x +$ _____

Solve the following equations by using the square root property.

19. $x^2 = 9$

20. $y^2 = 25$

21. $t^2 = 36$

22. $m^2 = 1$

23. $x^2 = 7$ **24.** $x^2 = \dfrac{9}{4}$

25. $y^2 = -49$ **26.** $(x + 6)^2 = 16$

27. $(y - 5)^2 = 9$ **28.** $(t + 8)^2 = 2$

29. $(2x - 3)^2 = 1$ **30.** $(3y + 2)^2 = 3$

31. $x^2 + 4x + 4 = 1$ **32.** $y^2 + 6y + 9 = 25$

33. $m^2 - 2m + 1 = 4$ **34.** $n^2 - 3n + \dfrac{9}{4} = \dfrac{1}{4}$

35. $t^2 - 4t + 4 = 5$ **36.** $x^2 + 2x + 1 = 11$

37. $x^2 + 6x + 9 = 13$ **38.** $y^2 - 8y + 16 = -2$

Solve these quadratic equations by completing the square.

39. $x^2 + 5x + 6 = 0$ **40.** $y^2 - 2y - 15 = 0$

41. $2x^2 - x - 3 = 0$ **42.** $m^2 + m - 2 = 0$

43. $2n^2 + n - 10 = 0$ **44.** $3x^2 - x - 4 = 0$

45. $-2x^2 - x + 21 = 0$ **46.** $m^2 + 3m + 1 = 0$

47. $y^2 + 4y + 5 = 0$ **48.** $-x^2 + 5x + 1 = 0$

49. $5t^2 - 2t - 3 = 0$ **50.** $y^2 - 3y + 2 = 0$

51. $x^2 + x - 1 = 0$ **52.** $m^2 - 5m + 3 = 0$

53. $2x^2 + 7x + 4 = 0$ **54.** $3y^2 - 7y + 3 = 0$

55. $m^2 - 6m + 1 = 0$ **56.** $3x^2 = 1 + 2x$

57. $3t^2 = 9t - 5$ **58.** $3r^2 - 5 = 6r$

59. $x^2 - 4x = -4$ **60.** $2y^2 - 5 = 4y$

THE QUADRATIC FORMULA

6.11

Every quadratic equation can be solved by completing the square, but the method is complicated and time consuming. Thus, in this section we derive the quadratic formula, which enables us to solve any quadratic equation easily by merely evaluating a general formula. To derive the quadratic formula, we begin with the general form of a quadratic equation: $ax^2 + bx + c = 0$ where a, b, c are constants and $a \neq 0$. Then we solve for x by completing the square, as follows:

Step (1) Obtain variables on one side and constants on the other.

$$ax^2 + bx + c = 0$$

$$ax^2 + bx \quad = -c$$

Step (2) Divide both sides by a. Then turn the left side into a perfect square trinomial.

$$ax^2 + bx \qquad = -c$$

$$x^2 + \frac{b}{a}x \qquad = -\frac{c}{a} \qquad \text{[divide both sides by } a\text{]}$$

$$x^2 + \frac{b}{a}x + \frac{b^2}{4a^2} = -\frac{c}{a} + \frac{b^2}{4a^2} \qquad \left[\text{add } \frac{b^2}{4a^2} \text{ to both sides}\right]$$

$$\left(\frac{1}{2} \cdot \frac{b}{a}\right)^2$$

$$x^2 + \frac{b}{a}x + \frac{b^2}{4a^2} = \frac{-4ac}{4a^2} + \frac{b^2}{4a^2} \qquad \text{[find LCD, which is } 4a^2\text{]}$$

$$x^2 + \frac{b}{a}x + \frac{b^2}{4a^2} = \frac{b^2 - 4ac}{4a^2} \qquad \text{[combine the fractions]}$$

Step (3) Write the left side as a binomial squared. Then use the square root property of equations.

$$x^2 + \frac{b}{a}x + \frac{b^2}{4a^2} = \frac{b^2 - 4ac}{4a^2}$$

$$\left(x + \frac{b}{2a}\right)^2 = \frac{b^2 - 4ac}{4a^2} \qquad \text{[write as a binomial squared]}$$

$$x + \frac{b}{2a} = \pm\sqrt{\frac{b^2 - 4ac}{4a^2}} \qquad \text{[use the square root property]}$$

$$x + \frac{b}{2a} = \frac{\pm\sqrt{b^2 - 4ac}}{2a} \qquad \text{[take the square root of } 4a^2\text{]}$$

$$x = \frac{-b}{2a} \pm \frac{\sqrt{b^2 - 4ac}}{2a} \qquad \text{[solve for } x\text{]}$$

$$x = \frac{-b \pm \sqrt{b^2 - 4ac}}{2a} \qquad \text{[combine the fractions]}$$

This derivation provides us with the quadratic formula. We will show how it can be used to solve any quadratic equation quickly and easily.

The Quadratic Formula

For quadratic equations in the form $ax^2 + bx + c = 0$,

$$x = \frac{-b \pm \sqrt{b^2 - 4ac}}{2a}$$

(Memorize this formula!)

Use the quadratic formula to solve $3x^2 + x = 2$ **Example 1**

Solution:

Step (1) Write the equation in standard quadratic form with all terms on one side. Then specify the values of a, b, and c.

$$3x^2 + x = 2$$

$$3x^2 + x - 2 = 0$$

$$3x^2 + 1x - 2 = 0$$

$$ax^2 + bx + c = 0$$

$$a = 3 \quad b = 1 \quad \text{and} \quad c = -2$$

Step (2) Substitute these values into the quadratic formula and evaluate.

$$a = 3 \quad b = 1 \quad c = -2$$

$$x = \frac{-b \pm \sqrt{b^2 - 4ac}}{2a}$$

$$x = \frac{-1 \pm \sqrt{1^2 - 4 \cdot 3 \cdot (-2)}}{2 \cdot 3}$$

$$x = \frac{-1 \pm \sqrt{1 + 24}}{6}$$

$$x = \frac{-1 \pm \sqrt{25}}{6}$$

$$x = \frac{-1 \pm 5}{6}$$

To separate the two solutions, first use the plus sign and then the minus sign.

$$x = \frac{-1 + 5}{6} = \frac{4}{6} = \frac{2}{3}$$

$$x = \frac{-1 - 5}{6} = \frac{-6}{6} = -1$$

The two solutions are $\frac{2}{3}$ and -1

Example 2   Use the quadratic formula to solve $x^2 - 4x + 2 = 0$

Solution: Determine the values of a, b, and c. Then substitute into the quadratic formula.

$$x^2 - 4x + 2 = 0 \qquad a = 1 \quad b = -4 \quad \text{and} \quad c = 2$$

$$x = \frac{-b \pm \sqrt{b^2 - 4ac}}{2a}$$

$$x = \frac{-(-4) \pm \sqrt{(-4)^2 - 4 \cdot 1 \cdot 2}}{2 \cdot 1}$$

$$x = \frac{4 \pm \sqrt{16 - 8}}{2}$$

$$x = \frac{4 \pm \sqrt{8}}{2}$$

Simplify the radical and reduce the resulting fraction.

$$x = \frac{4 \pm \sqrt{4 \cdot 2}}{2}$$

$$x = \frac{4 \pm 2\sqrt{2}}{2}$$

$$x = \frac{\cancel{2}(2 \pm \sqrt{2})}{\cancel{2}} \qquad \text{[factor out a 2; then cancel]}$$

$$x = 2 \pm \sqrt{2}$$

The two solutions are $2 + \sqrt{2}$ and $2 - \sqrt{2}$

In example 2 you found two solutions, $2 + \sqrt{2}$ and $2 - \sqrt{2}$. Both solutions involve a radical and are exact solutions. On some occasions you may be asked to approximate these solutions with decimals. If so, go to the Powers and Roots table to find that the approximate value of $\sqrt{2}$ is 1.414. Then convert the two solutions to decimal form as follows:

$$2 + \sqrt{2} \approx 2 + 1.414 \qquad\qquad 2 - \sqrt{2} \approx 2 - 1.414$$
$$\approx 3.414 \qquad\qquad\qquad\qquad \approx .586$$

Thus the approximate decimal solutions of $x^2 - 4x + 2 = 0$ are 3.414 and 0.586.

Use the quadratic formula to solve $2x^2 + x + 3 = 0$ **Example 3**

Solution: Determine the values of a, b, and c. Then substitute into the quadratic formula.

$$2x^2 + x + 3 = 0; \qquad a = 2 \quad b = 1 \quad \text{and} \quad c = 3$$

$$x = \frac{-b \pm \sqrt{b^2 - 4ac}}{2a}$$

$$x = \frac{-1 \pm \sqrt{1^2 - 4 \cdot 2 \cdot 3}}{2 \cdot 2}$$

$$x = \frac{-1 \pm \sqrt{1 - 24}}{4}$$

$$x = \frac{-1 \pm \sqrt{-23}}{4}$$

The square root of a negative number is not a real number. Therefore, $\sqrt{-23}$ shows the original equation has no real number solutions.

Example 4

Solve $\frac{1}{3}m^2 = \frac{2}{3}m + \frac{1}{6}$

Solution: Multiply both sides of the equation by the LCD, 6. Then use the quadratic formula.

$$\frac{1}{3}m^2 = \frac{2}{3}m + \frac{1}{6}$$

$$6 \cdot \frac{1}{3}m^2 = 6 \cdot \frac{2}{3}m + 6 \cdot \frac{1}{6}$$

$$2m^2 = 4m + 1$$

$$2m^2 - 4m - 1 = 0; \quad a = 2 \quad b = -4 \quad \text{and} \quad c = -1$$

$$m = \frac{-b \pm \sqrt{b^2 - 4ac}}{2a}$$

$$m = \frac{-(-4) \pm \sqrt{(-4)^2 - 4 \cdot 2 \cdot (-1)}}{2 \cdot 2}$$

$$m = \frac{4 \pm \sqrt{16 + 8}}{4}$$

$$m = \frac{4 \pm \sqrt{24}}{4}$$

$$m = \frac{4 \pm \sqrt{4 \cdot 6}}{4}$$

$$m = \frac{4 \pm 2\sqrt{6}}{4}$$

$$m = \frac{2(2 \pm \sqrt{6})}{2 \cdot 2} \quad \text{[factor and reduce]}$$

$$m = \frac{2 \pm \sqrt{6}}{2}$$

The two solutions are $\dfrac{2 + \sqrt{6}}{2}$ and $\dfrac{2 - \sqrt{6}}{2}$

Write each of the following quadratic equations in the form $ax^2 + bx + c = 0$ and specify the values of a, b, and c.

Exercise 6.11

1. $2x^2 + 5x + 1 = 0$ **2.** $x^2 + 2x - 1 = 0$

3. $4x^2 - 3x = 5$ **4.** $2x^2 + 3x = 6$

5. $x^2 - 3x = 7$ **6.** $3x^2 + 4x = 7$

7. $x^2 - 5x + 2 = 0$ **8.** $2x^2 + 3 = x$

9. $4x^2 + 5 = 0$ **10.** $3x^2 - x = 0$

Solve these equations using the quadratic formula. If the solution contains radicals, then also give the decimal approximations. Round answers to the nearest thousandth.

11. $x^2 + 5x + 6 = 0$ **12.** $x^2 + 7x + 12 = 0$

13. $y^2 - 5y + 6 = 0$ **14.** $2m^2 + 3m = 2$

15. $x^2 - 9 = 0$ **16.** $2t^2 = 5t$

17. $3n^2 = 8n + 3$ **18.** $2x^2 + x - 6 = 0$

19. $y^2 - 3y + 1 = 0$ **20.** $2x^2 + 4x + 3 = 0$

21. $3t^2 + 2t + 5 = 0$ **22.** $x^2 - 4x + 1 = 0$

23. $3m^2 + 2m - 5 = 0$ **24.** $2r^2 = 2r + 3$

25. $3x^2 - 5x = 10$ **26.** $2y^2 = 3y - 4$

27. $x^2 - 2x - 10 = 0$ **28.** $r^2 - 4r - 1 = 0$

29. $6m^2 + m = 2$ **30.** $x^2 - x + 1 = 0$

31. $2x^2 = 2x + 3$ **32.** $4y = 2y^2 - 21$

33. $n^2 = 20$ **34.** $3x^2 = 4$

The following are more difficult equations. Use the quadratic formula to solve for only the exact solutions.

35. $-x^2 + 5x - 7 = 0$

36. $(2x - 3)(3x - 1) = 5$

37. $(3y - 5)(7y + 1) = (2y + 7)(6y - 1)$

38. $2x^2 + \dfrac{2}{3}x = 1$

39. $3y^2 + 2 = \dfrac{1}{3}y$ **40.** $\dfrac{3x + 1}{2x - 1} = \dfrac{2x}{x - 2}$

41. $\dfrac{m + 1}{m} - \dfrac{13}{6} = \dfrac{-m}{m + 1}$ **42.** $\dfrac{t + 3}{t - 1} = \dfrac{t^2 - t + 3}{t^2 - 4t + 5}$

43. $(x - 8)(2x - 3) = 34$

44. $2 + \dfrac{5}{x} = \dfrac{12}{x^2}$

45. $\dfrac{2}{5}y^2 - \dfrac{3}{5}y = 1$

46. $x = \dfrac{3}{2}x^2 - \dfrac{4}{3}$

Solve, using the quadratic formula.

47. A rectangle whose length is 4 in. more than its width has an area of 45 in². Find the length and width.

48. If a square machine part is changed to a rectangular part by decreasing one dimension of the square by 3 mm (millimeters), the area becomes 21 mm². Find the dimensions of the square machine part.

49. If two positive numbers differ by 3 and their product is 54, find the numbers.

50. A certain number plus 4 times its reciprocal is $8\dfrac{1}{2}$. Find the number.

51. The base of a triangle is 4 ft less that its height. The area of the triangle is 48 ft². Find the length of the base.

52. The area of a border around a picture 10 in. long and 8 in. wide is half the area of the picture. What are the outside dimensions of the border?

Given the quadratic formula $x = \dfrac{-b \pm \sqrt{b^2 - 4ac}}{2a}$, $b^2 - 4ac$ is called the *discriminant*. What can be said about the real solutions of a quadratic equation when:

53. $b^2 < 4ac$; i.e., the discriminant represents a negative number.

54. $b^2 = 4ac$; i.e., the discriminant represents zero.

55. $b^2 > 4ac$; i.e., the discriminant represents a positive number.

Summary

SYMBOLS

$\sqrt{\ }$ Radical sign
$\sqrt[n]{b}$ Index is n; radicand is b
$\pm$ Plus or minus
$\approx$ Approximately equal to

DEFINITION OF NEGATIVE EXPONENTS

$x^{-n} = \dfrac{1}{x^n}$ where $x \neq 0$

SCIENTIFIC NOTATION

A number is written in scientific notation when it is expressed as the product of a number between 1 and 10 and a power of 10.
Examples: $2,350 = 2.35 \times 10^3$ and $0.000623 = 6.23 \times 10^{-4}$

(These properties are true for each even root greater than or equal to 2.)
(1) Each positive number has two even roots.
(2) Zero has only one even root.
(3) Negative numbers have no real even roots.

**PROPERTIES OF
EVEN ROOTS**

(These properties are true for each odd root greater than or equal to 3.)
(1) Every real number has exactly one odd root.
(2) An odd root of a positive number is a positive number.
(3) An odd root of a negative number is a negative number.
(4) An odd root of zero is zero.

**PROPERTIES OF
ODD ROOTS**

If m and n are integers where $n \geq 2$, then

$$x^{m/n} = \sqrt[n]{x^m} \quad \text{or} \quad (\sqrt[n]{x})^m$$

If n is even, then $x \geq 0$

**DEFINITION OF
RATIONAL EXPONENTS**

$$\sqrt[n]{a \cdot b} = \sqrt[n]{a} \cdot \sqrt[n]{b}$$

**THE BASIC PROPERTY
OF RADICALS**

where (1) n is even and $a, b \geq 0$
        (2) n is odd and a, b represent any real numbers

Like radicals can be added or subtracted by combining their numerical coefficients. Radicals must be simplified before you attempt to add or subtract.

**ADDITION AND
SUBTRACTION OF
RADICAL EXPRESSIONS**

Radicals having the same index can be multiplied or divided by taking the product or the quotient of the radicands. The result is then placed under the corresponding radical sign.

**MULTIPLICATION AND
DIVISION OF RADICALS**

To rationalize a monomial denominator, multiply the numerator and denominator by a radical, which will produce a perfect n power in the denominator. This will clear the denominator of radicals.

**RATIONALIZING A
MONOMIAL
DENOMINATOR**

Multiply the numerator and denominator by a binomial identical to the denominator except for its middle sign. Then use the "sum times a difference" shortcut.

**RATIONALIZING A
BINOMIAL
DENOMINATOR
CONTAINING SQUARE
ROOTS**

Squaring both sides of an equation will preserve every solution of the original equation, but may produce some extraneous solutions that will not check in the original equation.

**THE SQUARING
PROPERTY OF
EQUATIONS**

Step (1)   Isolate the term containing the radical on one side of the equation.

Step (2)   Square both sides of the equation to eliminate the radical. Then solve the resulting equation.

**SOLVING EQUATIONS
CONTAINING SQUARE
ROOTS**

Step (3)   Check all answers in the original equation to determine any extraneous solutions.

THE SQUART ROOT PROPERTY OF EQUATIONS

If $x^2 = k$, where x is the variable and k is a nonnegative constant, then $x = \pm\sqrt{k}$

SOLVING QUADRATIC EQUATIONS BY COMPLETING THE SQUARE

Step (1)   Write the equation so that the terms containing a variable are on one side and the constant is on the other.

Step (2)   Divide both sides of the equation by the coefficient of the squared variable. Then add the number to both sides that creates a perfect square trinomial. This number is found by squaring one-half the coefficient of the middle term.

Step (3)   Write the perfect square trinomial as the square of a binomial. Then use the square root property of equations.

THE QUADRATIC FORMULA

$$x = \frac{-b \pm \sqrt{b^2 - 4ac}}{2a} \quad \text{where } a \neq 0$$

Evaluate these expressions.

1. 3^{-2}

2. 10^{-1}

3. $\left(\dfrac{2}{3}\right)^{-2}$

4. $\sqrt{16}$

5. $-\sqrt{16}$

6. $4^{1/2}$

7. $8^{1/3}$

8. $9^{-1/2}$

9. $(-27)^{2/3}$

10. $\left(\dfrac{4}{9}\right)^{3/2}$

Simplify by converting negative exponents to positive exponents.

11. $\dfrac{3x^{-2}}{5y^{-3}}$

12. $\dfrac{2a^2b^{-3}c^{-1}}{7d^{-3}e^4}$

13. $\dfrac{5}{x^{-2}+y^{-2}}$

Using the properties of exponents, simplify. The final answer should not contain negative exponents.

14. $(3a^{-5})(2a^8)$

15. $\dfrac{12x^3y^{-2}}{4x^5y^{-1}}$

16. $(3m^3n^{-2})^{-2}$

Convert these expressions to scientific notation.

17. 26,700

18. 0.00000703

Convert these expressions to ordinary notation.

19. 5.13×10^6

20. 9.11×10^{-4}

Simplify as far as possible. Do not evaluate.

21. $2\sqrt{75}$

22. $5\sqrt[3]{54}$

Combine, using addition or subtraction.

23. $3\sqrt{2} - 5\sqrt{2}$

24. $2\sqrt[3]{7} + 3\sqrt{5} - 4\sqrt[3]{7} - 2\sqrt{5}$

25. $5\sqrt[3]{16} - 2\sqrt[3]{54} + 3\sqrt{2} - \sqrt{8}$

Multiply or divide the following expressions.

26. $(3\sqrt{5})(2\sqrt{3})$

27. $3\sqrt[3]{2}(2\sqrt[3]{7} - 4\sqrt[3]{3})$

28. $(2\sqrt{3} + \sqrt{5})(2\sqrt{3} - \sqrt{5})$

29. $(3\sqrt{2} + 1)(4\sqrt{2} - 3)$

30. $\dfrac{6\sqrt{10}}{2\sqrt{5}}$

31. $\sqrt{\dfrac{3}{4}}$

Rationalize the denominators. Do not evaluate.

32. $\dfrac{2}{\sqrt{7}}$

33. $\dfrac{3\sqrt{2}}{5\sqrt{3}}$

34. $\dfrac{2}{3\sqrt[3]{2}}$ **35.** $\sqrt{\dfrac{1}{2}}$

36. $\dfrac{4}{3\sqrt{2} + \sqrt{3}}$

Solve the following radical equations.

37. $\sqrt{2y} = 4$ **38.** $\sqrt{2x - 5} = 1$

39. $\sqrt{2m - 3} = m - 3$

Solve the following quadratic equations by completing the square.

40. $2x^2 + 3x = 20$ **41.** $3x^2 - x - 1 = 0$

Solve the following quadratic equations by using the quadratic formula. If the answer involves a radical, give the approximate solutions as well as the exact solutions.

42. $x^2 - 3x - 4 = 0$ **43.** $x^2 - 3x + 1 = 0$

44. $2y^2 + 4y + 3 = 0$ **45.** $m^2 - 4m + 2 = 0$

Solve the following applied problems. Give the equation as well as the solutions.

46. Two positive numbers differ by 4, and their product is 117. Find the numbers.

47. A rectangle whose length is 3 in. more than its width has an area of 40 in.² Find the length and the width.

Solutions to Self-Checking Exercise

1. $3^{-2} = \dfrac{1}{3^2} = \dfrac{1}{9}$ **2.** $10^{-1} = \dfrac{1}{10}$

3. $\left(\dfrac{2}{3}\right)^{-2} = \left(\dfrac{3}{2}\right)^2 = \dfrac{9}{4}$ **4.** $\sqrt{16} = 4$

5. $-\sqrt{16} = -4$ **6.** $4^{1/2} = \sqrt{4} = 2$

7. $8^{1/3} = \sqrt[3]{8} = 2$ **8.** $9^{-1/2} = \dfrac{1}{\sqrt{9}} = \dfrac{1}{3}$

9. $(-27)^{2/3} = (\sqrt[3]{-27})^2 = 9$ **10.** $\left(\dfrac{4}{9}\right)^{3/2} = \left(\sqrt{\dfrac{4}{9}}\right)^3 = \dfrac{8}{27}$

11. $\dfrac{3x^{-2}}{5y^{-3}} = \dfrac{3y^3}{5x^2}$ **12.** $\dfrac{2a^2b^{-3}c^{-1}}{7d^{-3}e^4} = \dfrac{2a^2d^3}{7b^3ce^4}$

13. $\dfrac{5}{x^{-2} + y^{-2}} = \dfrac{5}{\dfrac{1}{x^2} + \dfrac{1}{y^2}}$ **14.** $(3a^{-5})(2a^8) = 6a^3$

$\qquad\qquad = \dfrac{5x^2y^2}{x^2y^2\left(\dfrac{1}{x^2} + \dfrac{1}{y^2}\right)}$

$\qquad\qquad = \dfrac{5x^2y^2}{y^2 + x^2}$

15. $\dfrac{12x^3y^{-2}}{4x^5y^{-1}} = 3x^{-2}y^{-1} = \dfrac{3}{x^2y}$

16. $(3m^3n^{-2})^{-2} = 3^{-2}m^{-6}n^4$

$$= \dfrac{1}{9}\cdot\dfrac{1}{m^6}\cdot n^4$$

$$= \dfrac{n^4}{9m^6}$$

17. $26{,}700 = 2.67 \times 10^4$

18. $0.00000703 = 7.03 \times 10^{-6}$

19. $5.13 \times 10^6 = 5{,}130{,}000$

20. $9.11 \times 10^{-4} = 0.000911$

21. $2\sqrt{75} = 2\cdot\sqrt{25\cdot 3}$
$\qquad\quad = 2\cdot 5\cdot\sqrt{3}$
$\qquad\quad = 10\sqrt{3}$

22. $5\sqrt[3]{54} = 5\cdot\sqrt[3]{27\cdot 2}$
$\qquad\quad = 5\cdot 3\cdot\sqrt[3]{2}$
$\qquad\quad = 15\sqrt[3]{2}$

23. $3\sqrt{2} - 5\sqrt{2} = -2\sqrt{2}$

24. $2\sqrt[3]{7} + 3\sqrt{5} - 4\sqrt[3]{7} - 2\sqrt{5} = -2\sqrt[3]{7} + \sqrt{5}$

25. $5\sqrt[3]{16} - 2\sqrt[3]{54} + 3\sqrt{2} - \sqrt{8} = 5\cdot\sqrt[3]{8\cdot 2} - 2\sqrt[3]{27\cdot 2} + 3\cdot\sqrt{2} - \sqrt{4\cdot 2}$
$\qquad\qquad\qquad\qquad\qquad\qquad = 5\cdot 2\cdot\sqrt[3]{2} - 2\cdot 3\cdot\sqrt[3]{2} + 3\cdot\sqrt{2} - 2\cdot\sqrt{2}$
$\qquad\qquad\qquad\qquad\qquad\qquad = 10\sqrt[3]{2} - 6\sqrt[3]{2} + 3\sqrt{2} - 2\sqrt{2}$
$\qquad\qquad\qquad\qquad\qquad\qquad = 4\sqrt[3]{2} + \sqrt{2}$

26. $(3\sqrt{5})(2\sqrt{3}) = 6\sqrt{15}$

27. $3\sqrt[3]{2}(2\sqrt[3]{7} - 4\sqrt[3]{3}) = 6\sqrt[3]{14} - 12\sqrt[3]{6}$

28. $(2\sqrt{3} + \sqrt{5})(2\sqrt{3} - \sqrt{5}) = 4\sqrt{9} - \sqrt{25}$
$\qquad\qquad\qquad\qquad\qquad\quad = 4\cdot 3 - 5$
$\qquad\qquad\qquad\qquad\qquad\quad = 12 - 5$
$\qquad\qquad\qquad\qquad\qquad\quad = 7$

29. $\overset{\qquad\quad F\qquad\qquad O\qquad\qquad I\qquad\qquad L}{(3\sqrt{2} + 1)(4\sqrt{2} - 3) = (3\sqrt{2})(4\sqrt{2}) - 3(3\sqrt{2}) + 1(4\sqrt{2}) - 1\cdot 3}$
$\qquad\qquad\qquad\qquad\quad = \quad 12\sqrt{4}\quad - \quad 9\sqrt{2}\quad + \quad 4\sqrt{2}\quad - \quad 3$
$\qquad\qquad\qquad\qquad\quad = 24 - 5\sqrt{2} - 3$
$\qquad\qquad\qquad\qquad\quad = 21 - 5\sqrt{2}$

30. $\dfrac{6\sqrt{10}}{2\sqrt{5}} = 3\sqrt{2}$

31. $\sqrt{\dfrac{3}{4}} = \dfrac{\sqrt{3}}{\sqrt{4}} = \dfrac{\sqrt{3}}{2}$

32. $\dfrac{2}{\sqrt{7}} = \dfrac{2\cdot\sqrt{7}}{\sqrt{7}\cdot\sqrt{7}}$
$\qquad = \dfrac{2\sqrt{7}}{7}$

33. $\dfrac{3\sqrt{2}}{5\sqrt{3}} = \dfrac{3\sqrt{2}\cdot\sqrt{3}}{5\sqrt{3}\cdot\sqrt{3}}$
$\qquad = \dfrac{3\sqrt{6}}{5\cdot 3}$
$\qquad = \dfrac{\sqrt{6}}{5}$

34. $\dfrac{2}{3\sqrt[3]{2}} = \dfrac{2\cdot\sqrt[3]{4}}{3\sqrt[3]{2}\cdot\sqrt[3]{4}}$
$\qquad = \dfrac{2\sqrt[3]{4}}{3\cdot 2}$
$\qquad = \dfrac{\sqrt[3]{4}}{3}$

35. $\sqrt{\dfrac{1}{2}} = \dfrac{\sqrt{1}}{\sqrt{2}}$
$\qquad = \dfrac{1}{\sqrt{2}}$
$\qquad = \dfrac{1\cdot\sqrt{2}}{\sqrt{2}\cdot\sqrt{2}}$
$\qquad = \dfrac{\sqrt{2}}{2}$

36. $\dfrac{4}{3\sqrt{2} + \sqrt{3}} = \dfrac{4(3\sqrt{2} - \sqrt{3})}{(3\sqrt{2} + \sqrt{3})(3\sqrt{2} - \sqrt{3})}$

$\phantom{\dfrac{4}{3\sqrt{2} + \sqrt{3}}} = \dfrac{4(3\sqrt{2} - \sqrt{3})}{9\sqrt{4} - \sqrt{9}}$

$\phantom{\dfrac{4}{3\sqrt{2} + \sqrt{3}}} = \dfrac{4(3\sqrt{2} - \sqrt{3})}{18 - 3}$

$\phantom{\dfrac{4}{3\sqrt{2} + \sqrt{3}}} = \dfrac{12\sqrt{2} - 4\sqrt{3}}{15}$

37.
$$\sqrt{2y} = 4$$
$$(\sqrt{2y})^2 = 4^2$$
$$2y = 16$$
$$y = 8$$
(no extraneous solutions)

38.
$$\sqrt{2x - 5} = 1$$
$$(\sqrt{2x - 5})^2 = 1^2$$
$$2x - 5 = 1$$
$$2x = 6$$
$$x = 3$$
(no extraneous solutions)

39.
$$\sqrt{2m - 3} = m - 3$$
$$(\sqrt{2m - 3})^2 = (m - 3)^2$$
$$2m - 3 = m^2 - 6m + 9$$
$$0 = m^2 - 8m + 12$$
$$0 = (m - 6)(m - 2)$$
$$m = 6, \ m = 2$$
2 is extraneous; the only solution is 6

40.
$$2x^2 + 3x = 20$$
$$x^2 + \frac{3}{2}x = 10$$
$$x^2 + \frac{3}{2}x + \frac{9}{16} = 10 + \frac{9}{16}$$
$$x^2 + \frac{3}{2}x + \frac{9}{16} = \frac{169}{16}$$
$$\left(x + \frac{3}{4}\right)^2 = \frac{169}{16}$$
$$x + \frac{3}{4} = \pm\frac{13}{4}$$
$$x = -\frac{3}{4} \pm \frac{13}{4}$$
$$x = \frac{-3 \pm 13}{4}$$
$$x = \frac{5}{2} \ \text{ and } \ x = -4$$

41.
$$3x^2 - x - 1 = 0$$
$$3x^2 - x = 1$$
$$x^2 - \frac{1}{3}x = \frac{1}{3}$$
$$x^2 - \frac{1}{3}x + \frac{1}{36} = \frac{1}{3} + \frac{1}{36}$$
$$x^2 - \frac{1}{3}x + \frac{1}{36} = \frac{13}{36}$$
$$\left(x - \frac{1}{6}\right)^2 = \frac{13}{36}$$
$$x - \frac{1}{6} = \pm\frac{\sqrt{13}}{6}$$
$$x = \frac{1}{6} \pm \frac{\sqrt{13}}{6}$$
$$x = \frac{1 \pm \sqrt{13}}{6}$$

42. $x^2 - 3x - 4 = 0$
$$a = 1 \quad b = -3 \quad c = -4$$
$$x = \frac{3 \pm \sqrt{9 - 4 \cdot 1 \cdot (-4)}}{2 \cdot 1}$$
$$x = \frac{3 \pm \sqrt{25}}{2}$$
$$x = \frac{3 \pm 5}{2}$$
$$x = 4 \ \text{ and } \ x = -1$$

43. $x^2 - 3x + 1 = 0$

$a = 1 \quad b = -3 \quad c = 1$

$x = \dfrac{3 \pm \sqrt{9 - 4 \cdot 1 \cdot 1}}{2 \cdot 1}$

$x = \dfrac{3 \pm \sqrt{5}}{2}$ [exact solutions]

$x \approx \dfrac{3 \pm 2.236}{2}$

$x \approx 2.618 \quad \text{and} \quad x \approx 0.382$

44. $2y^2 + 4y + 3 = 0$

$a = 2 \quad b = 4 \quad c = 3$

$y = \dfrac{-4 \pm \sqrt{16 - 4 \cdot 2 \cdot 3}}{2 \cdot 2}$

$y = \dfrac{-4 \pm \sqrt{-8}}{4}$

no real solutions

45. $m^2 - 4m + 2 = 0$

$a = 1 \quad b = -4 \quad c = 2$

$m = \dfrac{4 \pm \sqrt{16 - 4 \cdot 1 \cdot 2}}{2 \cdot 1}$

$m = \dfrac{4 \pm \sqrt{8}}{2} = \dfrac{4 \pm 2\sqrt{2}}{2} = \dfrac{\cancel{2}(2 \pm \sqrt{2})}{\cancel{2}}$

$m = 2 \pm \sqrt{2}$ [exact solutions]

$m \approx 2 \pm 1.414$

$m \approx 3.414 \quad \text{and} \quad m \approx 0.586$

46. $x(x - 4) = 117$

        $x^2 - 4x = 117$

$x^2 - 4x - 117 = 0$

$a = 1 \quad b = -4 \quad c = -117$

$x = \dfrac{4 \pm \sqrt{16 - 4 \cdot 1 \cdot (-117)}}{2 \cdot 1}$

$x = \dfrac{4 \pm \sqrt{484}}{2}$

$x = \dfrac{4 \pm 22}{2}$

$x = 13 \quad \text{and} \quad x = -9$

13 is the only positive solution.

Thus, the numbers are 13 and 9.

47. $w(w + 3) = 40$

        $w^2 + 3w = 40$

$w^2 + 3w - 40 = 0$

$a = 1 \quad b = 3 \quad c = -40$

$w = \dfrac{-3 \pm \sqrt{9 - 4 \cdot 1 \cdot (-40)}}{2 \cdot 1}$

$w = \dfrac{-3 \pm \sqrt{169}}{2}$

$w = \dfrac{-3 \pm 13}{2}$

$w = 5 \quad \text{and} \quad w = -8$

5 is the only positive solution.

Thus, the width is 5 in. and the length is 8 in.

INTRODUCTION TO GRAPHING

7

EQUATIONS CONTAINING TWO VARIABLES

7.1

Thus far you have learned to solve several types of equations. These include first degree equations, such as $3x + 1 = 7$; fractional equations, such as $x + 2 = \dfrac{6}{x + 3}$; and quadratic equations, such as $2x^2 - 5x - 3 = 0$. These equations share a common characteristic—they each contain a single variable and each solution is composed of a single number.

Next we turn our attention to equations containing two variables, such as $y = 3x$ or $y = x^2 - 4$. You will find that equations of this type usually have an infinite number of solutions. And each solution will be composed of a pair of numbers, one for each variable.

Let us formulate an equation containing two variables by considering this illustration. A bank makes a monthly charge on checking accounts by charging 60¢ per month plus 10¢ per check. This can be symbolized by the equation $y = 10x + 60$, where y is the monthly charge in cents for x number of checks. For example, to find the monthly charge for writing 4 checks, we substitute 4 for x and solve for y, as follows:

$$y = 10x + 60$$
$$y = 10(4) + 60$$
$$y = 40 + 60$$
$$y = 100$$

If $x = 4$, then $y = 100$. Thus, the monthly charge for writing 4 checks is 100¢, or $1.00. Likewise, we find the monthly charge for writing 8 checks.

$$y = 10x + 60$$
$$y = 10(8) + 60$$
$$y = 80 + 60$$
$$y = 140$$

If $x = 8$, then $y = 140$. Thus, the monthly charge for writing 8 checks is 140¢, or $1.40. Continuing in the same manner, we find the monthly charge for writing 12 checks.

$$y = 10x + 60$$
$$y = 10(12) + 60$$
$$y = 120 + 60$$
$$y = 180$$

If $x = 12$, then $y = 180$. Therefore, the monthly charge for writing 12 checks is 180¢, or $1.80. Obviously, this process can be carried on indefinitely by substituting other values for x.

You should see that in the equation $y = 10x + 60$ the monthly charge y depends upon the number of checks x. In other words, the value of y is dependent upon the value chosen for x. Therefore, x is said to be the **independent variable** and y is said to be the **dependent variable**. Often we find it convenient to list the solutions in a table of values. The values of the independent variable are placed on the left, while the corresponding values of the dependent variable are placed on the right.

independent variable

dependent variable

Table of Values	
x	y
4	100
8	140
12	180

Since a solution consists of two numbers, it is written as a pair. For example, the solution represented by $x = 4$ and $y = 100$ is abbreviated (4, 100) Within parentheses the x-value is written first and the corresponding y-value is written second. A pair of numbers written in this form is called an **ordered pair**.

ordered pair

To summarize the preceding illustration we note that:

(1)   The equation of two variables is $y = 10x + 60$.
(2)   The independent variable is x.
(3)   The dependent variable is y.
(4)   A table of values, such as the following, gives rise to ordered pairs.

Table of Values		Ordered pairs	Meaning
x	y		
0	60	(0, 60)	If $x = 0$, then $y = 60$
2	80	(2, 80)	If $x = 2$, then $y = 80$
4	100	(4, 100)	If $x = 4$, then $y = 100$
8	140	(8, 140)	If $x = 8$, then $y = 140$
12	180	(12, 180)	If $x = 12$, then $y = 180$

With these techniques we can analyze more equations containing two variables.

Given $y = 2x$, find y when $x = -2$, $x = 0$, and $x = 3$. Give a table of values and also write the solutions as ordered pairs.

Example 1

Solution: Substitute each of the values of x into the equation $y = 2x$.
(a)   When $x = -2$, $y = 2(-2)$, which is -4
(b)   When $x = 0$, $y = 2(0)$, which is 0
(c)   When $x = 3$, $y = 2(3)$, which is 6

Table of Values		Ordered pairs
x	y	
-2	-4	$(-2, -4)$
0	0	$(0, 0)$
3	6	$(3, 6)$

Example 2   Given $y = x^2 + 5x + 6$, find y when $x = -2$, $x = -1$, $x = 0$, and $x = 3$. Give a table of values and write the solutions as ordered pairs.

Solution: Substitute each of the values of x into the equation $y = x^2 + 5x + 6$
(a)   When $x = -2$, $y = (-2)^2 + 5(-2) + 6$
$$y = 4 - 10 + 6$$
$$y = 0$$

(b)   When $x = -1$, $y = (-1)^2 + 5(-1) + 6$
$$y = 1 - 5 + 6$$
$$y = 2$$

(c)   When $x = 0$, $y = 0^2 + 5(0) + 6$
$$y = 0 + 0 + 6$$
$$y = 6$$

(d)   When $x = 3$, $y = 3^2 + 5(3) + 6$
$$y = 9 + 15 + 6$$
$$y = 30$$

Table of Values		Ordered pairs
x	y	
-2	0	$(-2, 0)$
-1	2	$(-1, 2)$
0	6	$(0, 6)$
3	30	$(3, 30)$

Exercise 7.1   Give a table of values illustrating each of the following equations. Also write each solution as an ordered pair.

1. Given $y = 5x$, find y when $x = -2$, $x = 0$, $x = 1$, and $x = 3$

2. Given $y = 2x + 3$, find y when $x = -3$, $x = -1$, $x = 0$, and $x = 2$

3. Given $y = 3x - 1$, find y when $x = -4$, $x = -2$, $x = 0$, and $x = 1$

4. Given $y = -x + 2$, find y when $x = -3$, $x = -\frac{1}{2}$, and $x = \frac{3}{2}$

5. Given $y = x^2$, find y when $x = -3$, $x = -1$, $x = 0$, and $x = 2$

6. Given $y = x^2 - 4$, find y when $x = -5$, $x = -2$, $x = 1$, and $x = 4$

7. Given $y = x^2 - 6x + 8$, find y when $x = -4$, $x = -1$, $x = 0$, and $x = 3$

8. Given $y = \frac{1}{x}$, find y when $x = -4$, $x = -2$, $x = \frac{1}{2}$, and $x = 1$

9. Given $y = \sqrt{9 - x^2}$, find y when $x = -3$, $x = -1$, $x = 0$, and $x = 2$

10. Given $2x + 3y = 6$, find y when $x = -3$, $x = -1$, $x = 0$, and $x = 3$

Complete the following tables.

11. $y = x$

x	y
-5	
-3	
-1	
0	
2	
4	
6	

12. $y = -x$

x	y
-5	
-3	
-1	
0	
2	
4	
6	

13. $y = 2x$

x	y
-6	
-4	
$-\frac{3}{2}$	
0	
$\frac{1}{2}$	
2	
4	

14. $y = 2x + 1$

x	y
-5	
-3	
-1	
$-\frac{1}{2}$	
0	
1	
2	

15. $y = 3x + 1$

x	y
-4	
-2	
-1	
0	
$\frac{1}{3}$	
$\frac{4}{3}$	
2	

16. $y = 2x + 4$

x	y
-3	
-1	
$-\frac{1}{2}$	
0	
1	
3	
5	

17. $y = -x - 1$

x	y
-3	
-2	
$-\dfrac{1}{2}$	
0	
2	
$\dfrac{5}{2}$	

18. $y = -2x - 3$

x	y
-4	
-2	
-1	
0	
1	
3	
5	

19. $y = \dfrac{1}{2}x + 2$

x	y
-4	
-2	
-1	
0	
2	
3	
4	

20. $x - y = 3$

x	y
-3	
-1	
0	
1	
2	
3	

21. $2x - y = 4$

x	y
-4	
-3	
-1	
0	
2	
4	

22. $x + 3y = 6$

x	y
	-3
	-1
	0
	$\dfrac{1}{3}$
	1
	2

23. $y = x^2 + 1$

x	y
-4	
-3	
-2	
-1	
0	
1	
2	
3	

24. $y = -x^2 + 2$

x	y
-4	
-3	
-2	
-1	
0	
1	
2	
3	

25. $y = 3x^2 + 1$

x	y
−3	
−2	
−1	
0	
1	
2	
3	

26. $y = x^2 - 2x - 3$

x	y
−3	
−2	
−1	
0	
1	
2	
3	
4	

27. $y = 2x^2 - 3x - 2$

x	y
−2	
−1	
$-\dfrac{1}{2}$	
0	
1	
2	
3	

28. $y = 3x^2 - x$

x	y
−2	
−1	
0	
$\dfrac{1}{3}$	
1	
2	
3	

29. $x = -2y^2$

x	y
	−3
	−2
	−1
	0
	1
	2
	3

30. $x = y^2 - 7y + 10$

x	y
	−1
	0
	1
	2
	3
	4
	5

31. $x = 3y^2 - 5y - 2$

x	y
	−2
	−1
	$-\dfrac{1}{3}$
	0
	1
	2
	3

32. $y = \dfrac{1}{x}$

x	y
−5	
−3	
−2	
−1	
$\dfrac{1}{2}$	
1	
2	
10	

33. $y = \dfrac{10}{x}$

x	y
−100	
−10	
−5	
−2	
$-\dfrac{1}{2}$	
1	
2	
10	

34. $y = \sqrt{25 - x^2}$

x	y
−5	
−3	
−2	
0	
1	
3	
5	

35. $y = |x|$

x	y
−10	
−5	
−2	
$-\dfrac{3}{2}$	
−1	
0	
3	

36. $s = \dfrac{1}{t^2}$

t	s
−3	
−2	
−1	
$\dfrac{1}{2}$	
1	
2	
3	

37. $V = \dfrac{100}{P}$

P	V
$\dfrac{1}{10}$	
$\dfrac{1}{5}$	
$\dfrac{1}{2}$	
1	
2	
5	
10	

Decide whether or not the given ordered pair is a solution of the given equation.

38. $y = -3x + 2;$ $(-1, 5)$

39. $x - 3y = 5;$ $(11, 2)$

40. $y = -x^2 - x;$ $(-2, 6)$

41. $x - 2y^2 = 0;$ $(0, -2)$

42. $x = 2y^2 - y + 1;$ $(0, 1)$

43. $y = \sqrt{16 - x^2};$ $(0, 4)$

44. $y = \dfrac{9}{x^2};$ $(-3, 1)$

45. $y = \dfrac{x + 3}{x - 5};$ $\left(0, -\dfrac{3}{5}\right)$

46. $y = \dfrac{1}{2}x^2;$ $(2, 1)$

47. $\dfrac{y}{x} = 3;$ $(3, 1)$

48. $x^2 + y^2 = 25;$ $(0, -5)$

49. $y = \dfrac{1}{\sqrt{x^2 + 5}};$ $\left(-2, \dfrac{1}{3}\right)$

50. $x^{1/2} + y^{1/2} = 5;$ $(4, 9)$

Describe each of these situations with an equation of two variables.

51. A car rental agency charges a daily rate of $10 plus 15¢ per mile.

52. For each trip, a taxi company charges 75¢ plus 30¢ per mile.

53. A bank makes a monthly charge on checking accounts of 75¢ plus 15¢ per check.

54. A plumber charges $15 per trip plus $20 per hour.

THE RECTANGULAR COORDINATE SYSTEM

7.2

In section 7.1 we discussed equations of two variables and found that their solutions consist of ordered pairs. In this section you will learn how ordered pairs are used to denote points in a plane. A **plane** is a flat level surface having no thickness and extending infinitely in two dimensions. A **point** is a location in the plane. It has no size, only location.

plane

point

The method we use to name points in a plane was first invented by a French philosopher and mathematician named René Descartes (1596 – 1650). He found, as shown in figure 7.1, that a plane can be partitioned into a coordinate grid by constructing two perpendicular number lines called the **x-axis** and the **y-axis** Their point of intersection is called the **origin** and is labeled point O. Note that on the x-axis the positive direc-

x-axis   y-axis

origin

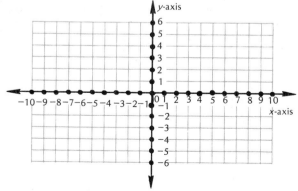

FIGURE 7.1

abscissa

ordinate

tion is to the right, while on the y-axis the positive direction is up. This configuration is called the *rectangular coordinate system*. In honor of its inventor, it is also referred to as the *Cartesian coordinate system*.

To identify a point in the rectangular coordinate system, we use an ordered pair of the form (x, y). The x-coordinate is called the **abscissa** and measures the directed distance to the point along the x-axis. The y-coordinate is called the **ordinate** and measures the directed distance to the point along the y-axis. To illustrate, study figure 7.2, where we have located two points, (7, 4) and (−5, −3). In the ordered pair (7, 4), the abscissa 7 tells us that the point is located 7 units to the right of the y-axis. The ordinate 4 tells us that the point is located 4 units above the x-axis. Likewise, in the ordered pair (−5, −3) the abscissa −5 tells us that the point is located 5 units to the left of the y-axis, while the ordinate −3 tells us the point is located 3 units below the x-axis.

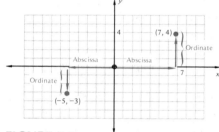

FIGURE 7.2

Example 1

On a rectangular coordinate system plot the points corresponding to these ordered pairs:

(a) (0, 0)   (b) (3, 0)
(c) (0, 4)   (d) (9, 5)
(e) (−6, 2)   (f) (−8, 0)
(g) (−3, −1)   (h) (0, −5)
(i) (7, −4)

Solution: Construct a rectangular coordinate system as seen in figure 7.3. Remember, the abscissa tells how far the point is located to the left or right of the y-axis. The ordinate tells how far the point is located above or below the x-axis.

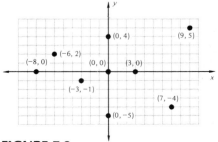

FIGURE 7.3

As you have learned, ordered pairs are used to identify points in the plane. The following summary reviews the important details.

Summary of Ordered Pairs

Given an ordered pair (x, y):

(1)   The abscissa x gives the distance the point is located to the right or left of the y-axis.

    (a) If $x > 0$, the point is to the right of the y-axis
    (b) If $x < 0$, the point is to the left of the y-axis
    (c) If $x = 0$, the point is located on the y-axis

(2)   The ordinate y gives the distance the point is located above or below the x-axis.

    (a) If $y > 0$, the point is above the x-axis
    (b) If $y < 0$, the point is below the x-axis
    (c) If $y = 0$, the point is located on the x-axis

(3)   The ordered pair $(0, 0)$ names the origin. It is the point where the x and y-axis intersect.

The x- and y-axes divide the plane into four regions called **quadrants**. Note in figure 7.4 that the quadrants are numbered I through IV, proceeding counterclockwise. The axes are considered to be boundary lines and are not part of any quadrant. As illustrated by the figure, each point in quadrant I has a positive abscissa and a positive ordinate; each point in quadrant II has a negative abscissa and a positive ordinate; in quadrant III each point has a negative abscissa and a negative ordinate; and, in quadrant IV each point has a positive abscissa and a negative ordinate.

quadrants

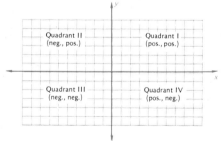

FIGURE 7.4

Without plotting, give the quadrant in which each of these points is located.

(a) $(-2, 6)$
(c) $(0, 3)$
(e) $(-2, 0)$

(b) $(-5, -1)$
(d) $(4, -2)$
(f) $(1, 3)$

Example 2

Solution: (a)   (−2, 6) is in quadrant II since $x < 0$ and $y > 0$

(b)   (−5, −1) is in quadrant III since $x < 0$ and $y < 0$

(c)   (0, 3) is not in any quadrant. It is located on the y-axis since $x = 0$

(d)   (4, −2) is in quadrant IV since $x > 0$ and $y < 0$

(e)   (−2, 0) is not in any quadrant. It is located on the x-axis since $y = 0$

(f)   (1, 3) is in quadrant I since $x > 0$ and $y > 0$

Exercise 7.2

Using ordered pairs, identify each of the points in figure 7.5.

1. A	**2.** B	**3.** C
4. D	**5.** E	**6.** F
7. G	**8.** H	**9.** I
10. J	**11.** K	**12.** L
13. M	**14.** N	**15.** O
16. P	**17.** Q	**18.** R
19. S	**20.** T	

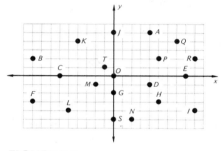

FIGURE 7.5

Write an ordered pair to fit each situation.

21. The abscissa is −6 and the ordinate is 3

22. The abscissa is 3 and the ordinate is −6

23. $x = -1$ and $y = 4$

24. $x = 4$ and $y = -1$

25. It names the origin.

26. It names a point which is 5 units below the x-axis and 3 units to the right of the y-axis.

Specify the abscissa and the ordinate for each of these ordered pairs.

27. (−5, 1)   **28.** (0, 3)

29. (3, 0)   **30.** (2, 3)

31. (2, −1)   **32.** $\left(-\dfrac{1}{2}, \dfrac{4}{5}\right)$

33. $\left(\dfrac{7}{3}, -\dfrac{1}{6}\right)$ **34.** $(\sqrt{3}, \sqrt{7})$

Give the quadrant in which each of these points is located.

35. (2, 4)   **36.** (−3, 5)

37. (−10, 12)   **38.** (1, −6)

39. (0, −6)   **40.** (0, 0)

41. (1, 1)   **42.** (−5, −10)

43. $\left(-\dfrac{1}{2}, -4\right)$ **44.** $\left(\dfrac{1}{3}, \dfrac{4}{5}\right)$

45. $(\sqrt{2}, -\sqrt[3]{5})$ **46.** (a, b) where a > 0 and b < 0

Fill in the blanks to make each statement true.

47. The x- and y-axes intersect at the point having coordinates (__, __), and this point is called the _____.

48. Any point located on the x-axis has an ordinate of _____.

49. Any point located on the y-axis has an abscissa of _____.

50. The point named by (−2, −6) is located __ units below the x-axis and __ units to the left of the y-axis.

GRAPHING EQUATIONS OF TWO VARIABLES

7.3

A graph is a picture illustrating a relationship between two variables. Each day you probably see various graphs in newspapers and magazines. Articles on many different subjects are illustrated with graphs because they are an excellent way of clarifying complex relationships. They show immediately any changes within the variables. Figure 7.6 illustrates the relationship between the speed of a full-size automobile and the distance it will travel after the driver decides to apply the brakes. The speed of the automobile is represented on the x-axis, while the corresponding stopping distance is shown on the y-axis.

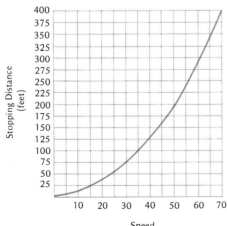

FIGURE 7.6

Selecting specific examples, the graph shows that a car traveling 25 mph will travel 50 ft after the driver decides to apply the brakes. At a speed of 50 mph, the car would travel 200 ft. And, if the car were traveling 70 mph, it would travel an incredible 400 ft before the driver is able to bring it to a halt. This graph also illustrates visually that as the speed of the automobile increases, the stopping distance increases even more dramatically.

graph

For the rest of this section we will use our knowledge of the rectangular coordinate system and learn to construct graphs of equations containing two variables. In mathematics, a **graph** is defined to be the set of points whose coordinates satisfy the given condition. Thus, to construct a graph we follow this procedure:

Constructing a Graph

Step (1)   Make a table of values by selecting values for one of the variables. This will provide you with ordered pairs. Obtain as many as necessary to determine a pattern of points.

Step (2)   Plot the points corresponding to the ordered pairs. Remember, if you need more points, go back to the table of values and obtain more ordered pairs.

Step (3)   Look for a pattern. Then connect the points with a smooth line or curve.

Example 1

Graph $y = 2x$

Solution:

Step (1)   Make a table of values. Select any values for x that you wish. However, it is best to choose some positive values as well as some negative values. The compute the corresponding values of y. This provides several ordered pairs.

Table of Values		Ordered pairs	Computation
x	y		$y = 2x$
−3	−6	(−3, −6)	If $x = -3$, $y = 2(-3) = -6$
−2	−4	(−2, −4)	If $x = -2$, $y = 2(-2) = -4$
−1	−2	(−1, −2)	If $x = -1$, $y = 2(-1) = -2$
0	0	(0, 0)	If $x = 0$, $y = 2(0) = 0$
1	2	(1, 2)	If $x = 1$, $y = 2(1) = 2$
2	4	(2, 4)	If $x = 2$, $y = 2(2) = 4$
3	6	(3, 6)	If $x = 3$, $y = 2(3) = 6$

Step (2) Plot the points corresponding to the ordered pairs (see figure 7.7).

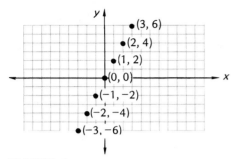

FIGURE 7.7

Step (3) Connect the points with a smooth line (see figure 7.8). The arrows at each end indicate that the graph continues on indefinitely.

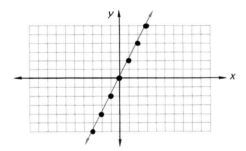

FIGURE 7.8

Graph $y = x + 2$ Example 2

Solution:

Step (1) Make a table of values. Select values for x and compute the corresponding values of y.

Table of Values		Ordered pairs	Computation $y = x + 2$
x	y		
-4	-2	$(-4, -2)$	If $x = -4$, $y = -4 + 2 = -2$
-3	-1	$(-3, -1)$	If $x = -3$, $y = -3 + 2 = -1$
-2	0	$(-2, 0)$	If $x = -2$, $y = -2 + 2 = 0$
-1	1	$(-1, 1)$	If $x = -1$, $y = -1 + 2 = 1$
0	2	$(0, 2)$	If $x = 0$, $y = 0 + 2 = 2$
1	3	$(1, 3)$	If $x = 1$, $y = 1 + 2 = 3$
2	4	$(2, 4)$	If $x = 2$, $y = 2 + 2 = 4$

Step (2) Plot each of the points on a rectangular coordinate system (see figure 7.9).

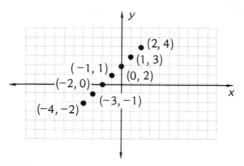

FIGURE 7.9

> **Step (3)** Connect the points with a smooth line to obtain a straight line (see figure 7.10).

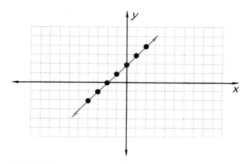

FIGURE 7.10

Example 3   Graph $y = x^2$

Solution:

> **Step (1)** Make a table of values. Select values for x and compute the corresponding values of y.

Table of Values		Ordered pairs	Computation $y = x^2$
x	y		
-3	9	$(-3, 9)$	If $x = -3$, then $y = (-3)^2 = 9$
-2	4	$(-2, 4)$	If $x = -2$, then $y = (-2)^2 = 4$
-1	1	$(-1, 1)$	If $x = -1$, then $y = (-1)^2 = 1$
0	0	$(0, 0)$	If $x = 0$, then $y = 0^2 = 0$
1	1	$(1, 1)$	If $x = 1$, then $y = 1^2 = 1$
2	4	$(2, 4)$	If $x = 2$, then $y = 2^2 = 4$
3	9	$(3, 9)$	If $x = 3$, then $y = 3^2 = 9$

> **Step (2)** Plot each of the points (see figure 7.11).

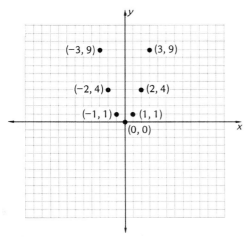

FIGURE 7.11

Step (3) Connect the points with a smooth curve (see figure 7.12). This U-shaped graph is called a *parabola*. It will be studied in more detail in section 7.8.

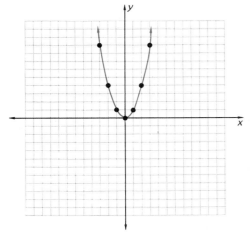

FIGURE 7.12

Graph $x = 2y^2 - 1$ **Example 4**

Solution:

Step (1) Make a table of values. For this equation it is easier to select values for y and then compute the corresponding values for x.

Table of Values		Ordered pairs	Computation $x = 2y^2 - 1$
x	y		
7	−2	(7, −2)	If $y = -2$, then $x = 2(-2)^2 - 1 = 7$
1	−1	(1, −1)	If $y = -1$, then $x = 2(-1)^2 - 1 = 1$
−1	0	(−1, 0)	If $y = 0$, then $x = 2 \cdot 0^2 - 1 = -1$
1	1	(1, 1)	If $y = 1$, then $x = 2 \cdot 1^2 - 1 = 1$
7	2	(7, 2)	If $y = 2$, then $x = 2 \cdot 2^2 - 1 = 7$

Step (2)   Plot each of the points (see figure 7.13).

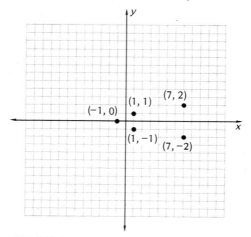

FIGURE 7.13

Step (3)   Connect the points with a smooth curve to obtain a parabola.

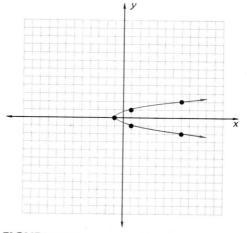

FIGURE 7.14

The formula $s = 16t^2$ gives the distance in feet that a dropped object will fall in t seconds due to the force of gravity. Construct the graph of this relationship.

Example 5

Solution:

Step (1) Make a table of values. t is the independent variable and represents time. Therefore, we will select nonnegative numbers for t and we will write the ordered pairs in the form (t, s).

Table of Values		Ordered pairs	Computation $s = 16t^2$
t	s		
0	0	$(0, 0)$	If $t = 0$ sec, then $s = 16 \cdot 0^2$ or 0 ft.
1	16	$(1, 16)$	If $t = 1$ sec, then $s = 16 \cdot 1^2$ or 16 ft.
2	64	$(2, 64)$	If $t = 2$ sec, then $s = 16 \cdot 2^2$ or 64 ft.
3	144	$(3, 144)$	If $t = 3$ sec, then $s = 16 \cdot 3^2$ or 144 ft.
4	256	$(4, 256)$	If $t = 4$ sec, then $s = 16 \cdot 4^2$ or 256 ft.
5	400	$(5, 400)$	If $t = 5$ sec, then $s = 16 \cdot 5^2$ or 400 ft.

Step (2) Plot each of the points on a rectangular coordinate system (see figure 7.15). The independent variable is t, so it is assigned to the horizontal axis. Since neither variable represents a negative number, we use only the first quadrant. For convenience, the vertical s-axis is given a smaller unit of measure.

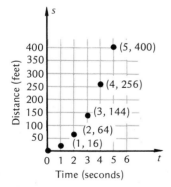

FIGURE 7.15

Step (3) Connect the points with a smooth curve, getting half of a parabola. Figure 7.16 gives a visual representation of the formula $s = 16t^2$.

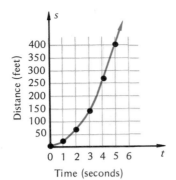

FIGURE 7.16

Exercise 7.3

Graph the following equations.

1. $y = x$ 2. $y = -x$
3. $y = 3x$ 4. $y = -3x$
5. $y = x + 3$ 6. $y = x - 3$
7. $y = 2x + 1$ 8. $y = 2x - 1$
9. $y = -2x + 1$ 10. $y = 3x + 2$
11. $y = 3x - 2$ 12. $y = -3x - 2$
13. $y = -x^2$ 14. $y = x^2 + 1$
15. $y = x^2 - 1$ 16. $y = x^2 - 4$
17. $y = -x^2 + 4$ 18. $x = 3y^2$
19. $x = y^2 - 2$ 20. $y = x^2 + x - 6$
21. $x = 2y^2 - 3y - 2$ 22. $y = \dfrac{1}{x}$

23. $y = \dfrac{12}{x}$ 24. $x = -\dfrac{1}{y}$

Graph the following formulas.

25. $s = 50t$ (For a vehicle traveling 50 mph, this formula gives the distance traveled in miles after t hours.)

26. $P = \dfrac{100}{V}$ (This formula gives the volume in cubic inches that a particular gas occupies at a given pressure P in pounds per square inch.)

27. $F = \dfrac{9}{5}C + 32$ [This formula converts Celsius (°C) temperature to Fahrenheit (°F).]

28. $A = S^2$.   (This formula gives the area of a square where the length of each side is S. If S is given in inches, then the area A is in square inches.)

Figure 7.17 graphs the relationship between pressure and volume of a gas at a constant temperature. Pressure is measured in pounds per square inch and volume is measured in cubic inches. Use the graph to answer exercises 29–34.

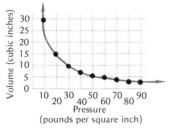

FIGURE 7.17

29. When a pressure of 30 lbs/in.² is applied, what is the resulting volume?
30. When a pressure of 40 lbs/in.² is applied, what is the resulting volume?
31. What pressure is necessary to give a volume of 30 in.³?
32. What pressure is necessary to give a volume of 5 in.³?
33. As the pressure increases, does the volume increase or decrease?
34. As the pressure decreases, does the volume increase or decrease?

Figure 7.18 graphs the relationship between voltage and current in a circuit where the resistance is constant. Voltage is measured in volts (V) and current is measured in amperes (A). Use the figure to answer exercises 35–39.

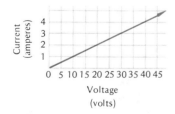

FIGURE 7.18

35. If 35 V are applied to the circuit, what is the resulting current?
36. If 2 A of current are desired, how much voltage must be applied to the circuit?
37. If 10 V are applied to the circuit, what is the resulting current?
38. If 3 A of current are desired, how much voltage must be applied to the circuit?
39. As the voltage increases, does the current increase or decrease?

Figure 7.19 illustrates the relationship between the speed of a full-size automobile and the distance it will travel after the driver decided to apply the brakes. Use the figure to answer exercises 40–43.

40. If the speed of an automobile is 25 mph, how far will it travel after the driver decides to apply the brakes?

41. If the speed of an automobile is 55 mph, how far will it travel after the driver decides to apply the brakes?

42. If a driver wishes to stop the car within a distance of 100 ft, he must keep his speed beneath _____ mph.

43. As the speed of an automobile increases, does the stopping distance increase slightly or increase radically?

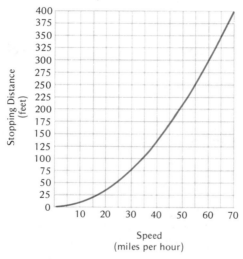

FIGURE 7.19

GRAPHING LINEAR EQUATIONS

7.4   **linear equation**   A **linear equation** in two variables is any equation that can be written in the form $y = mx + b$ where m and b are constants. For example, $y = 2x + 3$ is a linear equation where $m = 2$ and $b = 3$. In a linear equation both variables are raised to the first power.

Example 1   Here are some illustrations of linear equations.

(a) $y = -x + 5$ is a linear equation where $m = -1$ and $b = 5$

(b) $3x + y = 4$ is a linear equation. Solving for y produces the equation $y = -3x + 4$. Thus, $m = -3$ and $b = 4$

(c) $y = 6$ is a linear equation. It can be written as $y = 0x + 6$. Hence, $m = 0$ and $b = 6$

(d) $2x + 3y = -6$ is a linear equation. Solving for y produces the following:

$$2x + 3y = -6$$
$$3y = -2x - 6$$
$$y = \frac{-2x - 6}{3}$$
$$y = -\frac{2}{3}x - 2$$

Thus, $m = -\frac{2}{3}$ and $b = -2$

A linear equation can be graphed using the techniques developed in section 7.2. Construct a table of values and plot the corresponding ordered pairs on a rectangular coordinate system. Then connect the points with a smooth line extending infinitely far in both directions.

Graph the linear equation $y = 2x + 1$ **Example 2**

Solution:

Step (1)   Make a table of values and form the corresponding ordered pairs.

Table of Values		Ordered pairs	Meaning
x	y		
−3	−5	(−3, −5)	If $x = -3$, then $y = 2(-3) + 1$ or −5
−2	−3	(−2, −3)	If $x = -2$, then $y = 2(-2) + 1$ or −3
−1	−1	(−1, −1)	If $x = -1$, then $y = 2(-1) + 1$ or −1
0	1	(0, 1)	If $x = 0$, then $y = 2 \cdot 0 + 1$ or 1
1	3	(1, 3)	If $x = 1$, then $y = 2 \cdot 1 + 1$ or 3
2	5	(2, 5)	If $x = 2$, then $y = 2 \cdot 2 + 1$ or 5
3	7	(3, 7)	If $x = 3$, then $y = 2 \cdot 3 + 1$ or 7

Step (2)   Plot the points on a rectangular coordinate system (see figure 7.20).

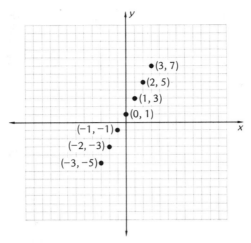

FIGURE 7.20

Step (3)   Form the graph of $y = 2x + 1$ by connecting the points with a smooth line (see figure 7.21). The arrows indicate that the graph continues infinitely far in both directions.

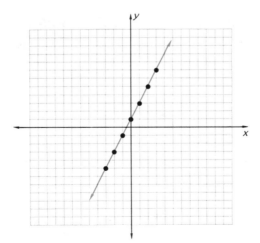

FIGURE 7.21

As seen in Example 2, the graph of a linear equation in two variables is a straight line. A straight line is completely determined by two points. Therefore, every linear equation can be graphed by plotting only two ordered pairs. However, it is best to obtain a third ordered pair to serve as a check.

Example 3   Graph the linear equation $y = x + 5$

Solution: Make a table of values consisting of three ordered pairs. Then plot the corresponding points and connect them with a straight line (see figure 7.22).

Table of Values	
x	y
−6	−1
−2	3
0	5

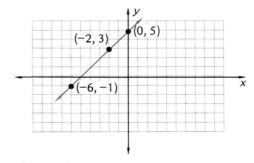

FIGURE 7.22

Graph the linear equation $y = -2x - 1$

Example 4

Solution: Make a table of values consisting of three ordered pairs. Then plot the corresponding points and connect them with a straight line (see figure 7.23).

Table of Values	
x	y
−3	5
0	−1
2	−5

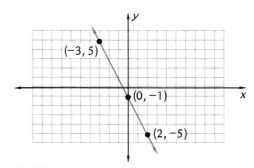

FIGURE 7.23

Graph the linear equation $3x + 2y = 6$

Example 5

Solution: When both variables appear on one side of the equation, it is convenient to obtain two of the ordered pairs by letting $x = 0$, and then by letting $y = 0$. The third ordered pair is obtained by letting x or y equal some other number. (The graph is shown in figure 7.24.)

(a) Let $x = 0$; $3x + 2y = 6$

$$3 \cdot 0 + 2y = 6$$

$$2y = 6$$

$$y = 3$$

(b) Let $y = 0$; $3x + 2y = 6$

$$3x + 2 \cdot 0 = 6$$

$$3x = 6$$

$$x = 2$$

(c) Let $x = -2$; $3x + 2y = 6$

$$3(-2) + 2y = 6$$

$$-6 + 2y = 6$$

$$2y = 12$$

$$y = 6$$

Table of Values	
x	y
0	3
2	0
-2	6

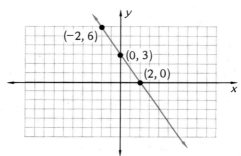

FIGURE 7.24

Next we examine two special types of equations whose graphs are also straight lines. These equations are of the form $y = k$ and $x = k$ where k is a constant. For example, consider the equation $y = 4$. This equation states that y is always 4, no matter what number x represents. In other words, the equation $y = 4$ produces ordered pairs where the ordinate is always 4. Therefore, we arrive at the following table of values and the corresponding graph, figure 7.25.

Table of Values for y = 4	
x	y
-5	4
-3	4
-1	4
0	4
2	4
4	4

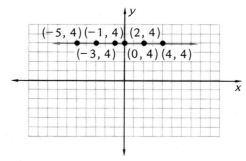

FIGURE 7.25

As shown, the graph of the equation $y = 4$ is a horizontal straight line, 4 units above the x-axis. Thus, we make the following generalization:

Horizontal Line

The graph of an equation of the form $y = k$, where k is a constant, is a horizontal straight line k units from the x-axis.

Next consider the equation $x = 3$. This equation states that x is always 3 no matter what number y represents. Notice below that the equation produces ordered pairs where the abscissa is always 3. The graph, illustrated in figure 7.26, is a vertical straight line 3 units to the right of the y-axis.

Table of Values for $x = 3$	
x	y
3	−4
3	−2
3	0
3	1
3	3

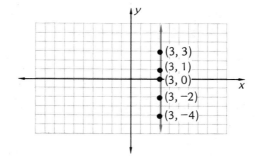

FIGURE 7.26

From this example we arrive at the following generalization:

Vertical Line

The graph of an equation of the form $x = k$, where k is a constant, is a vertical straight line k units from the y-axis.

Write each linear equation in the form $y = mx + b$. Then identify the values of m and b.

Exercise 7.4

1. $x + y = 6$ **2.** $x - y = 1$

3. $2x + y = 7$ **4.** $3x - y = 4$

5. $-5x + y = 1$ **6.** $-3x - y = 2$

7. $2x + y = 0$ **8.** $x + 2y = 6$

9. $x - 3y = 9$ **10.** $2x + 5y = 10$

11. $3x - 4y = 12$ **12.** $2x + 7y = 5$

13. $y = 5$ **14.** $y + 2 = 0$

Graph these linear equations by plotting at least three points.

15. $y = x$ **16.** $y = -x$

17. $y = x + 2$ **18.** $y = x - 2$

19. $y = 2x$ **20.** $y = -2x$

21. $y = 2x + 1$ **22.** $y = 2x - 1$

23. $y = -2x + 1$ **24.** $y = -2x - 1$

25. $y = 3x + 2$ **26.** $y = \frac{1}{2}x - 1$

27. $x + y = 3$ **28.** $x - y = 1$

29. $x - 2y = 0$ **30.** $y - 4x = 0$

31. $x = 2y + 6$ **32.** $2x + y = 5$

33. $x + 3y = 6$ **34.** $x - 2y = 4$

35. $2x - 7y = 14$ **36.** $3x + 5y = 15$

37. $2x - 5y = 10$ **38.** $3x - 4y = 12$

Graph the following special cases whose graphs are straight lines.

39. $y = 1$ **40.** $x = 4$

41. $x = -3$ **42.** $y = -2$

43. $x = -1$ **44.** $y = -5$

45. $x = 0$ **46.** $y = 0$

47. $x + 2 = 0$ **48.** $y - 6 = 0$

49. $2x - 1 = 0$ **50.** $2y = 5$

Translate each of these statements into an equation. Then plot the graph.

51. The y-value is 3 times the x-value.

52. The y-value is 1 more than 3 times the x-value.

53. Twice the x-value is 6 more than 3 times the y-value.

54. The ordinate is always twice the abscissa.

55. The abscissa is always 3 more than $\frac{1}{2}$ the ordinate.

56. The abscissa is always -2.

57. The ordinate is always 3.

THE SLOPE OF A STRAIGHT LINE

The slope of a straight line is a measure of its slant. Before computing slope, we must discuss the rise and run between two given points on a straight line. Given two points on a straight line, the **run** is the directed horizontal distance between the points, and the **rise** is the directed vertical distance between the two points.

Figure 7.27 is the graph of the linear equation $y = 2x - 4$, which contains the points $(1, -2)$ and $(3, 2)$. To determine the rise and the run, start at the point $(1, -2)$ and proceed 2 units to the right, stopping directly beneath the point $(3, 2)$. The continue up 4 units, arriving at the point $(3, 2)$. This procedure forms a right triangle. The run is $+2$ and the rise is $+4$.

7.5

run
rise

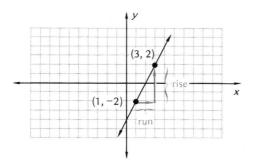

FIGURE 7.27

Given the linear equation $y = -\frac{1}{3}x + 1$ and the two points $(-3, 2)$ and $(6, -1)$, determine the rise and the run by starting at the point $(-3, 2)$

Example 1

Solution: Graph the straight line (see figure 7.28). Then, starting at the point $(-3, 2)$, construct a right triangle terminating at the point $(6, -1)$. Be sure to record the directed horizontal and vertical distances. The run is 9 units to the right, $+9$. The rise is 3 units down, -3.

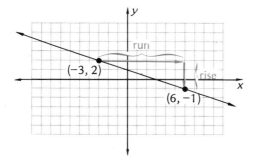

FIGURE 7.28

The slope of a straight line is a measure of its slant. Slope is defined to be the rise divided by the run.

slope

$$\text{slope} = \frac{\text{rise}}{\text{run}}$$

The slope tells how far a straight line rises or falls as it proceeds one unit to the right. The slope of a straight line can be determined by following these steps:

Step (1)   Graph the straight line.

Step (2)   Select two points. Then determine the rise and the run.

Step (3)   Divide the rise by the run. This ratio will be the same no matter which two points you select.

Example 2   Find the slope of the straight line whose equation is $y = 2x - 4$

Solution: Graph the straight line (see figure 7.29). Select any two points on the line and determine the rise and the run. Compute the slope by dividing the rise by the run.

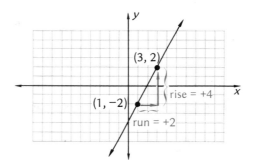

FIGURE 7.29

$$\text{slope} = \frac{\text{rise}}{\text{run}}$$

$$\text{slope} = \frac{+4}{+2}$$

$$\text{slope} = 2$$

A slope of 2 tells us the straight line rises 2 units for every 1 unit it goes to the right. Notice that a straight line having a positive slope will slant from lower left to upper right.

The slope of a straight line is constant. That is, it will not change if you select different points on the line. You should choose two other points for the line in Example 2, say $(-1, -6)$ and $(4, 4)$. Then show that the slope is still 2.

Example 3

Find the slope of the straight line whose equation is $y = -\frac{1}{3}x + 1$

Solution: Graph the straight line (see figure 7.30). Select any two points on the line and determine the rise and the run. Compute the slope by dividing the rise by the run.

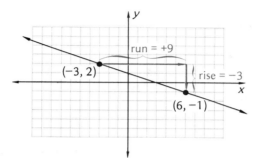

FIGURE 7.30

$$\text{slope} = \frac{\text{rise}}{\text{run}}$$

$$\text{slope} = \frac{-3}{+9}$$

$$\text{slope} = -\frac{1}{3}$$

A slope of $-\frac{1}{3}$ tells us the straight line falls $\frac{1}{3}$ of a unit for every 1 unit it goes to the right. Notice that a straight line having a negative slope will slant from upper left to lower right.

Find the slope of a straight line whose equation is $y = 3$

Example 4

Solution: $y = 3$ is the equation of a horizontal line 3 units above the x-axis (see figure 7.31). Select any two points, say $(-1, 3)$ and $(4, 3)$. The run is $+5$, but the rise is 0.

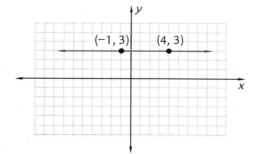

FIGURE 7.31

$$\text{slope} = \frac{\text{rise}}{\text{run}} \qquad \text{slope} = \frac{0}{+5} \qquad \text{slope} = 0$$

A slope of zero tells us the straight line is horizontal. It neither rises nor falls.

Example 5   Find the slope of a straight line whose equation is $x = -4$

Solution: $x = -4$ is the equation of a vertical line 4 units to the left of the y-axis (see figure 7.32). Select any two points, say $(-4, -3)$ and $(-4, 4)$. The run is 0, while the rise is 7.

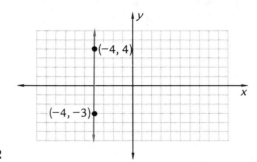

FIGURE 7.32

$$\text{slope} = \frac{\text{rise}}{\text{run}}$$

$$\text{slope} = \frac{7}{0}$$

Division by zero is not possible. Therefore, a vertical line has no slope.

The preceding examples suggest the following properties of slope.

Important Properties of Slope

(1) Slope $= \dfrac{\text{rise}}{\text{run}}$

(2) Straight lines having a positive slope will slant from lower left to upper right.

(3) Straight lines having a negative slope will slant from upper left to lower right.

(4) A horizontal line has a slope of zero.

(5) A vertical line has no slope whatsoever.

(6) As you may have observed in Examples 2 and 3, if the equation of a straight line is in the form $y = mx + b$, then m will equal the slope.

Exercises 1–14 each describe two points, labeled A and B. Graph the points and connect them with a straight line. Then compute the rise and the run starting with point A and terminating with point B.

Exercise 7.5

1. A (2, 3); B (5, 9)
2. A (1, 1); B (3, 5)
3. A (0, 0); B (2, 4)
4. A (1, 0); B (3, 2)
5. A (0, 2); B (3, 1)
6. A (−2, 1); B (1, 5)
7. A (−3, −4); B (0, 2)
8. A (1, −2); B (4, 5)
9. A (−3, −7); B (−1, 2)
10. A (0, −3); B (4, 0)
11. A (−6, −1); B (2, 4)
12. A (−3, −5); B (1, −5)
13. A (1, −6); B (1, 2)
14. A (2, −3); B (−1, 4)

Graph the following linear equations. Then compute the slope by using any two points on the straight line.

15. $y = x$
16. $y = 2x$
17. $y = -2x$
18. $y = x + 3$
19. $y = -x + 2$
20. $y = 3x - 1$
21. $y = -3x + 2$
22. $y = -2$
23. $x = 4$
24. $x + y = 3$
25. $x - y = 2$
26. $2x + y = 4$
27. $x - 3y = 6$
28. $3x - 2y = 12$
29. $5x + 2y = 10$
30. $3x - 5y = 15$

Place each linear equation in the form $y = mx + b$. Graph the corresponding straight line and compute its slope. Then, show that the slope is equal to the value of m.

31. $x - y = 2$
32. $3x + y = 1$
33. $2x - y = 3$
34. $4x + 2y = 3$
35. $x - 2y = 4$
36. $8x = 2y + 5$
37. $x - 2y = 6$
38. $x + 3y = 1$
39. $5x + y = 0$
40. $y + 3 = 0$

True or false.

41. A horizontal line has no slope.

42. A vertical line has a slope of zero.

43. If a straight line has a slope of -2, it will slant from upper left to lower right.

44. If a straight line has a slope of 3, it will rise 3 units for every 1 unit it goes to the right.

45. If a straight line has a slope of $-\dfrac{1}{2}$, it will fall 2 units for every 1 unit it goes to the right.

THE INTERCEPTS OF A STRAIGHT LINE

7.6

Every straight line which is not vertical or horizontal will intersect both the x-axis and the y-axis. For example, figure 7.33 graphs the linear equation $y = 2x + 4$. Notice the straight line intersects the x-axis at the point $(-2, 0)$ and it intersects the y-axis at the point $(0, 4)$.

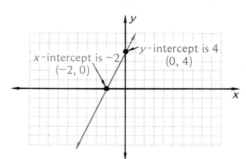

FIGURE 7.33

x-intercept

y-intercept

The word "intercept" means to intersect. Thus, the **x-intercept** is defined to be the x-value (abscissa) of the point where the line intersects the x-axis. Similarly, the **y-intercept** is defined to be the y-value (ordinate) of the point where the line intersects the y-axis. In figure 7.33, the line intersects the x-axis at $(-2, 0)$ and the y-axis at $(0, 4)$. Thus, the x-intercept is -2, while the y-intercept is 4. The x- and y-intercepts can be found graphically by simply locating the points where the line crosses the x- and y-axes.

Example 1

Graph the equation $y = -2x + 4$ and find the x- and y-intercepts.

Solution:

Table of Values	
x	y
−1	6
1	2
3	−2

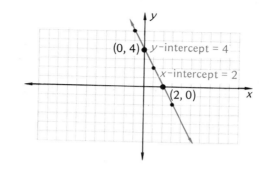

FIGURE 7.34

As seen in figure 7.34, the coordinates at the x-intercept are (2, 0). Thus, the x-intercept is 2. The coordinates at the y-intercept are (0, 4). Therefore, the y-intercept is 4.

The x- and y-intercepts can also be found algebraically by remembering that the ordinate at the x-intercept is zero and the abscissa at the y-intercept is zero.

Finding the x- and y-Intercepts Algebraically

(1) To find the x-intercept, substitute 0 for y, and then solve the equation for x.

(2) To find the y-intercept, substitute 0 for x, and then solve the equation for y.

Algebraically find the x- and y-intercepts of the straight line whose equation is $3x - 4y = 12$

Example 2

Solution:

Step (1)   To find the x-intercept, substitute 0 for y

$$3x - 4y = 12$$
$$3x - 4 \cdot 0 = 12$$
$$3x - 0 = 12$$
$$3x = 12$$
$$x = 4$$

Step (2)   To find the y-intercept, substitute 0 for x

$$3x - 4y = 12$$
$$3 \cdot 0 - 4y = 12$$

$$0 - 4y = 12$$
$$-4y = 12$$
$$y = -3$$

Thus, the x-intercept is 4, while the y-intercept is −3

Linear equations can be graphed very rapidly by finding the x- and y-intercepts algebraically. These two points are plotted on their respective axes and connected with a straight line.

Example 3

Graph the linear equation $2x + 3y = 6$ by determining its x- and y-intercepts.

Solution:

Step (1) Determine the x-intercept.

$$2x + 3y = 6$$
$$2x + 3 \cdot 0 = 6$$
$$2x = 6$$
$$x = 3$$

Step (2) Determine the y-intercept.

$$2x + 3y = 6$$
$$2 \cdot 0 + 3y = 6$$
$$3y = 6$$
$$y = 2$$

Step (3) Plot the two points and construct the graph (figure 7.35).

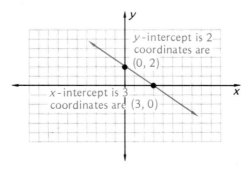

FIGURE 7.35

We can now prove this property:

> If a linear equation is in the form
> $$y = mx + b,$$
> then b is the y-intercept.

To prove, 0 is substituted for x, and the equation is solved for y:

$$y = mx + b$$
$$y = m \cdot 0 + b$$
$$y = 0 + b$$
$$y = b$$

Thus, b is the y-intercept.

Each of these linear equations is in the form $y = mx + b$, where b represents the y-intercept.
(a) $y = 3x + 2$; the y-intercept is 2
(b) $y = 2x - 5$; the y-intercept is -5
(c) $y = -x + 1$; the y-intercept is 1

Example 4

Before using this method to determine the y-intercept, remember that the linear equation must be in the form $y = mx + b$. For example, if $2y = 8x - 6$, the y-intercept is not -6. You must solve for y by dividing both sides by 2. Thus, $y = 4x - 3$ and the y-intercept is -3.

In summary, $y = mx + b$ is often called the *slope-intercept* form of a linear equation. m, as you learned in section 7.5, represents the slope, while b represents the y-intercept.

Write the linear equation $4x + 3y = 12$ in the slope-intercept form. Then determine the slope and the y-intercept.

Example 5

Solution: $4x + 3y = 12$

$$3y = -4x + 12$$
$$y = \frac{-4}{3}x + \frac{12}{3}$$
$$y = -\frac{4}{3}x + 4$$

The slope is $-\frac{4}{3}$; the y-intercept is 4

Exercise 7.6

For each of these linear equations, determine the x- and y-intercepts algebraically. Then construct the graph.

1. $x + y = 1$

2. $x + y = 2$

3. $x + y = 5$

4. $x - y = 1$

5. $x - y = 3$

6. $x - y = 6$

7. $x + 2y = 4$

8. $x - 2y = 6$

9. $x + 3y = 6$

10. $x - 5y = 10$

11. $2x + y = 4$

12. $3x - y = 6$

13. $5x + y = 10$

14. $2x + y = 3$

15. $x - 3y = 5$

16. $x + 4y = 6$

17. $2x - 3y = 6$

18. $8x + 2y = 12$

19. $5x - 2y = 10$

20. $4x + 5y = 20$

21. $2x - 7y = 14$

22. $3x + 8y = 24$

23. $3x - 2y = 8$

24. $5x + 2y = 12$

25. $4x - y = 2$

26. $x + 3y \doteq 7$

27. $y = 2x - 4$

28. $y = 3x + 9$

29. $y = 4x + 3$

30. $y = 5x + 8$

31. $y = x + 4$

32. $x = 2y - 6$

33. $x = -3y + 1$

34. $x = 4y + 7$

Write each of these linear equations in the form $y = mx + b$. Then determine the slope and the y-intercept.

35. $3y = 6x + 9$

36. $2y = 8x - 2$

37. $x = 2y + 4$

38. $x = 3y - 6$

39. $x + y = 3$

40. $x - y = 2$

41. $2x + 4y = 1$

42. $3x - 6y = 1$

43. $x - 5y = 10$

44. $y - 3x = 0$

45. $4x - y = 2$

46. $3x - 2y = 8$

True or false.

47. If the x-intercept is -3, then the coordinates of that point are $(-3, 0)$.

48. If the y-intercept is 2, then the coordinates of that point are $(2, 0)$.

49. If the x- and y-intercepts of a straight line are both zero, then the line passes through the origin.

50. If the equation of a straight line is $2x + 5y = 10$, then the y-intercept is 10.

51. If a straight line passes through the point $(0, -2)$, then its x-intercept is zero.

52. If a straight line passes through the point $(-4, 1)$, then its x-intercept is -4 and its y-intercept is 1.

53. If a straight line passes through the point $(5, 0)$, then its x-intercept is 5.

54. A straight line having an equation of $y = -1$ will not have an x-intercept.

55. A straight line having an equation of $x = 2$ will not have a y-intercept.

GRAPHING LINEAR INEQUALITIES

You know how to graph linear equations, such as $x + 2y = 6$. In this section you will learn to graph linear inequalities, such as $x + 2y < 6$. Linear inequalities are similar to linear equations except the equal sign has been replaced with one of the following inequality symbols:

$<$ read "is less than"
$\leq$ read "is less than or equal to"
$>$ read "is greater than"
$\geq$ read "is greater than or equal to"

7.7

Each of the following is a linear inequality.

Example 1

(a) $x + 2y < 6$
(b) $2x + 5y \leq 10$
(c) $y > 2x + 3$
(d) $y \geq 3x - 4$

To help us graph a linear inequality, we need to recall that the graph of a linear equality is a straight line. Figure 7.36 shows the graph of $x + 2y = 6$. Observe that the straight line, not being vertical, divides the plane into an upper region and a lower region. The upper region contains all of the points above the line, while the lower region contains all the points below the line. The line itself is called the *boundary* and is not contained in either the upper or the lower region. For this reason, the boundary is illustrated by a broken line. It turns out that the graph

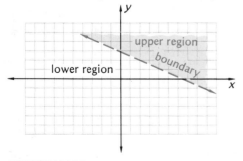

FIGURE 7.36

of a linear inequality will consist of either the upper region or the lower region, but not both. The boundary line will be part of the graph if the given inequality includes an equal sign, i.e., $\leq$ or $\geq$. Otherwise, the boundary is not part of the graph.

Thus, to construct the graph of a linear inequality, we first graph the boundary line. If the boundary is to be included, it is drawn solid; otherwise it is designated by a broken line. Then a test point is selected from either the upper or lower region. The coordinates of the test point are substituted into the original inequality. If the resulting statement is true, then the graph includes that region. But if the statement is false, the graph will include the other region.

Example 2   Graph $x + 2y \leq 6$

Solution:

Step (1)   Graph the boundary given by the equation $x + 2y = 6$. The original inequality contains an equal sign. Therefore, the graph will include the boundary, drawn as a solid line (see figure 7.37a).

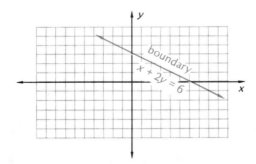

FIGURE 7.37a

Step (2)   Select a test point from either region. Do not choose a point on the boundary. The origin $(0, 0)$ is a convenient point located in the lower region. Substitute the coordinates into the original inequality and determine whether the resulting statement is true or false.

$$x + 2y \leq 6$$

$$0 + 2 \cdot 0 \leq 6$$

$$0 \leq 6$$

Since the statement is true, the graph includes the lower region containing the test point $(0, 0)$. And, as noted in Step

(1), it will also include the boundary. The graph is the shaded region, together with the solid line shown in figure 7.37b.

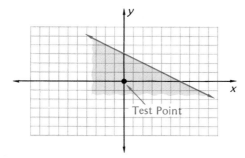

FIGURE 7.37b

Graph $2x - 5y > 10$

Exampie 3

Solution:

Step (1) Graph the boundary given by the equation $2x - 5y = 10$ (see figure 7.38a). The original inequality contains no equal sign. Therefore, the graph will not include the boundary, drawn as a broken line.

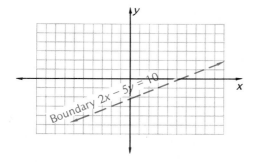

FIGURE 7.38a

Step (2) For our test point we select the origin (0, 0) and substitute the coordinates into the original inequality.

$$2x - 5y > 10$$
$$2 \cdot 0 - 5 \cdot 0 > 10$$
$$0 > 10 \quad \text{[false]}$$

Since the statement is false, the graph of the inequality will not include the upper region, containing the test point (0, 0). Therefore, as illustrated in figure 7.38b, the graph will contain the lower region, but not the boundary.

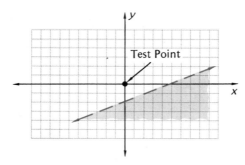

FIGURE 7.38b

Example 4

Graph $y \geq 3x$

Solution:

Step (1) Graph the boundary, given by the equation $y = 3x$ (see figure 7.39a). Since the original inequality contains an equal sign, the graph will include the boundary. Thus, it is drawn as a solid line.

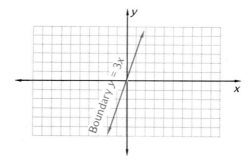

FIGURE 7.39a

Step (2) For the test point we can select any point not on the boundary. Therefore, we cannot choose the point (0, 0), because it lies on the line $y = 3x$. Instead, we arbitrarily select the point (0, 2), lying in the upper region. Substituting 0 for x and 2 for y in the original inequality, we arrive at the following true statement:

$$y \geq 3x$$

$$2 \geq 3 \cdot 0$$

$$2 \geq 0 \qquad \text{[true]}$$

Thus, as illustrated by the shaded portion of figure 7.39b, the graph includes the upper region as well as the boundary.

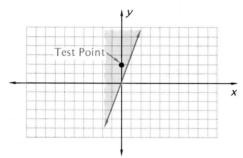

FIGURE 7.39b

Graph $x < 2$

Example 5

Solution:

Step (1) Graph the boundary given by the equation $x = 2$. This is a vertical line dividing the plane into a left region and a right region (see figure 7.40a). Since there is no equal sign in the original inequality, the graph will not include the boundary. Therefore, it is pictured as a broken line.

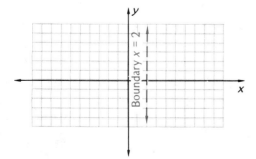

FIGURE 7.40a

Step (2) For the test point we select $(0, 0)$ and substitute the coordinates into the original inequality.

$x < 2$

$0 < 2$ [true]

Since the statement is true, the graph includes the region containing the test point. Thus, as shown in figure 7.40b, the graph includes the region to the left of the boundary.

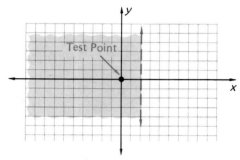

FIGURE 7.40b

Exercise 7.7   In each of the following figures, the required boundary has been drawn. Finish the graph for each inequality by shading the correct region.

1. $2x + y \geq 4$

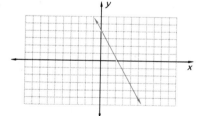

2. $6y - x < 6$

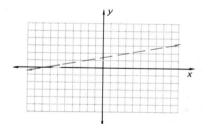

3. $y \geq 2x - 4$

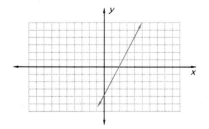

4. $3x - 2y < -6$

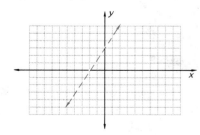

5. $y + x > 0$

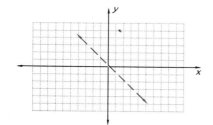

6. $y \leq -x - 2$

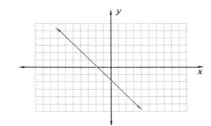

7. $x + 2y + 6 > 0$

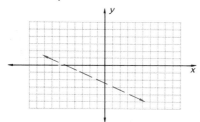

8. $y > -2$

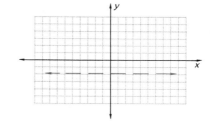

9. $x + 4 \geq 0$

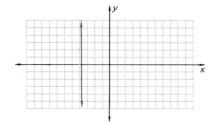

10. $x < 5$

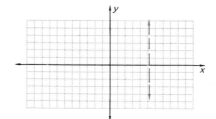

Graph the following linear inequalities.

11. $x + y \geq 3$ **12.** $x + y \leq 1$

13. $x + y < 5$ **14.** $x + y > 6$

15. $x - y < 1$ **16.** $x - y \geq 4$

17. $x - y \leq 0$ **18.** $x - y > 3$

19. $x + 2y < 6$ **20.** $x - 2y > 6$

21. $2x - y > 4$ **22.** $2x + y \leq 2$

23. $x - 3y \geq 9$ **24.** $y - 3x \leq 6$

25. $x + 5y < 10$ **26.** $y - 4x \geq 4$

27. $2x + 3y < 6$ **28.** $3x - 4y \leq 12$

29. $3x + 7y < 21$ **30.** $4x - 5y \leq 20$

31. $5x - 3y \geq 15$ **32.** $3x + 5y < 15$

33. $3x + 8y < -24$ **34.** $2x - 5y + 10 \geq 0$

35. $x \geq 2y$ **36.** $x \leq 3y$

37. $x - 5y > 5$ **38.** $x \geq -2y$

39. $y + 3x \leq 0$ **40.** $x - 4y < 8$

41. $y \geq 3x + 2$ **42.** $y \leq 2x + 5$

43. $y > x - 2$ **44.** $y \geq \frac{1}{2}x + 7$

45. $y \geq -3x - 5$ **46.** $y > \frac{2}{3}x + 1$

47. $y < 3$ **48.** $y + 2 \geq 0$

49. $y - 5 \leq 0$ **50.** $x > -1$

51. $x \leq 0$ **52.** $x - 1 < 0$

GRAPHING QUADRATIC EQUATIONS IN TWO VARIABLES

7.8

You have learned the basic techniques of graphing both linear equations and linear inequalities. In this section, we continue our discussion of graphing by examining quadratic equations in two variables of the form

$$y = ax^2 + bx + c$$

where a, b, c are constants and $a \neq 0$.

Example 1

Each of the following is a quadratic equation in two variables.
(a) $y = 2x^2 - 3x - 5$; $a = 2, b = -3$, and $c = -5$
(b) $y = x^2 - 4$; $a = 1, b = 0$, and $c = -4$
(c) $y = x^2$; $a = 1, b = 0$, and $c = 0$

The simplest quadratic equation is $y = x^2$. To graph this equation, a table of values is constructed giving some ordered pairs which satisfy the original equation. Then, each of the corresponding points is plotted on a coordinate system. Finally, when a pattern is evident, the points are connected with a smooth curve.

Example 2

Graph $y = x^2$

Solution:

Step (1)   Make a table of values. Select values for x and compute the corresponding values of y.

Table of Values		Ordered pairs	Computation $y = x^2$
x	y		
-3	9	$(-3, 9)$	If $x = -3$, then $y = (-3)^2 = 9$
-2	4	$(-2, 4)$	If $x = -2$, then $y = (-2)^2 = 4$
-1	1	$(-1, 1)$	If $x = -1$, then $y = (-1)^2 = 1$
0	0	$(0, 0)$	If $x = 0$, then $y = 0^2 = 0$
1	1	$(1, 1)$	If $x = 1$, then $y = 1^2 = 1$
2	4	$(2, 4)$	If $x = 2$, then $y = 2^2 = 4$
3	9	$(3, 9)$	If $x = 3$, then $y = 3^2 = 9$

Step (2) Plot each of the points and connect them with a smooth curve (see figure 7.41).

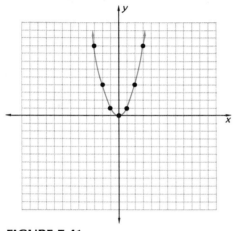

FIGURE 7.41

The graph of $y = x^2$ shown in figure 7.41 is called a **parabola**. It is a U-shaped curve continually becoming wider. This particular parabola is said to open upward. Every quadratic equation of the form $y = ax^2 + bx + c$ has a parabola for its graph. Parabolas are very useful curves and have many real-life applications. The cross sections of spotlights and floodlights form parabolas. Many radar antennae are parabolic in shape. Disregarding air resistance, the path of a moving projectile is a parabola. And the cables supporting a suspension bridge hang in the form of a parabola.

parabola

Graph $y = -x^2 + 9$

Example 3

Solution:

Step (1) Make a table of values.

Table of Values		Ordered pairs	Computation $y = -x^2 + 9$
x	y		
−4	−7	(−4, −7)	If $x = -4$, then $y = -(-4)^2 + 9 = -7$
−3	0	(−3, 0)	If $x = -3$, then $y = -(-3)^2 + 9 = 0$
−2	5	(−2, 5)	If $x = -2$, then $y = -(-2)^2 + 9 = 5$
−1	8	(−1, 8)	If $x = -1$, then $y = -(-1)^2 + 9 = 8$
0	9	(0, 9)	If $x = 0$, then $y = -0^2 + 9 = 9$
1	8	(1, 8)	If $x = 1$, then $y = -1^2 + 9 = 8$
2	5	(2, 5)	If $x = 2$, then $y = -2^2 + 9 = 5$
3	0	(3, 0)	If $x = 3$, then $y = -3^2 + 9 = 0$
4	−7	(4, −7)	If $x = 4$, then $y = -4^2 + 9 = -7$

Step (2) Plot each of the points and draw the parabola (see figure 7.42).

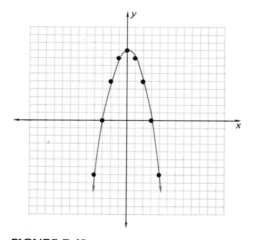

FIGURE 7.42

The graph of $y = -x^2 + 9$ shown in figure 7.42 is a parabola which opens downward. Comparing Example 2 with Example 3 shows that the sign of a (the numerical coefficient of x^2) determines whether the parabola opens upward or downward.

(1) In the equation $y = x^2$, $a = +1$. As seen in Example 2, the parabola opens upward.

(2) In the equation $y = -x^2 - 9$, $a = -1$. As seen in Example 3, the parabola opens downward.

Example 4

Graph $y = x^2 + 2x - 3$

Solution: Since $a = +1$, we expect the parabola to open upward.

Step (1) Make a table of values. Find several points on both sides of the parabola.

Table of Values		Ordered pairs	Computation $y = x^2 + 2x - 3$
x	y		
−4	5	(−4, 5)	If $x = -4$, then $y = (-4)^2 + 2(-4) - 3 = 5$
−3	0	(−3, 0)	If $x = -3$, then $y = (-3)^2 + 2(-3) - 3 = 0$
−2	−3	(−2, −3)	If $x = -2$, then $y = (-2)^2 + 2(-2) - 3 = -3$
−1	−4	(−1, −4)	If $x = -1$, then $y = (-1)^2 + 2(-1) - 3 = -4$
0	−3	(0, −3)	If $x = 0$, then $y = 0^2 + 2(0) - 3 = -3$
1	0	(1, 0)	If $x = 1$, then $y = 1^2 + 2(1) - 3 = 0$
2	5	(2, 5)	If $x = 2$, then $y = 2^2 + 2(2) - 3 = 5$

Step (2) Plot each of the points and draw the parabola (see figure 7.43).

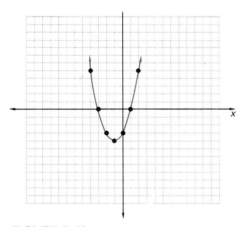

FIGURE 7.43

Graph $y = -2x^2 - 2x + 4$ Example 5

Solution: Since $a = -2$, we expect the parabola to open downward.

Step (1) Make a table of values. Be sure to find several points on each side of the parabola.

Table of Values		Ordered pairs	Computation $y = -2x^2 - 2x + 4$
x	y		
−3	−8	(−3, −8)	If $x = -3$, then $y = -2(-3)^2 - 2(-3) + 4 = -8$
−2	0	(−2, 0)	If $x = -2$, then $y = -2(-2)^2 - 2(-2) + 4 = 0$
−1	4	(−1, 4)	If $x = -1$, then $y = -2(-1)^2 - 2(-1) + 4 = 4$
0	4	(0, 4)	If $x = 0$, then $y = -2(0)^2 - 2(0) + 4 = 4$
1	0	(1, 0)	If $x = 1$, then $y = -2(1)^2 - 2(1) + 4 = 0$
2	−8	(2, −8)	If $x = 2$, then $y = -2(2)^2 - 2(2) + 4 = -8$

Step (2)   Plot each point; then draw the parabola (see figure 7.44). We have to approximate the actual turning point because its coordinates were not included in the table of values.

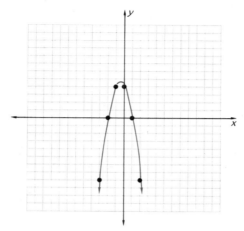

FIGURE 7.44

Exercise 7.8

For each quadratic equation, determine the value of a and state whether the graph will open upward or downward.

1. $y = x^2$ **2.** $y = -x^2$

3. $y = -3x^2$ **4.** $y = 2x^2$

5. $y = x^2 + 4$ **6.** $y = -x^2 + 1$

7. $y = x^2 + x - 1$ **8.** $y = -x^2 + 2x + 1$

9. $y = -3x^2 + x - 1$ **10.** $y = 2x^2 - x$

11. $y = 4 - 3x^2$ **12.** $y = 3 - 2x + x^2$

Sketch the graph of each quadratic equation.

13. $y = x^2$ **14.** $y = -x^2$

15. $y = -3x^2$ **16.** $y = 3x^2$

17. $y = \frac{1}{2}x^2$ **18.** $y = -\frac{1}{3}x^2$

19. $y = x^2 - 1$ **20.** $y = x^2 + 1$

21. $y = -x^2 - 1$ **22.** $y = -x^2 + 1$

23. $y = x^2 + 4$ **24.** $y = -x^2 + 4$

25. $y = 2x^2 - 1$ **26.** $y = -3x^2 + 5$

27. $y = x^2 + x$ **28.** $y = x^2 - x$

29. $y = x^2 + 2x - 8$ **30.** $y = x^2 + 2x - 3$

31. $y = -x^2 - 2x + 3$ **32.** $y = x^2 + 7x + 12$

33. $y = -x^2 + 3x + 10$ **34.** $y = 2x^2 + x - 6$

35. $y = 3x^2 + x - 3$ **36.** $y = -3x^2 + 4x + 7$

A condition describing a correspondence between two variables is called a **relation**. For example, the quadratic equation $y = x^2 + 1$ is a relation between the variables x and y. It states that y is always equal to one more than the square of x. Or, in terms of ordered pairs, the ordinate is always one more than the square of the abscissa. Some of the ordered pairs making up this relation are $(-2, 5)$, $(-1, 2)$, $(0, 1)$, $\left(\frac{1}{2}, \frac{5}{4}\right)$, and $(3, 10)$. Thus, a relation is also considered to be a nonempty set of ordered pairs. Using set notation this relation is symbolized as follows:

$$\{(x, y) \mid y = x^2 + 1\}$$

It reads "the set of ordered pairs (x, y) such that $y = x^2 + 1$." Note that the single vertical line is read "such that."

Here are some more illustrations of relations.

Equation or inequality	Set notation	Meaning
(a) $y = 3x - 2$	$\{(x, y) \mid y = 3x - 2\}$	The set of ordered pairs (x, y) such that $y = 3x - 2$
(b) $2x - 3y = 6$	$\{(x, y) \mid 2x - 3y = 6\}$	The set of ordered pairs (x, y) such that $2x - 3y = 6$
(c) $y = 2x^2 - x + 3$	$\{(x, y) \mid y = 2x^2 - x + 3\}$	The set of ordered pairs (x, y) such that $y = 2x^2 - x + 3$
(d) $3x + 4y \leq 12$	$\{(x, y) \mid 3x + 4y \leq 12\}$	The set of ordered pairs (x, y) such that $3x + 4y \leq 12$

Associated with each relation are two sets of numbers called the *domain* and *range*. The **domain** is the set of all x-values; that is, all of the values that can legitimately be substituted for x. The **range** is the set of all y-values. For example, in the relation defined by the equation $y = x^2$, we are free to substitute any value for x. Thus, the domain consists of all real numbers and is symbolized below:

$$D = \{x \mid x \text{ is any real number}\}$$

It reads "the domain is the set of all x-values such that x is any real number." The range of this relation is the set of all y-values given by the equation $y = x^2$. Each y-value is equal to the square of x. Since the square of a number is never negative, the range consists of all nonnegative real numbers. This is symbolized in set notation as follows:

$$R = \{y \mid y \geq 0\}$$

7.9

relation

Example 1

domain

range

It reads "the range is the set of all y-values such that y is greater than or equal to zero."

Example 2

Find the domain and range of the relation $\{(x, y) \mid y = \sqrt{x}\}$.

Solution: The equation $y = \sqrt{x}$ places a restriction on the x-values. We cannot take the square root of negative numbers. Therefore, only nonnegative numbers can be substituted for x. This gives a domain of $D = \{x \mid x \geq 0\}$. To determine the range, we note that y equals the principal square root of x. Since principal roots are never negative, the range is $R = \{y \mid y \geq 0\}$.

Example 3

Study each of these relations, being sure that you can obtain the indicated domain and range.

(a) $\{(x, y) \mid y = 2x + 3\}$ $D = \{x \mid x \text{ is any real number}\}$
$R = \{y \mid y \text{ is any real number}\}$

(b) $\{(x, y) \mid y = x^2 + 1\}$ $D = \{x \mid x \text{ is any real number}\}$
$R = \{y \mid y \geq 1\}$

(c) $\{(x, y) \mid y = \sqrt{x - 1}\}$ $D = \{x \mid x \geq 1\}$
$R = \{y \mid y \geq 0\}$

(d) $\{(x, y) \mid x = y^2\}$ $D = \{x \mid x \geq 0\}$
$R = \{y \mid y \text{ is any real number}\}$

The graph of a relation is the set of points whose coordinates satisfy the given equation or inequality. Relations are graphed using the techniques discussed earlier.

Example 4

Graph the relation $\{(x, y) \mid y = \sqrt{x}\}$

Solution: Make a table of values and plot the corresponding points. Then, connect them with a smooth curve (see figure 7.45).

Table of Values		Ordered pairs	Computation
x	y		
0	0	(0, 0)	$y = \sqrt{0} = 0$
1	1	(1, 1)	$y = \sqrt{1} = 1$
4	2	(4, 2)	$y = \sqrt{4} = 2$
9	3	(9, 3)	$y = \sqrt{9} = 3$

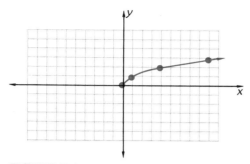

FIGURE 7.45

The domain and range of a relation can also be found graphically. The domain, being the set of all possible x-values, can be found by projecting the graph onto the x-axis. This projection is sometimes called the **horizontal sweep** of the graph. Similarly, the range, being the set of all possible y-values, can be found by projecting the graph onto the y-axis. This projection is sometimes called the **vertical sweep** of the graph.

horizontal sweep

vertical sweep

Graph the relation $y = \sqrt{x}$. Then find its domain and range.

Example 5

Solution: From Example 4 we have figure 7.46a.

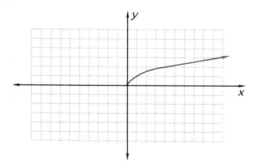

FIGURE 7.46a

Step (1)   To find the domain, project the graph onto the x-axis as in figure 7.46b.

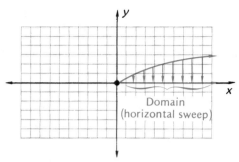

FIGURE 7.46b

Thus, the domain is $\{x \mid x \geq 0\}$.

Step (2) To find the range, project the graph onto the y-axis as shown in figure 7.46c.

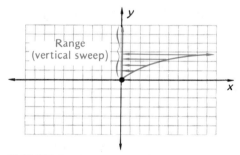

FIGURE 7.46c

Thus, the range is $\{y \mid y \geq 0\}$.

Example 6

Find the domain and range of the relation whose graph is the semicircle shown in figure 7.47a.

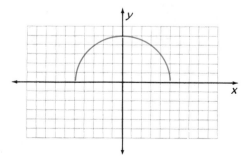

FIGURE 7.47a

Solution:

Step (1)   Find the domain by projecting the graph onto the x-axis (see figure 7.47b).

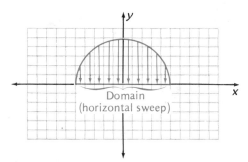

FIGURE 7.47b

Thus, the domain is $\{x \mid -5 \leq x \leq 5\}$.

Step (2)   Find the range by projecting the graph onto the y-axis (see figure 7.47c).

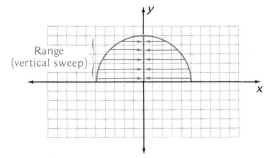

FIGURE 7.47c

Thus, the range is $\{y \mid 0 \leq y \leq 5\}$.

Symbolize each of these relations using set notation.

Exercise 7.9

$y = 2x - 3$ can be written as $\{(x, y) \mid y = 2x - 3\}$.

Example

1. $y = 3x - 4$ **2.** $x + 2y \leq 6$

3. $y = x^2 - 1$ **4.** $x = y^2 + 3$

5. $y = \sqrt{x}$ **6.** $x - \sqrt{y} = 0$

7. $y = \dfrac{1}{\sqrt{x}}$ **8.** $x^2 + y^2 = 4$

9. $3y^2 - x^2 = 0$ **10.** $x = \dfrac{1}{y}$

Find the domain and range for each of these relations.

11. $y = x$

12. $\{(x, y) \mid y = x - 3\}$

13. $x + 2y = 4$

14. $\{(x, y) \mid y = x^2\}$

15. $x = y^2$

16. $\{(x, y) \mid y = x^2 - 3\}$

17. $y = -x^2$

18. $\{(x, y) \mid x = y^2 + 1\}$

19. $y = \sqrt{x}$

20. $\{(x, y) \mid y = \sqrt{x + 4}\}$

21. $y = \sqrt{x - 1}$

22. $\left\{(x, y) \mid y = \dfrac{1}{x}\right\}$

23. $y = \dfrac{1}{x^2}$

24. $\left\{(x, y) \mid y = \dfrac{1}{\sqrt{x}}\right\}$

25. $x = \sqrt{y - 2}$

26. $\{(x, y) \mid y = |x|\}$

Find the domain and range for each of the relations whose graph follows. The equation is written beside each graph simply as a point of interest.

27. $y = 2x - 3$

28. $y = -2x - 3$

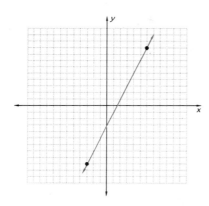
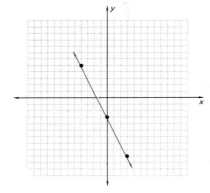

29. $x = 3$

30. $y = -6$

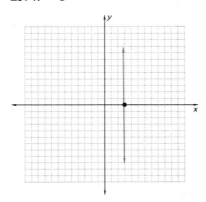
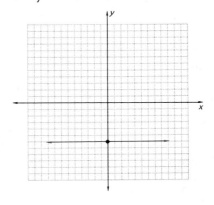

31. $y = x^2 + 1$

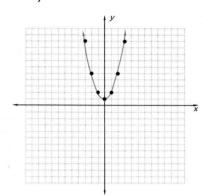

32. $x = y^2 + 3$

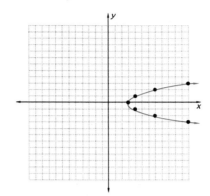

33. $y = -x^2$

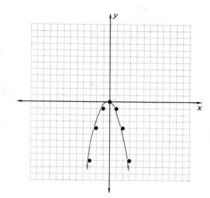

34. $x = -y^2 + 3$

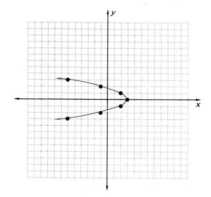

35. $y = x^2 + 2x - 8$

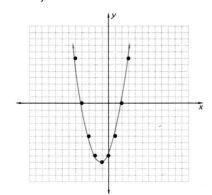

36. $y = |x - 2|$

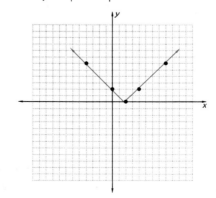

37. $|x| + |y| = 6$

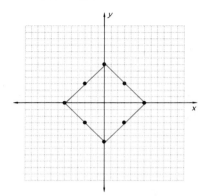

38. $x^2 + y^2 = 16$

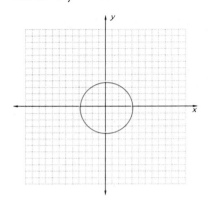

39. $y = \sqrt{9 - x^2}$

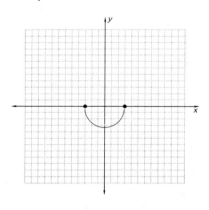

40. $y = \sqrt{x + 5}$

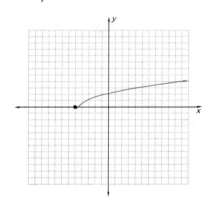

41. $x^2 + 4y^2 = 64$

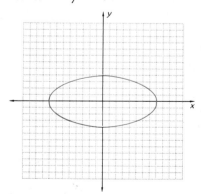

42. $y = 2^x$

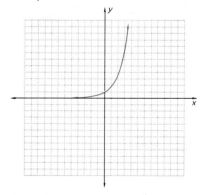

Graph each of these relations. Then state the domain and range.

43. $y = 3x - 1$ **44.** $x + 3y \le 6$

45. $y = x^2 - 2$ **46.** $y = -x^2 + 1$

47. $y = \sqrt{x}$ **48.** $x = \sqrt{y}$

49. $x = y^2$ **50.** $x = y^2 - 1$

FUNCTIONS AND FUNCTIONAL NOTATION

7.10

function

An important concept in mathematics is that of a function. In this section we will study this new concept and discuss some of the associated symbolism, called *functional notation*. A function is defined to be a relation where each x-value in the domain produces exactly one corresponding y-value.

For example, the relation $y = 2x$ is a function. It states that each x-value will produce a y-value equal to twice x. Thus, each x-value will produce only one y-value. If x is 3, then y is $2 \cdot 3$ or 6. And if x is -4, then y is $2 \cdot (-4)$ or -8. On the other hand, the relation $y^2 = x$ is not a function. Some x-values will produce more than one y-value. To illustrate, let $x = 4$. Then, $y^2 = 4$. Solving for y we obtain two values; $y = 2$ and $y = -2$. Remember, a relation is not a function if any x-value produces two or more y-values.

Example 1

Each of the following relations is also a function because each x-value in the domain will produce exactly one y-value. This has been illustrated for each case with one value of x. Verify further by substituting other values for x.

(a) $y = 2x - 5$;   if $x = 3$, then y can only be 1

(b) $y = x^2 - x + 2$;   if $x = 0$, then y can only be 2

(c) $y = \sqrt{x}$;   if $x = 9$, then y can only be 3

Example 2

None of the following relations is a function. In each case, some values of x will produce more than one corresponding value of y.

(a) $y < x + 2$;   if $x = 4$, then y represents every value less than 6

(b) $3y^2 = x$;   if $x = 3$, then $y = 1$ and $y = -1$

(c) $x = |y|$;   if $x = 5$, then $y = 5$ and $y = -5$

Again, a functional relationship will exist only if each x in the domain produces exactly one corresponding value of y. Graphically, each x-value of the domain will produce exactly one point on the graph. For example, figure 7.48 shows the graph of the function $y = x^2$. As indicated by the arrows, each x-value of the domain produces exactly one point on the curve. In other words, no part of the curve doubles back over itself.

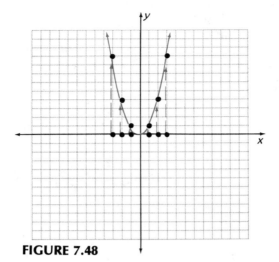

FIGURE 7.48

On the other hand, if the relation is not a function, then for at least one x-value of the domain there are two or more different points on the graph. This is illustrated in figure 7.49, the graph of the relation $x = y^2 - 2$. Since this is not a function, a portion of the curve doubles back over itself, placing at least two points along the same vertical line.

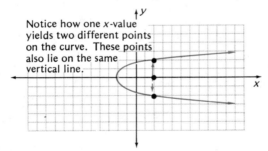

Notice how one x-value yields two different points on the curve. These points also lie on the same vertical line.

FIGURE 7.49

As we have seen, a curve is not the graph of a function if it contains two or more different points for the same value of x. Or, stated another way, a vertical line drawn through that portion of the graph would cut across it more than once. This concept enables us to state the following test for a function. It is called the *vertical line test*.

The Vertical Line Test for a Function

Imagine a set of vertical lines passing through a graph. If any one of these lines cuts the graph more than once, then the relation is not a function.

Use the vertical line test to determine whether figure 7.50a is a graph of a function.

Example 3

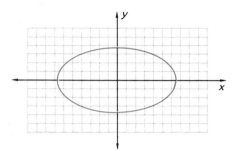

FIGURE 7.50a

Solution: Imagine a set of vertical lines passing through the graph, as shown in figure 7.50b. See whether any of these lines cut the graph more than once.

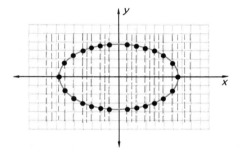

FIGURE 7.50b

Many vertical lines cut through the graph at more than one point. Therefore, it is not the graph of a function.

Use the vertical line test to determine whether figure 7.51a is a graph of a function.

Example 4

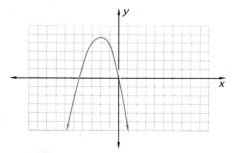

FIGURE 7.51a

Solution: Imagine a set of vertical lines passing through the graph, as shown in figure 7.51b. See whether any of these lines cut the graph more than once.

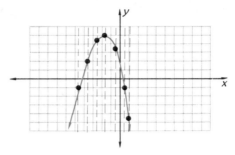

FIGURE 7.51b

No vertical line cuts through the graph at more than one point. Therefore, it is the graph of a function.

It is common to use lowercase letters such as f, g, and h to name functions. For example, using set notation and the letter f, the function defined by the equation $y = 2x + 3$ would be written as follows:

$$f = \{(x, y) \mid y = 2x + 3\}$$

functional notation

Given a function named f, another way to symbolize y is $f(x)$, read "f of x." This symbolism is called **functional notation** and does not mean "f times x." Since $f(x)$ is another name for y, the preceding function could also be written as:

$$f = \{(x, f(x)) \mid f(x) = 2x + 3\}$$

Often you will see the symbol $f(x)$ where x has been replaced with a number, such as $f(4)$. This is read "f of 4" and instructs us to go to the function named f and find the y-value when $x = 4$. Thus, in this case, $f(4) = 2(4) + 3$, or 11. Remember, the notation $f(4) = 11$ simply means that when $x = 4$, $y = 11$.

We use the following functions to illustrate functional notation in examples 5 through 8.

$$f = \{(x, f(x)) \mid f(x) = 2x + 3\}$$
$$g = \{(x, g(x)) \mid g(x) = 2x^2 - 3x + 1\}$$
$$h = \{(x, h(x)) \mid h(x) = \sqrt{x}\}$$

Evaluate $f(-1)$ **Example 5**

Solution: $f(x) = 2x + 3$
$f(-1) = 2(-1) + 3$
$f(-1) = 1$

Evaluate $g(3)$ **Example 6**

Solution: $g(x) = 2x^2 - 3x + 1$

$g(3) = 2(3)^2 - 3(3) + 1$

$g(3) = 18 - 9 + 1$

$g(3) = 10$

Evaluate $h(16)$ **Example 7**

Solution: $h(x) = \sqrt{x}$

$h(16) = \sqrt{16}$

$h(16) = 4$

Evaluate $g(0)$ **Example 8**

Solution: $g(x) = 2x^2 - 3x + 1$

$g(0) = 2(0)^2 - 3(0) + 1$

$g(0) = 0 - 0 + 1$

$g(0) = 1$

State whether or not each of these relations is a function.

1. $y = 2x$ **2.** $y > x + 1$

3. $x + y = 6$ **4.** $2x - 5y \leq 10$

5. $x = y^2$ **6.** $y = x^2$

7. $y = 2x^2 - 1$ **8.** $x = y^2 + 3$

9. $y = \sqrt{x}$ **10.** $x = |y|$

Use the vertical line test to determine which of the following are graphs of functions.

11.

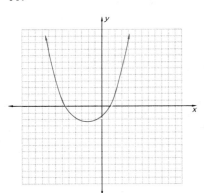

12.

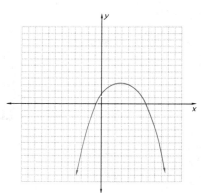

13.

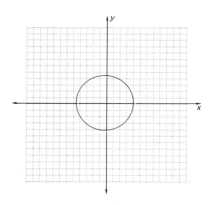

14.

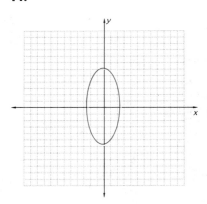

15.

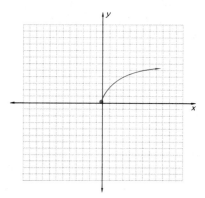

16.

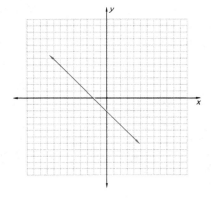

17.

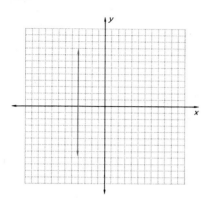

18.

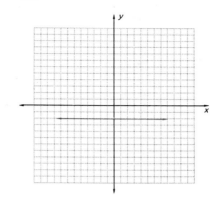

19.

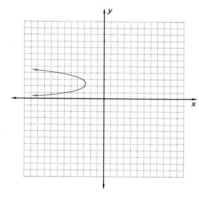

20.

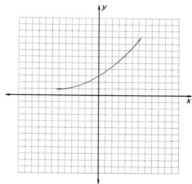

21.

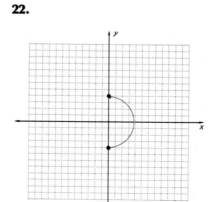

22.

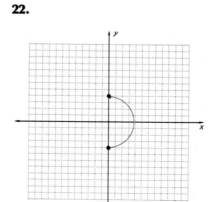

23.

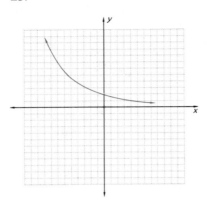

24.

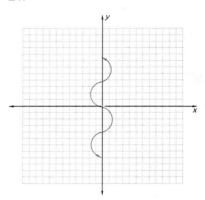

25.

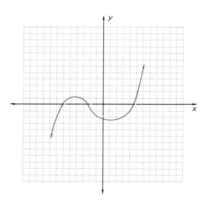

26.

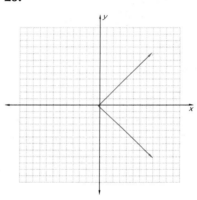

27.

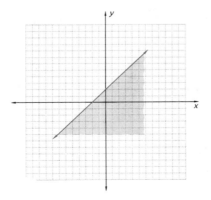

28.

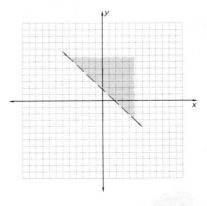

29.

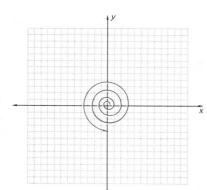

30.

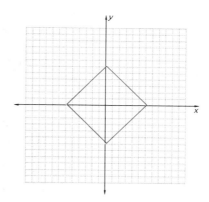

31.

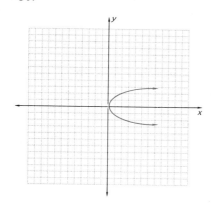

32.

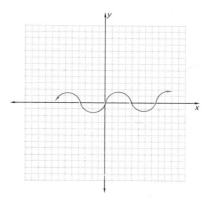

Given these functions, evaluate the following expressions.

$f = \{(x, f(x)) \mid f(x) = 3x - 1\}$

$g = \{(x, g(x)) \mid g(x) = 2x^2 - 5x - 3\}$

$h = \{(x, h(x)) \mid h(x) = \sqrt{x + 3}\}$

33. $f(0)$

34. $g(-1)$

35. $h(1)$

36. $f(-1)$

37. $g(-2)$

38. $h(-3)$

39. $f(-5)$

40. $g(0)$

41. $h(6)$

42. $f(2)$

43. $g(1)$

44. $h(13)$

45. $f(3)$

46. $g(3)$

47. $h(0)$

48. $f\left(\dfrac{4}{3}\right)$

49. $g\left(\dfrac{1}{2}\right)$

50. $h\left(\dfrac{22}{9}\right)$

Summary

SYMBOLS

(x, y)	Ordered pair where x represents the abscissa and y represents the ordinate.
$y = mx + b$	Equation of a straight line where m represents the slope and b represents the y-intercept.
$<$	Less than.
$\leq$	Less than or equal to.
$>$	Greater than.
$\geq$	Greater than or equal to.
$\{(x, y) \mid y = x^2\}$	Set notation for a relation.
$\{(x, f(x)) \mid f(x) = 3x\}$	Set notation for a function.
$f(x)$	The y-value of a function named f. (Read "f of x.")

SUMMARY OF ORDERED PAIRS

Given an ordered pair (x, y):
(1) The abscissa x gives the distance the point is located to the right or left of the y-axis.
(2) The ordinate y gives the distance the point is located above or below the x-axis.
(3) The ordered pair $(0, 0)$ names the origin. It is the point where the x- and y-axes intersect.

TO CONSTRUCT THE GRAPH OF AN EQUATION

(1) Make a table of values by selecting values for one of the variables. This will provide you with ordered pairs. Find as many as necessary to determine a pattern of points.
(2) Plot the points corresponding to the ordered pairs. Remember, if you need more points, go back to the table of values and find more ordered pairs.
(3) Look for a pattern. Then connect the points with a smooth line or curve.

GRAPHING A STRAIGHT LINE

(1) The graph of an equation of the form $y = mx + b$ is a straight line. To graph the equation, find three ordered pairs. Plot the corresponding points and connect them with a straight line.
(2) The graph of an equation of the form $y = k$, where k is a constant, is a horizontal straight line k units from the x-axis.
(3) The graph of an equation of the form $x = k$, where k is a constant, is a vertical straight line k units from the y-axis.
(4) The slope of a straight line is a measure of its slant and is defined to be the rise divided by the run.
(5) The x-intercept is the x-value (abscissa) of the point where the line intersects the x-axis. It can be found by substituting zero for y, and then solving the equation for x.
(6) The y-intercept is the y-value (ordinate) of the point where the line intersects the y-axis. It can be found by substituting zero for x, and then solving the equation for y.

(7) $y = mx + b$ is called the slope-intercept form of a linear equation. In this form, m represents the slope while b represents the y-intercept.

(1)   Graph the boundary line. If the boundary is to be included, draw the line solid; otherwise draw it dashed.

(2)   Select a test point from either the upper or lower region. Substitute the coordinates into the original inequality. If the resulting statement is true, the graph includes that region. If the statement is false, the graph includes the other region.

(1)   The graph of a quadratic equation of the form $y = ax^2 + bx + c$, where $a \neq 0$, is a U-shaped curve called a *parabola*. It will either open upwards or downwards.

(2)   In the equation $y = ax^2 + bx + c$, if $a > 0$, the parabola will open upward. If $a < 0$, the parabola will open downward.

(3)   When graphing a parabola, be sure to obtain several points on both sides of the U-shaped curve.

(1)   A relation is a condition that describes a correspondence between two variables.

(2)   The domain of a relation is the set of all x-values. It gives the horizontal sweep of the graph.

(3)   The range of a relation is the set of all y-values. It gives the vertical sweep of the graph.

(1)   A function is a relation where each x-value in the domain produces exactly one corresponding y-value.

(2)   To determine whether a graph is from a function use the vertical line test. Imagine a set of vertical lines passing through the graph. If any one of these lines cuts the graph more than once, then the relation is not a function.

(3)   Functions are named using lowercase letters such as f, g, or h. The y-value of a function named g is $g(x)$. This is read "g of x."

Self-Checking Exercise

Complete the following tables.

1. $y = 3x - 2$

x	y
-4	
-3	
-2	
$-\frac{1}{3}$	
0	
$\frac{4}{3}$	
2	
$\frac{5}{2}$	
3	
7	

2. $y = 2x^2 - 3x + 1$

x	y
-3	
-2	
-1	
0	
$\frac{1}{2}$	
1	
2	
$\frac{5}{2}$	
3	
4	
5	

3. $y = \dfrac{10}{x}$

x	y
-10	
-5	
-2	
-1	
$-\frac{1}{2}$	
$-\frac{1}{10}$	
$\frac{1}{100}$	
1	
2	
10	
20	
100	

Write an ordered pair such that:

4. The abscissa is -5 and the ordinate is 2

5. $x = \dfrac{3}{4}$ and $y = -\dfrac{1}{2}$

6. It names the origin.

Name the quadrant in which the following points are located.

7. $(3, 8)$

8. $(-2, 3)$

9. $(-3, -2)$

10. $\left(3, -\dfrac{1}{3}\right)$

Figure 7.52 is a graph of the relationship between voltage and current in a circuit where the resistance is constant. Voltage is measured in volts (V) and current is measured in amperes (A).

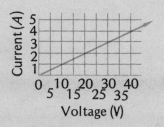

FIGURE 7.52

11. If 40 V are applied to the circuit, what is the resulting current?

12. If 25 V are applied to the circuit, what is the resulting current?

13. What voltage is necessary to produce a current of 1 A?

14. What voltage is necessary to produce a current of 3.5 A?

Graph the following equations.

15. $y = 2x - 3$ **16.** $2x - 3y = 6$

17. $y = -6$ **18.** $x = 5$

19. $y = -x^2 + 9$ **20.** $y = x^2 + 2x - 8$

Graph the following linear equations. Then compute the slope by choosing any two points on the straight line.

21. $2x + y = 4$ **22.** $y = x - 3$

23. $y = 5$

Determine the x- and y-intercepts of these linear equations algebraically.

24. $3x + y = 9$ **25.** $y = 2x - 3$

26. $5x - 2y = 10$

Find the domain and range of these relations.

27. $\{(x, y) \mid y = 2x + 1\}$ **28.** $\{(x, y) \mid y = x^2\}$

29. $\{(x, y) \mid y = \sqrt{x - 1}\}$

Given these functions, evaluate the following expressions.

$f = \{(x, f(x)) \mid f(x) = 3x - 5\}$

$g = \{(x, g(x)) \mid g(x) = -x^2 + 3x\}$

30. $f(0)$ **31.** $g(0)$

32. $f(2)$ **33.** $g(-2)$

34. $f\left(\dfrac{1}{3}\right)$ **35.** $g\left(\dfrac{1}{2}\right)$

36. $f(-3)$ **37.** $g(1)$

Use the vertical line test to determine which of the following are graphs of functions.

38.

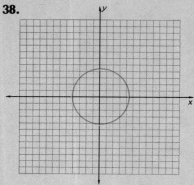

39.

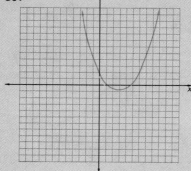

40.

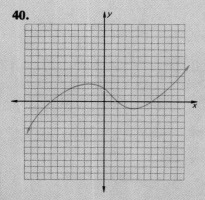

41.

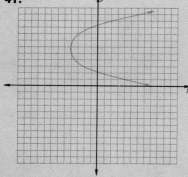

Solutions to Self-Checking Exercise

1. $y = 3x - 2$

x	y
−4	−14
−3	−11
−2	−8
$-\dfrac{1}{3}$	−3
0	−2
$\dfrac{4}{3}$	2
2	4
$\dfrac{5}{2}$	$\dfrac{11}{2}$
3	7
7	19

2. $y = 2x^2 - 3x + 1$

x	y
−3	28
−2	15
−1	6
0	1
$\dfrac{1}{2}$	0
1	0
2	3
$\dfrac{5}{2}$	6
3	10
4	21
5	36

3. $y = \dfrac{10}{x}$

x	y
−10	−1
−5	−2
−2	−5
−1	−10
$-\dfrac{1}{2}$	−20
$-\dfrac{1}{10}$	−100
$\dfrac{1}{100}$	1000
1	10
2	5
10	1
20	$\dfrac{1}{2}$
100	$\dfrac{1}{10}$

4. $(-5, 2)$

5. $\left(\dfrac{3}{4}, -\dfrac{1}{2}\right)$

6. $(0, 0)$

7. Quadrant I.

8. Quadrant II.

9. Quadrant III.

10. Quadrant IV.

11. 4 A

12. 2.5 A

13. 10 V

14. 35 V

15. $y = 2x - 3$

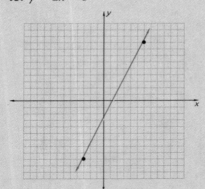

16. $2x - 3y = 6$

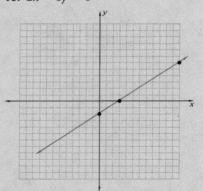

17. $y = -6$

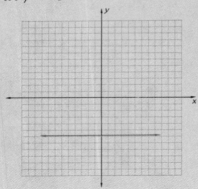

18. $x = 5$

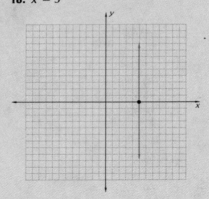

19. $y = -x^2 + 9$

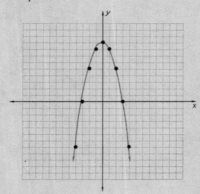

20. $y = x^2 + 2x - 8$

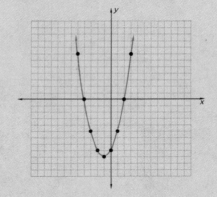

21. $2x + y = 4$
     slope $= -2$

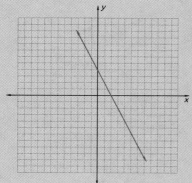

22. $y = x - 3$
     slope $= 1$

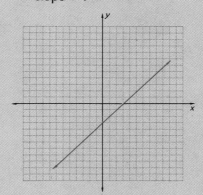

23. $y = 5$
     slope $= 0$

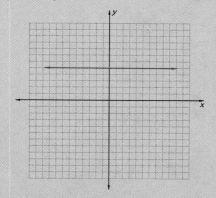

24. x-intercept

$3x + y = 9$
$3x + 0 = 9$
$3x = 9$
$x = 3$

y-intercept

$3x + y = 9$
$3(0) + y = 9$
$0 + y$
$y = 9$

25. x-intercept

$y = 2x - 3$
$0 = 2x - 3$
$3 = 2x$
$\dfrac{3}{2} = x$

y-intercept

$y = 2x - 3$
$y = 2(0) - 3$
$y = 0 - 3$
$y = -3$

26. x-intercept

$5x - 2y = 10$
$5x - 2(0) = 10$
$5x - 0 = 10$
$5x = 10$
$x = 2$

y-intercept

$5x - 2y = 10$
$5(0) - 2y = 10$
$0 - 2y = 10$
$-2y = 10$
$y = -5$

27. $\{(x, y) \mid y = 2x + 1\}$
$D = \{x \mid x \text{ is any real number}\}$
$R = \{y \mid y \text{ is any real number}\}$

28. $\{(x, y) \mid y = x^2\}$
$D = \{x \mid x \text{ is any real number}\}$
$R = \{y \mid y \geq 0\}$

29. $\{(x, y) \mid y = \sqrt{x - 1}\}$
$D = \{x \mid x \geq 1\}$
$R = \{y \mid y \geq 0\}$

30. $f(0) = -5$

31. $g(0) = 0$

32. $f(2) = 1$

33. $g(-2) = -10$

34. $f\left(\dfrac{1}{3}\right) = -4$

35. $g\left(\dfrac{1}{2}\right) = \dfrac{5}{4}$

36. $f(-3) = -14$

37. $g(1) = 2$

38. Is not a function.

39. Is a function.

40. Is a function.

41. Is not a function.

SYSTEMS OF LINEAR EQUATIONS AND LINEAR INEQUALITIES

8

SYSTEMS OF LINEAR EQUATIONS

8.1

A system of linear equations in two variables consists of two linear equations containing the same two variables.

Example 1

Each of the following is a system of linear equations.

(a) $x + y = 5$
    $x - y = 1$

(b) $m + 2n = 5$
    $2m + 3n = 7$

(c) $2s + 4t = 5$
    $s = 3t$

(d) $y = 3x - 4$
    $y = 5x - 6$

To solve a system of linear equations, we must find all of the ordered pairs that satisfy both equations at the same time. Such an ordered pair is said to be a **simultaneous solution** of the system. In later sections of this chapter you will learn three techniques for solving a system of linear equations: (1) graphing, (2) addition, and (3) substitution. However, before studying these techniques, you must learn how to check a simultaneous solution.

simultaneous solution

To check whether an ordered pair is a simultaneous solution of a system, simply substitute the ordered pair into both equations. If it makes both equations true, it is a solution. Otherwise it is not.

Example 2

Check to see whether the ordered pair (3, 2) is a simultaneous solution of this system:

$x + y = 5$

$x - y = 1$

Solution: Substitute 3 for x and 2 for y in both equations.

Check 1
$x + y = 5$
$3 + 2 = 5$
$5 = 5$
True

Check 2
$x - y = 1$
$3 - 2 = 1$
$1 = 1$
True

Since the ordered pair (3, 2) satisfies both equations, it is a simultaneous solution of the system.

Check to see whether the ordered pair (7, −1) is a simultaneous solution   **Example 3**
of this system:

$m + 2n = 5$

$2m + 3n = 7$

Solution: Unless specified otherwise, ordered pairs are in alphabetical
order. Thus, $m = 7$ and $n = −1$. Substitute these values into
both equations.

Check 1
$m + 2n = 5$
$7 + 2(−1) = 5$
$7 − 2 = 5$
$5 = 5$
True

Check 2
$2m + 3n = 7$
$2(7) + 3(−1) = 7$
$14 − 3 = 7$
$11 = 7$
False

Since the ordered pair (7, −1) does not satisfy both
equations, it is not a simultaneous solution.

You already know that the graph of a linear equation is a straight line.
Thus the graph of a system of linear equations must contain two straight
lines. And a simultaneous solution, since it satisfies both equations at
the same time, must represent a point where the two lines cross. This is
called a **point of intersection**.   **point of intersection**

In Example 2, we showed that the ordered pair (3, 2) is a simultaneous
solution of this system:

$$x + y = 5$$

$$x − y = 1$$

Figure 8.1 shows the graph of both equations and illustrates that the
simultaneous solution (3, 2) is indeed the point of intersection.

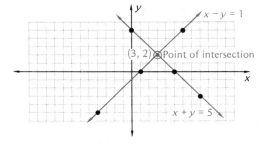

FIGURE 8.1

Since a simultaneous solution represents a point where two straight lines intersect, each system of linear equations will fit into exactly one of these three categories:

(1) The graphs will intersect at exactly one point, as in figure 8.2. A system of this type has exactly one simultaneous solution. It is **consistent system**   called a **consistent system**.

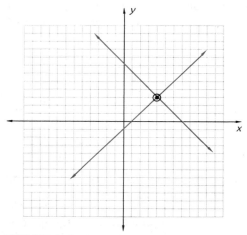

FIGURE 8.2

(2) The graphs will be parallel and not intersect at all, as in figure 8.3. A system of this type has no simultaneous solutions whatsoever. It is **inconsistent system**   called an **inconsistent system**.

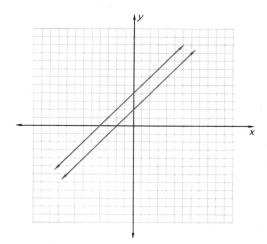

FIGURE 8.3

(3) The two equations will be equivalent and have the same graph, giving the appearance of one straight line, as in figure 8.4. The

graphs would then intersect at every point along the line. A system of this type has infinitely many simultaneous solutions, representing every point of the common line. It is called a **dependent system.**

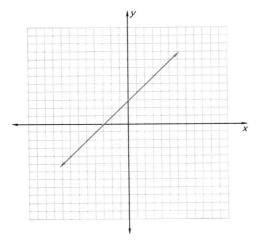

FIGURE 8.4

Determine whether or not the listed ordered pair is a simultaneous solution of the given system of linear equations.

1. $(4, 2)$ $x + y = 6$
 $x - y = 2$

2. $(3, 1)$ $2x + y = 7$
 $x - y = 2$

3. $(2, 0)$ $3x - y = 6$
 $2x - y = 4$

4. $(3, 1)$ $a + b = 4$
 $2a + 3b = 1$

5. $(3, -6)$ $2m + n = 0$
 $5m + 3n = 3$

6. $(2, 13)$ $2x - y = -9$
 $x + y = 3$

7. $(9, 3)$ $4x + 3y = 45$
 $4x + 5y = 51$

8. $(4, 2)$ $3a - 5b = 2$
 $2a + 7b = 22$

9. $\left(\dfrac{3}{5}, -\dfrac{2}{5}\right)$ $3m - 8n = 5$
 $5m + 5n = 1$

10. $(5, 0)$ $2r - 4t = 10$
 $3r + 2t = 17$

11. $\left(\dfrac{1}{3}, \dfrac{2}{3}\right)$ $3x + 9y = 7$
 $5x - 4y = -1$

12. $(1, -1)$ $5x - y = 4$
 $x - 5y = 4$

13. $(6, 2)$ $7a - 3b = 36$
 $2a + 5b = 22$

14. $(0, 2)$ $5m + 3n = 6$
 $2m + 4n = 8$

15. $(1, -3)$ $x - 2y = 7$
 $y = 3x$

16. $\left(1, \dfrac{1}{2}\right)$ $r = 2t$
 $r + 4t = 3$

17. $(0, 0)$ $2x = 3y$
 $x - 4y = 0$

18. $(-1, 1)$ $y = 3x - 4$
 $y = 5x - 6$

19. $(-1, 1)$ $2x - y + 3 = 0$
                      $3x + 2y - 5 = 0$

20. $(2, -5)$ $3a + 2b + 4 = 0$
                      $5a - 3b - 20 = 0$

21. $(0, 4)$ $3x + 2y - 12 = 0$
                      $x - 3y + 12 = 0$

22. $\left(\dfrac{1}{2}, -\dfrac{3}{2}\right)$ $4m + 6n + 7 = 0$
                      $2m - 8n - 13 = 0$

23. $\left(\dfrac{6}{7}, \dfrac{17}{7}\right)$ $3r + t = 5$
                      $2r + 3t = 9$

24. $(2, 3)$ $3x + y = 9$
                      $y = 3$

25. $(6, -5)$ $a - 6 = 0$
                      $b + 5 = 0$

26. $(2, -5)$ $m + 2 = 0$
                      $n - 5 = 0$

27. $(-1, 2)$ $x = 2$
                      $y = -1$

28. $\left(\dfrac{34}{27}, \dfrac{1}{27}\right)$ $3x - 3y - 4 = 0$
                      $5x + 4y - 6 = 0$

29. $\left(\dfrac{-6}{7}, \dfrac{15}{7}\right)$ $3r + 4s - 6 = 0$
                      $2r + 5s - 9 = 0$

30. $(2, 4)$ $2x + 3y - 3 = 0$
                      $x - 2y - 4 = 0$

Each of the following is a graph of a system of linear equations. Classify each system as consistent, inconsistent, or dependent.

31.

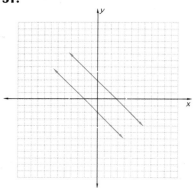

32.

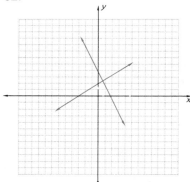

33.

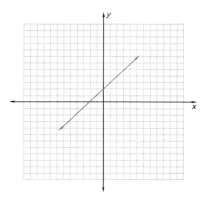

34.

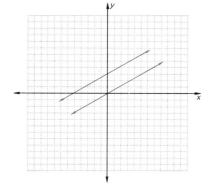

35.

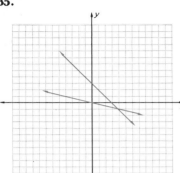

36.

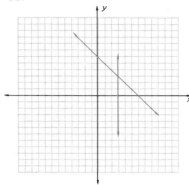

Fill in the blanks so that each statement is true.

37. The graph of an inconsistent system of linear equations consists of two _____ straight lines.

38. If the graph of a system of linear equations produces only one straight line, the system is said to be _____.

39. If the graph of a system of linear equations produces two straight lines intersecting at exactly one point, the system is said to be _____.

40. In a _____ system of linear equations, the two equations are equivalent.

SOLVING SYSTEMS OF LINEAR EQUATIONS BY GRAPHING

8.2

You know from previous experience that the graph of a linear equation is a straight line. Thus, to solve a system of linear equations by graphing, simply construct both straight lines on the same coordinate system. Then determine any points of intersection. The ordered pair representing an intersection point is a simultaneous solution of the system. Recall from section 8.1 that each system of linear equations will fit into only one of these three categories:

(1) A consistent system. The two straight lines intersect at exactly one point. The system has exactly one simultaneous solution.

(2) An inconsistent system. The two straight lines are parallel and do not intersect. Hence, the system has no simultaneous solutions.

(3) A dependent system. The two equations are equivalent and produce the same line. Consequently, every point on the common line is a simultaneous solution.

Solve this system by graphing.

Example 1

$2x + y = -2$

$x + 2y = 5$

Solution: Graph both equations. Then determine the coordinates of the intersection point. (See figure 8.5.) This system is consistent and the solution is $(-3, 4)$. Remember, the solution should be checked by substituting the coordinates into both of the original equations as follows:

Check 1

$$2x + y = -2$$
$$2(-3) + 4 = -2$$
$$-6 + 4 = -2$$
$$-2 = -2$$

True

Check 2

$$x + 2y = 5$$
$$-3 + 2(4) = 5$$
$$-3 + 8 = 5$$
$$5 = 5$$

True

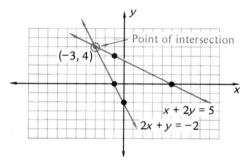

FIGURE 8.5

Example 2   Solve this system by graphing.

$$x + y = 3$$
$$x + y = 5$$

Solution: Graph both equations on the same coordinate system. Then look for an intersection point. (See figure 8.6.) The lines are parallel. This is an inconsistent system. It has no solutions.

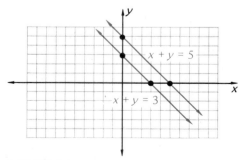

FIGURE 8.6

Solve this system by graphing.   Example 3

$x + y = 3$

$2x + 2y = 6$

Solution: Graph the equations. (See figure 8.7.) Both equations produce the same line. This system is dependent. Thus, every
point on the common line is a simultaneous solution. When
this occurs we will write "same line."

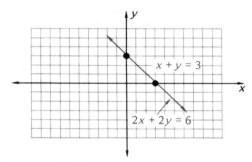

FIGURE 8.7

Solve this system by graphing.   Example 4

$x + 2y = 5$

$3x + 3y = 5$

Solution: Graph both equations. (See figure 8.8.) This system is consistent. x is somewhere between -1 and -2. y is somewhere
between 3 and 4. The best we can do is an approximation. So,
the solution appears to be about $\left(-1\frac{1}{3}, 3\frac{1}{3}\right)$

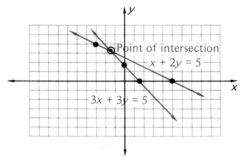

FIGURE 8.8

Example 4 illustrates a severe limitation of the graphing method. It can obviously be inaccurate, especially if the simultaneous solution involves fractions or decimals instead of integers. Therefore, in the next two sections we will learn two algebraic methods, called addition and substitution. These methods will not depend upon the inaccuracies of graphing and will give exact solutions.

Exercise 8.2

Solve the following systems of linear equations by graphing. Also, classify each system as consistent, inconsistent, or dependent.

1. $x + y = 6$
$x - y = 2$

2. $x + y = 4$
$x - y = 2$

3. $x + y = -5$
$x - y = -7$

4. $x + y = 3$
$x + y = 5$

5. $x + y = 2$
$3x + 3y = 6$

6. $x - y = 3$
$x + y = -1$

7. $x + y = 1$
$x - y = 5$

8. $x + y = 1$
$2x + y = -1$

9. $x + 2y = 3$
$2x + 4y = 7$

10. $3x - y = 4$
$5x - y = 6$

11. $2x + 3y = 2$
$4x + 6y = 4$

12. $x + 2y = 1$
$3x + 6y = 30$

13. $3x + 2y = 6$
$x + 2y = 6$

14. $2x + y = 4$
$x + y = 1$

15. $2x + y = -2$
$x + 2y = 5$

16. $x + y = 2$
$y = x$

17. $y = 8 - x$
$y = 3x$

18. $x + 2y = 0$
$2x + y = 3$

19. $y = 3 - x$
$x = 3 + y$

20. $y = x - 3$
$2x - 2y = 6$

21. $x + y = 5$
$x = 3 - y$

22. $y = x + 1$
$x = y + 1$

23. $y = x - 2$
$y = 2x + 1$

24. $y = x - 1$
$y = 5 - x$

25. $2x + y = 3$
$y = -x$

26. $y = 2 + x$
$4y = 3x$

27. $x + 3y - 5 = 0$
$2x + 6y - 5 = 0$

28. $2x - 3y = 6$
$4x - 6y = 12$

29. $2x - 3y = 6$
$4x - 6y = 24$

30. $2x + y - 5 = 0$
$x + 2y - 1 = 0$

31. $2x - 3y = -15$
$3x + 4y = 20$

32. $5x + 2y - 20 = 0$
$x + y - 10 = 0$

33. $x - 3y = -11$
$\quad\;\; 3x + 2y = 11$

34. $\quad\quad\; y = 3$
$\quad\;\; x + 2y = 8$

35. $2x - 3y + 5 = 0$
$\quad\quad\;\; y + 1 = 0$

36. $5x - 2y = -9$
$\quad\;\; 3x - 4y = 3$

Solve the following systems by graphing. The solutions do not contain integers. Therefore, estimate the coordinates of the intersection point as accurately as you can.

37. $x + y = 5$
$\quad\; x - y = 8$

38. $2x - y = -1$
$\quad\;\; x + 2y = 8$

39. $\;\; x - 2y = 9$
$\quad\; 3x + 2y = -2$

40. $\;\; x - y = -5$
$\quad\; 2x + y = 7$

SOLVING SYSTEMS OF LINEAR
EQUATIONS BY ADDITION

8.3

In section 8.2 you learned that solving a system of linear equations by graphing can be inaccurate, especially if the simultaneous solution contains fractions or decimals. Thus we now study an algebraic method which does not depend upon the inaccuracies and approximations of graphing. This method uses the addition property of equivalent equations to eliminate one of the variables from the system. It is called the addition method or the elimination method of solving a system of linear equations.

The addition property of equivalent equations, studied in section 2.9, permits the same number to be added to each side of an equation. Since an equation states that two expressions represent the same number, it follows that two equations can be added together to form a third equation. Under this interpretation the addition property of equivalent equations can be rewritten as follows.

Addition of Equations

Given expressions A, B, C, and D,

$\quad\quad$ If $A = B$ and $C = D$, then $A + C = B + D$

Solve this system by adding the equations.   **Example 1**

$2x - y = 3$

$\;\; x + y = 3$

Solution:

Step (1)   Add the two equations, thereby eliminating the *y*-terms.

$$2x - y = 3$$
$$\underline{x + y = 3}$$
$$3x + 0 = 6$$

Step (2)   Solve the resulting equation for *x*

$$3x + 0 = 6$$
$$3x = 6$$
$$x = 2$$

Thus, 2 is the *x*-value of the simultaneous solution.

Step (3)   To find the *y*-value, substitute 2 for *x* in either of the two equations forming the system. Choosing the second equation produces the following:

$$x + y = 3$$
$$2 + y = 3$$
$$y = 1$$

Hence, 1 is the *y*-value of the simultaneous solution. The solution of the system is (2, 1).

To check, we substitute into the other equation of the system, $2x - y = 3$

Check $2x - y = 3$

$$2 \cdot 2 - 1 = 3$$
$$3 = 3 \qquad \text{[True]}$$

Since the result is true, the simultaneous solution of the given system is (2, 1).

Example 2   Solve this system of equations.

$$3x + 4y = 6$$
$$x + 3y = 7$$

Solution: If we add the two given equations, we obtain $4x + 7y = 13$. This does not help solve the system because neither variable is eliminated. Therefore, before adding the equations, multiply each side of the second equation by -3. This will eliminate the x-terms from the system when the equations are added together.

$$3x + 4y = 6$$
$$x + 3y = 7 \quad \xrightarrow{\text{mult. by } -3} \quad \begin{array}{l} 3x + 4y = 6 \\ -3(x + 3y) = -3 \cdot 7 \end{array} \quad \xrightarrow{\text{producing}} \quad \begin{array}{l} 3x + 4y = 6 \\ -3x - 9y = -21 \end{array}$$

Now add the two equations, eliminating the x-terms. The resulting equation is then solved for y.

$$
\begin{array}{rcr}
3x + 4y & = & 6 \\
-3x - 9y & = & -21 \\
\hline
0 - 5y & = & -15 \\
-5y & = & -15 \\
y & = & 3
\end{array}
$$

Next substitute 3 for y into either of the two original equations. Choosing the second equation, we have the following:

$$x + 3y = 7$$
$$x + 3 \cdot 3 = 7$$
$$x + 9 = 7$$
$$x = -2$$

Therefore, the simultaneous solution of this system is $(-2, 3)$. It can be checked by substituting into the other equation, $3x + 4y = 6$

As stated earlier, this method of solving a system of linear equations is called the addition method or the elimination method. It provides a systematic way of eliminating one of the variables, leading to the exact simultaneous solution. It is not subject to the approximations of graphing. This method is summarized in the following steps.

Step (1)   Write each equation in the form $ax + by = c$. Then decide which variable to eliminate.

Step (2)   Multiply one or both equations by appropriate numbers so that the coefficients of the variables are opposites.

Step (3)   Add the two equations, obtaining an equation having only one variable.

Step (4)   Solve the equation obtained in Step (3).

Step (5) Substitute this value into either of the original two equations and solve for the remaining variable. You now have the simultaneous solution, which can be checked by substituting into the other equation.

Example 3

Use the addition method to solve this system.

$$5x = 2y - 20$$
$$2x + 3y = 11$$

Solution:

Step (1) Write each equation in the form $ax + by = c$. Arbitrarily we decide to eliminate the y-terms.

$$5x - 2y = -20$$
$$2x + 3y = 11$$

Step (2) Transform the coefficients of the y-terms into -6 and 6, by multiplying the first equation by 3 and the second equation by 2.

$$5x - 2y = -20 \xrightarrow{\text{multiply by 3}} 15x - 6y = -60$$
$$2x + 3y = 11 \xrightarrow{\text{multiply by 2}} 4x + 6y = 22$$

Step (3) Add the two resulting equations, thereby eliminating the y-terms.

$$15x - 6y = -60$$
$$\underline{4x + 6y = 22}$$
$$19x + 0 = -38$$

Step (4) Solve the equation for x.

$$19x + 0 = -38$$
$$19x = -38$$
$$x = -2$$

Step (5) Substitute -2 for x into either of the original two equations. Choosing the second equation, we substitute, and then solve for y

$$2x + 3y = 11$$
$$2(-2) + 3y = 11$$
$$-4 + 3y = 11$$
$$3y = 15$$
$$y = 5$$

Thus the simultaneous solution is $(-2, 5)$ and the system is consistent. Remember, this solution can be checked by substituting it into the other equation, $5x = 2y - 20$

Use the addition method to solve this problem.

Example 4

$$4x + 9y = 17$$
$$6x = 13 - 6y$$

Solution:

Step (1) Write each equation in the form $ax + by = c$. We shall eliminate the x-terms.

$$4x + 9y = 17$$
$$6x + 6y = 13$$

Step (2) The coefficients of the x-terms can be transformed into 12 and -12, respectively. Multiply the first equation by 3 and the second equation by -2.

$$4x + 9y = 17 \xrightarrow{\text{multiply by 3}} 12x + 27y = 51$$
$$6x + 6y = 13 \xrightarrow{\text{multiply by } -2} -12x - 12y = -26$$

Step (3) Add the two resulting equations, thereby eliminating the x-terms.

$$\begin{array}{r} 12x + 27y = 51 \\ -12x - 12y = -26 \\ \hline 0 + 15y = 25 \end{array}$$

Step (4) Solve the equation for y

$$0 + 15y = 25$$
$$15y = 25$$
$$y = \frac{25}{15}$$
$$y = \frac{5}{3}$$

Step (5) Substitute $\frac{5}{3}$ for y into either of the original two equations. Choosing the second one, we substitute and solve for x

$$6x = 13 - 6y$$

$$6x = 13 - 6 \cdot \frac{5}{3}$$

$$6x = 13 - 2 \cdot 5$$

$$6x = 3$$

$$x = \frac{3}{6}$$

$$x = \frac{1}{2}$$

Thus, the simultaneous solution is $\left(\frac{1}{2}, \frac{5}{3}\right)$ and the system is consistent. You should check this solution by substituting it into the other equation, $4x + 9y = 17$

The next two examples show what happens if the addition method is applied to either an inconsistent or dependent system of linear equations.

Example 5

Use the addition method to solve this system.

$$3x + 6y = 30$$
$$x = 1 - 2y$$

Solution:

Step (1) Write in the form $ax + by = c$. We shall eliminate the x-terms.

$$3x + 6y = 30$$
$$x + 2y = 1$$

Step (2) Multiply the second equation by -3. Then the coefficients of x will be 3 and -3 respectively.

$$3x + 6y = 30 \qquad\qquad\qquad 3x + 6y = 30$$
$$x + 2y = 1 \xrightarrow{\text{multiply by } -3} -3x - 6y = -3$$

Step (3) Add the resulting equations.

$$3x + 6y = 30$$
$$-3x - 6y = -3$$
$$\overline{0\ +\ 0\ =\ 27}$$
$$0 = 27 \qquad \text{[False]}$$

Notice that both variables were eliminated and a false equation resulted. When this occurs, the addition method is telling us that the system is inconsistent and there are no simultaneous solutions. If graphed, the two lines would be parallel.

Use the addition method to solve this system.

Example 6

$$2x + 3y = 2$$

$$4x + 6y = 4$$

Solution:

Step (1) Both equations are in the form $ax + by = c$. We will eliminate the x-terms.

$$2x + 3y = 2$$

$$4x + 6y = 4$$

Step (2) Multiply the first equation by -2. Then the coefficients of x will be -4 and 4 respectively.

$$2x + 3y = 2 \xrightarrow{\text{multiply by } -2} -4x - 6y = -4$$

$$4x + 6y = 4 \qquad\qquad\qquad 4x + 6y = \ \ 4$$

Step (3) Add the resulting equations.

$$-4x - 6y = -4$$
$$4x + 6y = \ \ 4$$
$$\overline{0\ +\ 0\ =\ \ 0}$$
$$0 = \ \ 0 \qquad \text{[True]}$$

Notice that both variables were eliminated and a true equation resulted. When this occurs, the additon method is telling us that the two equations are equivalent. Thus, we have a dependent system and the solutions lie all along the common line.

Exercise 8.3　　Use the addition method to solve the following systems. Also, classify each system as consistent, inconsistent, or dependent.

1. $x - y = 2$
$x + y = 4$

2. $x + y = 9$
$x - y = 1$

3. $a + b = 5$
$a - b = 7$

4. $m + n = 2$
$m - n = 1$

5. $2x - y = 1$
$x + y = 11$

6. $3x + y = 16$
$x - 2y = 4$

7. $m + n = 5$
$m - 2n = 1$

8. $x + 3y = 1$
$2x + 5y = 1$

9. $5m + 3n = 17$
$m + 3n = 1$

10. $2x = 1 + 5y$
$6x - 15y = 3$

11. $3r + 4s = 6$
$r + s = 2$

12. $5x = 6 + 2y$
$3x - 12 = 4y$

13. $a = 4 - b$
$b = -5a$

14. $x = 3 - y$
$y = 2x + 9$

15. $6x + 7y = 0$
$2x - 3y = 32$

16. $a + 2b = 3$
$3a = 9 - 6b$

17. $y = 3x - 4$
$y = 5x - 6$

18. $m + 2n = 3$
$2m + 4n = 7$

19. $a - b = 2$
$2a + b = 8$

20. $a - b = 1$
$a = b + 5$

21. $y = 3x - 1$
$6x - 2y = 1$

22. $x + y = 3$
$2x - y = -9$

23. $3x - 2y = 1$
$2x + 5y = 7$

24. $7m + 3n = 4$
$9m + 5n = 8$

25. $a - 2b = 3$
$4a - 8b = 12$

26. $2x + y + 2 = 0$
$x + 2y - 5 = 0$

27. $2u - v = -6$
$u - v = -5$

28. $2r + 3t = 5$
$3r - 2t = -12$

29. $x - y = 2$
$2x + y = 7$

30. $3a + b = 5$
$2a - 3b = 9$

31. $2x + 3y = 1$
$x + y = 4$

32. $3x + 5y = 1$
$4x + 7y = 2$

33. $2x = 6 - 2y$
$x + y - 4 = 0$

34. $2y = 6 - 2x$
$x + y = 3$

35. $2r - 3t = 1$
$5r + 2t = -26$

36. $6u - 2v = 7$
$-3u + v = 5$

37. $x + y = 5$
$x - y = 8$

38. $-x + 2y = 3$
$3x - 6y = -9$

39. $4x + y = -18$
$5x + 3y = -19$

40. $2x + 5y = -20$
$x - 2y = 8$

41. $a + 7b = 21$
$a - 5b = -15$

42. $7m + 3n = -12$
$3m - 4n = 16$

43. $6u + 12v = 13$
$3u + 8v = 9$

44. $4x - 5y = 1$
$5x - 7y = 8$

45. $5p - q = 7$
$3p - q = 5$

46. $3x + y - 7 = 0$
$2x - y - 8 = 0$

47. $5a + 2b = 11$
$5a - 2b = 19$

48. $8m + 5n = 7$
$12m - 6n = 11$

49. $4x - 5y = -12$
$6x - 3y = -9$

50. $7x - 5y = -17$
$3x + 4y = 5$

SOLVING SYSTEMS OF LINEAR EQUATIONS BY SUBSTITUTION

8.4

A second algebraic method for solving systems of linear equations is called the substitution method. This method uses the fact that any quantity may be substituted for its equal. It works particularly well when one of the equations has been solved for one of its variables.

Solve this system.

Example 1

$$y = 2x$$
$$x + 3y = 14$$

Solution: The first equation is solved for y. It tells us that y equals $2x$. Using this fact we can substitute $2x$ for y in the second equation.

$$x + 3y = 14$$
$$x + 3(2x) = 14$$

We then solve the resulting equation for x

$$x + 3(2x) = 14$$
$$x + 6x = 14$$
$$7x = 14$$
$$x = 2$$

Since $x = 2$, we go to the first equation $y = 2x$, and find y

$$y = 2x$$
$$y = 2 \cdot 2$$
$$y = 4$$

Therefore, the simultaneous solution is (2, 4) and the system is consistent. To check a solution obtained by the substitution method, you should substitute into both of the original equations as follows.

Check 1
$y = 2x$
$4 = 2 \cdot 2$
$4 = 4$
True

Check 2
$x + 3y = 14$
$2 + 3 \cdot 4 = 14$
$14 = 14$
True

Example 2

Solve this system using the substitution method.

$$2x + 3y = 3$$
$$x = 4 - 2y$$

Solution: The second equation is solved for x. It tells us that x equals $4 - 2y$. Therefore, we can substitute $4 - 2y$ for x in the first equation.

$$2x + 3y = 3$$
$$2(4 - 2y) + 3y = 3$$

The resulting equation is then solved for y

$$2(4 - 2y) + 3y = 3$$
$$8 - 4y + 3y = 3$$
$$8 - y = 3$$
$$-y = -5$$
$$y = 5$$

Since $y = 5$, we go back to the second equation $x = 4 - 2y$, and find x

$x = 4 - 2y$

$x = 4 - 2 \cdot 5$

$x = 4 - 10$

$x = -6$

Therefore, the simultaneous solution is $(-6, 5)$. Remember, this solution can be checked by substituting it into both of the original equations.

Use the substitution method to solve this system.   **Example 3**

$x + 6y = 1$

$3x + 4y = -4$

Solution: Solve the first equation for x

$x + 6y = 1$

$x = 1 - 6y$

Substitute $1 - 6y$ for x in the second equation.

$3x + 4y = -4$

$3(1 - 6y) + 4y = -4$

Solve the resulting equation for y

$3(1 - 6y) + 4y = -4$

$3 - 18y + 4y = -4$

$-14y = -7$

$y = \dfrac{1}{2}$

Go back to the equation $x = 1 - 6y$ and find x

$x = 1 - 6y$

$x = 1 - 6 \cdot \dfrac{1}{2}$

$x = 1 - 3$

$x = -2$

The simultaneous solution is $\left(-2, \frac{1}{2}\right)$. You should check this solution by substituting it in the original two equations.

Examples 4 and 5 illustrate what happens when the substitution method is applied to either an inconsistent or dependent system.

Example 4

Solve this system using the substitution method.

$$x + 2y = 1$$
$$3x + 6y = 5$$

Solution: Solve the first equation for x

$$x + 2y = 1$$
$$x = 1 - 2y$$

Substitute $1 - 2y$ for x in the second equation.

$$3x + 6y = 5$$
$$3(1 - 2y) + 6y = 5$$

Solve the resulting equation for y

$$3(1 - 2y) + 6y = 5$$
$$3 - 6y + 6y = 5$$
$$3 + \quad 0 \quad = 5$$
$$3 \quad = 5 \qquad \text{[False]}$$

Both variables were eliminated, producing a false equation. When this occurs, the substitution method is telling us that the system is inconsistent and there are no simultaneous solutions. If graphed, the two lines would be parallel.

Example 5

Solve this system using the substitution method.

$$6x - 4y = 14$$
$$2y = 3x - 7$$

Solution: Solve the second equation for y

$$2y = 3x - 7$$
$$y = \frac{3x - 7}{2}$$

Substitute $\dfrac{3x - 7}{2}$ for y in the first equation.

$$6x - 4y = 14$$

$$6x - 4\left(\frac{3x - 7}{2}\right) = 14$$

Solve the resulting equation for x

$$6x - 4\left(\frac{3x - 7}{2}\right) = 14$$

$$6x - 2(3x - 7) = 14$$

$$\underbrace{6x - 6x} + 14 = 14$$

$$0 \quad + 14 = 14$$

$$14 = 14 \qquad \text{[True]}$$

Both variables were eliminated, producing a true equation. When this happens, the substitution method is telling us that the two equations are equivalent. Thus, the system is dependent and the solutions lie all along the common line.

You now have at your command three methods for solving a system of linear equations. They are the graphing method, the addition method, and the substitution method. Since there are three ways of solving a given system, it is important to be able to select the easiest method. The following guidelines should prove helpful.

(1) The graphing method should be used when you want a visual interpretation. Recall that this method can be inaccurate.

(2) The substitution method is difficult to use unless one of the equations is easily solved for x or y without producing a fraction. Thus, substitution works well when one of the equations has a numerical coefficient of 1. For example, the substitution method would work well on the following system because the second equation is easily solved for x.

$$4x + 3y = 7$$

$$x + 2y = 4$$

(3) It is best to use the addition method when neither of the equations contains a numerical coefficient of 1; that is, when neither equation

can be solved for x or y without producing a fraction. Hence, the addition method is the best choice for solving this system.

$$2x + 5y = 1$$

$$3x - 6y = 7$$

Exercise 8.4

Use the substitution method to solve the following systems. Also, classify each system as consistent, inconsistent, or dependent.

1. $x + y = 8$
$\quad\ y = 3x$

2. $x + 3y = 7$
$\qquad\ y = 2x$

3. $2a + b = 2$
$\qquad b = -3a$

4. $3m - 4n = 38$
$\qquad\ m = -5n$

5. $2x - 3y = 10$
$\qquad\ x = -y$

6. $3x - 2y = 5$
$\qquad\ x = 4y$

7. $5p - 3q = 10$
$\qquad\ q = -5p$

8. $3x - 4y = -5$
$\qquad\ y = 2x$

9. $x + y = 8$
$\qquad y = x + 6$

10. $x + 2y = 8$
$\qquad\ x = y + 2$

11. $m - 3n = -5$
$\qquad\ m = n + 3$

12. $2x + 5y = -9$
$\qquad\ x = 3y + 1$

13. $5x + y = 3$
$\qquad\ y = 2x + 10$

14. $9x + 4y = -7$
$\qquad\ x = 2y - 13$

15. $3x - y = 5$
$\qquad\ y = 3x - 6$

16. $2x - y = -1$
$\qquad\ y = 2x + 1$

17. $m - n = 2$
$\quad\ 2m + n = 8$

18. $x - 4y = 0$
$\qquad 2x = 3y$

19. $a - b = 1$
$\qquad\ a = b + 5$

20. $\qquad r = 2s$
$\quad 2r - 3s = 1$

21. $x + y = 3$
$\qquad\ y = 2x + 9$

22. $\qquad x = 1 - 4y$
$\quad 2x - 8y = 1$

23. $x - 4y = 5$
$\quad 2x - 8y = 7$

24. $a + b = 5$
$\quad\ a - 2b = 1$

25. $x + 3y = 1$
$\quad 2x + 5y = 1$

26. $m + 3n = 1$
$\quad 5m + 3n = 1$

27. $x = 3 - y$
$\quad\ y = 2x + 9$

28. $p = 4 - q$
$\quad\ q = -5p$

29. $2x + y = 5$
$\qquad x + 3y = 0$

30. $a + b = 6$
$\quad 2a - 3b = 12$

31. $2r + 5t = 14$
$\qquad\ r = 2t - 11$

32. $2x - 3y = 14$
$\qquad\ x + 2y = 0$

33. $5x - 2y = 1$
$\qquad y = 2$

34. $3m - 7n = 1$
$\qquad m - 5 = 0$

35. $a - 3b = -4$
$\quad 2a + 6b = 5$

36. $3x - 5y = 8$
$\quad x + 2y = 1$

37. $2a - b = 8$
$\quad 3a - 2b = 11$

38. $x - 3y = 1$
$\qquad 3x = 9y + 3$

39. $2x - y = 2$
$\qquad 2y = 4x - 4$

40. $m - 2n = 0$
$\quad 4m - 3n = 15$

41. $3r - 4t = 5$
$\quad r + 7t = 10$

42. $2x + y = 1$
$\quad 10x + 15y = 4$

Solve the following systems using either the addition method or the substitution method. Also, classify each system as consistent, inconsistent, or dependent.

43. $2x + y = 9$
$\qquad x - y = 11$

44. $3x + 2y = -1$
$\quad 5x + 3y = -2$

45. $7m + 5n = 2$
$\quad 8m - 9n = 17$

46. $r - 4s = -5$
$\quad 2r + 3s = 12$

47. $7x - 3y = -32$
$\quad 2x - 5y = -5$

48. $a = 4 + 2b$
$\quad 3a + 4b = 2$

49. $3x + 2y = 4$
$\quad 2x + 2y = 3$

50. $y = 2x - 1$
$\quad y = 3x + 2$

51. $m = 3 - 2n$
$\quad 2m + 4n = 6$

52. $3x = 1 - 2y$
$\quad 4y = 1 - 6x$

For the following systems, simplify each equation and clear it of fractions. Then, solve by using either the addition method or the substitution method.

To clear $\dfrac{x}{2} + \dfrac{y}{3} = 1$ of fractions, multiply both sides by the LCD, 6   **Example**

Solution:
$$\frac{x}{2} + \frac{y}{3} = 1$$

$$6\left[\frac{x}{2} + \frac{y}{3}\right] = 6 \cdot 1$$

$$6 \cdot \frac{x}{2} + 6 \cdot \frac{y}{3} = 6 \cdot 1$$

$$3x + 2y = 6$$

53. $\dfrac{x}{2} + \dfrac{y}{3} = 1$

$\quad \dfrac{x}{4} + \dfrac{y}{5} = \dfrac{1}{2}$

54. $m - \dfrac{n}{2} = 4$

$\qquad m + n = 7$

55. $\dfrac{a}{5} + \dfrac{b}{3} = \dfrac{14}{15}$

$\dfrac{5a}{2} - 2b = -13$

56. $\dfrac{x+4}{4} + \dfrac{3y-2}{2} = 4$

$\dfrac{3x+2}{5} + \dfrac{2y+1}{7} = \dfrac{1}{5}$

57. $\dfrac{3x}{4} - \dfrac{1}{2} = x + \dfrac{y}{4}$

$\dfrac{x}{2} + 19 = 2x - \dfrac{3y}{2} + 4$

58. $\dfrac{x}{2} + \dfrac{y}{4} = x + \dfrac{3y}{4} - 15$

$x - \dfrac{5}{2} = \dfrac{5x}{4} - \dfrac{y}{2} - 4$

USING SYSTEMS OF LINEAR EQUATIONS TO SOLVE APPLIED PROBLEMS

8.5

When solving applied problems it is often easier to translate statements into equations using two variables than equations using one variable. Consequently, in this section, we will see how systems of linear equations can be used to solve many types of applied problems. To do this, as illustrated by the following examples, we use two variables and then translate the data into two equations. These two equations form a system of linear equations which we then solve to obtain the solution for the problem.

Example 1

The sum of two numbers is 45. Their difference is 23. Find both numbers.

Solution: Let x represent one number and y the other. Then write two equations describing the given information. This gives a system of linear equations.

$x + y = 45$ [the sum of two numbers is 45]

$x - y = 23$ [their difference is 23]

Solve the system using the addition method.

$$\begin{array}{r} x + y = 45 \\ \underline{x - y = 23} \\ 2x = 68 \\ x = 34 \end{array}$$

Substitute 34 for x in the first equation, $x + y = 45$, and solve for y

$$x + y = 45$$

$$34 + y = 45$$

$$y = 11$$

Hence, the two numbers are 34 and 11

Example 2

The sum of two electric voltages is 110 volts. If the larger voltage is tripled and the smaller voltage is doubled, the sum becomes 290 volts. What are the resulting voltages?

Solution: Let x represent the larger voltage and y the smaller. Set up a system by writing two equations.

$$x + y = 110 \qquad \text{(the sum of two voltages is 110 V)}$$

$$3x + 2y = 290 \qquad \text{(triple larger, double smaller; sum is 290 V)}$$

Solve the system using the addition method.

$$x + y = 110 \xrightarrow{\text{multiply by } -3} -3x - 3y = -330$$

$$\underline{3x + 2y = 290} \qquad\qquad \underline{3x + 2y = 290}$$

$$-y = -40$$

$$y = 40$$

Substitute 40 for y in the first equation, $x + y = 110$, and solve for x

$$x + y = 110$$

$$x + 40 = 110$$

$$x = 70$$

Thus, the larger voltage is 70 V and the smaller is 40 V.

Example 3

A rectangular steel plate has a perimeter of 60 centimeters. Find the dimensions of the plate if the length is 3 centimeters less than twice its width.

Solution: Let l represent the length of the plate and let w represent its width. Set up a system by writing two equations.

$$2l + 2w = 60 \qquad \text{(the perimeter is 60 cm)}$$

$$l = 2w - 3 \qquad \text{(the length is 3 cm less than twice its width)}$$

Solve the system using the substitution method. Substitute $2w - 3$ for l in the first equation.

$$2l + 2w = 60$$

$$2(2w - 3) + 2w = 60$$

$$4w - 6 + 2w = 60$$

$$6w - 6 = 60$$

$$6w = 66$$

$$w = 11$$

Substitute 11 for w in the second equation, $l = 2w - 3$, and solve for l

$$l = 2w - 3$$

$$l = 2(11) - 3$$

$$l = 19$$

Therefore, the length of the plate is 19 cm and the width is 11 cm.

Example 4

A man invests $6000. Part is invested at an annual rate of 5%, while the remainder is invested at 7%. How much is invested at each interest rate if the total annual interest income is $350?

Solution:　Let x represent the portion invested at 5% and let y represent the remainder invested at 7%. Set up a system by forming two equations.

$$x + y = 6000 \qquad \text{(the sum of both portions is \$6000)}$$

$$.05x + .07y = 350 \qquad \text{(the interest income is \$350)}$$

Let's solve this system using the addition method.

$$x + y = 6000 \xrightarrow{\text{mult. by } -5} -5x - 5y = -30000$$

$$\underline{.05x + .07y = 350} \xrightarrow{\text{mult. by } 100} \underline{5x + 7y = 35000}$$

$$0 + 2y = 5000$$

$$y = 2500$$

Substitute 2500 for y in the first equation, $x + y = 6000$, and solve for x

$$x + y = 6000$$
$$x + 2500 = 6000$$
$$x = 3500$$

We see that $3500 is invested at 5% while $2500 is invested at 7%

A medical assistant needs to make 16 cc (cubic centimeters) of a 50% saline (salt) solution. She has in stock a 20% saline solution and a 70% saline solution. How much of each solution should be mixed together to produce the required 50% saline solution?

Example 5

Solution: Let x represent the amount of 20% solution, and let y represent the amount of 70% solution. Then form a system of two equations by studying this diagram.

The two equations are:

$$x + y = 16$$
$$.20x + .70y = (.50)(16)$$

Let's solve this system by substitution. Solve the first equation, $x + y = 16$, for x

$$x + y = 16$$
$$x = 16 - y$$

Substitute $16 - y$ for x into the second equation.

$$.20x + .70y = (.50)(16)$$
$$.20(16 - y) + .70y = 8$$

Clear the equation of decimals by multiplying both sides by 10. Then solve for y

$$.20(16 - y) + .70y = 8$$
$$10(.20)(16 - y) + 10(.70y) = 10(8)$$
$$2(16 - y) + 7y = 80$$
$$32 - 2y + 7y = 80$$
$$32 + 5y = 80$$
$$5y = 48$$
$$y = \frac{48}{5} \quad \text{or} \quad 9.6$$

Substitute 9.6 for y in the first equation, $x + y = 16$, and solve for y

$$x + y = 16$$
$$x + 9.6 = 16$$
$$x = 6.4$$

The medical assistant should mix 6.4 cc of the 20% solution with 9.6 cc of the 70% solution.

Exercise 8.5

Write a system of linear equations for each problem. Then solve the system, using either the addition method or the substitution method.

1. The sum of two numbers is 62. Their difference is 26. Find the numbers.

2. The sum of two numbers is 63. If the larger number is doubled and the smaller number is tripled, the difference is 41. Find the numbers.

3. The sum of two numbers is 76. One of the numbers is 3 times larger than the other. Find the numbers.

4. The sum of two numbers is 53. If the larger number is doubled and the smaller number is tripled, the sum is 121. Find the numbers.

5. The sum of two voltages is 120 V. If the larger voltage is tripled and the smaller voltage is quadrupled, the sum becomes 400 V. What are the voltages?

6. A rectangular steel plate has a perimeter of 28 in. Find the dimensions of the plate if the length is 2 in. less than 3 times its width.

7. A rectangular field has a perimeter of 176 m. Find the dimensions of the field if the length is 4 m more than twice the width.

8. A rectangular field has a perimeter of 130 yd. Find the dimensions of the field if the length is 10 yd less than twice its width.

9. Find the dimensions of a rectangle of perimeter 180 in. whose length is $1\frac{1}{2}$ times the width.

10. Find the dimensions of a rectangle of perimeter 126 in. whose length is $2\frac{1}{2}$ times the width.

11. The length of a rectangle is 6 ft greater than its width. The perimeter is 44 ft. Find the dimensions.

12. The length of a rectangle is 4 ft more than its width. If the perimeter is 56 ft, find the dimensions.

13. A woman invests $10,000. Part of it is invested at 5% and the remainder is invested at 6%. How much is invested at each interest rate if the total annual interest is $540?

14. A man invests $4000. Part of it is invested at 7% and the remainder is invested at 8%. How much is invested at each interest rate if the total annual interest is $295?

15. A man invests $4000. Part of it is invested at 6% and the remainder is invested at 8%. How much is invested at each interest rate if the total annual interest is $270?

16. One sum of money invested at 6% and another at 5% yield a total of $750. If the investments were interchanged, their income would increase by $40. Find the sums.

17. How many liters of a 15% solution of alcohol should be added to 30 ℓ (liters) of a 50% solution of alcohol to produce a 40% solution?

18. How many pints of a 20% saline solution should be added to 30 pt of a 50% solution to obtain a 40% saline solution?

19. How many ounces of a 30% acid solution should be added to 20 oz of a 60% acid solution to obtain a 50% acid solution?

20. How many ounces of an alloy containing 10% titanium must be melted with an alloy containing 25% titanium to produce 40 oz of an alloy containing 15% titanium?

21. How many pounds of walnuts at 49¢ per pound should a grocer mix with 20 lb of pecans at 58¢ a pound to give a mixture worth 54¢ a pound?

22. A grocer wishes to mix candy selling for 70¢ a pound with candy selling for 95¢ a pound to obtain 50 lb of a mixture selling for 85¢ a pound. How many pounds of each type of candy should he mix together?

23. A 54-ft rope is cut into two pieces so that one piece is 20 ft longer than the other. Find the length of each piece.

24. A 20-ft board is cut into two pieces, one of which is 2 ft longer than the other. How long is each piece?

25. A board 96 in. long is cut into two pieces so that one piece is 12 in. longer than the other. Find the length of each piece.

26. A boat can travel 6 mi downstream in 40 min. The return trip requires 1 hr. Find the rate of the boat in still water and the rate of the current.

27. Two cars make a trip of 168 miles each. The faster car averages 14 mph faster and makes the trip in 1 hr less time than the slower car. Find the average rate of each car.

28. A speedboat goes 20 mi downstream in 1 hr. The return trip against the current takes 2 hr. Find the boat's speed in still water.

29. A jet plane can travel 500 mph into the wind and 650 mph with the wind. Find the speed of the plane in still air.

30. A store owner sells 15-cent and 30-cent candy bars. In one day, she sold 109 candy bars for $28.20. How many of each kind were sold?

31. A man has 45 coins consisting of nickels and dimes. The total value is 350 cents. How many of each are there?

32. A woman has $4.55 in change, consisting of 5 more quarters than nickels. How many nickels and quarters does she have?

33. A bank teller has 85 bills of $5 and $10 denominations. The total value is $535. How many bills of each denomination does he have?

34. A collection of quarters and half-dollars has a value of $8.75. How many quarters and half-dollars are in the collection if there are 8 fewer halves than quarters?

35. Tom is 3 times as old as his son. The sum of their ages is 52 years. How old is Tom and how old is his son?

36. Maria is twice as old as her sister. The sum of their ages is 27 years. How old is Maria and how old is her sister?

37. A woman is now 8 times as old as her daughter. In 10 years the sum of their ages will be 56. How old are they now?

GRAPHING SYSTEMS OF LINEAR INEQUALITIES

8.6

system of linear inequalities

A **system of linear inequalities** is composed of two or more linear inequalities. We will be concerned only with systems containing two inequalities, such as:

$$2x + 3y > 6$$

$$3x - 5y > 15$$

Such a system can be solved by graphing each inequality and locating the overlapping region. Recall from section 7.7 how to construct the graph of a linear inequality. First we graph the boundary line. If the boundary is to be included it is drawn solid; otherwise, it is broken. Next, we select a test point from either the upper or lower region. The coordinates of the test point are substituted into the original inequality. If the resulting statement is true, the graph includes that region. But if the statement is false, the graph includes the opposite region. To review this procedure, consider Example 1.

Graph $2x + 3y > 6$ **Example 1**

Solution:

Step (1)   Graph the boundary line given by the equation $2x + 3y = 6$. The original inequality does not include an equal sign. Therefore, the graph will not include the boundary. It is pictured by a broken line. (See figure 8.9a.)

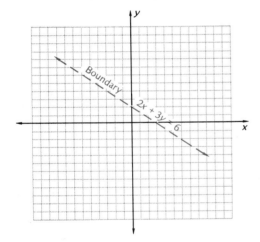

FIGURE 8.9a

Step (2)   For the test point we select $(0, 0)$ and substitute the coordinates into the original inequality.

$$2x + 3y > 6$$
$$2 \cdot 0 + 3 \cdot 0 > 6$$
$$0 > 6 \quad \text{[False]}$$

Since the statement is false, the graph includes the opposite region. As shown in figure 8.9b, the graph includes the upper region.

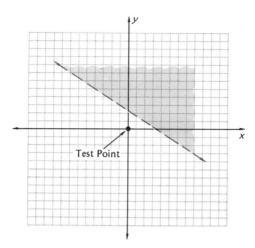

FIGURE 8.9b

We use this same graphing technique to solve systems of linear inequalities, as illustrated by the following examples.

Example 2   Graph the solution of this system.

$2x + 3y > 6$

$3x - 5y < 15$

Solution: Graph both inequalities on the same coordinate system. The solution of the system is the overlap of the two regions (see figure 8.10). The boundary lines are not included.

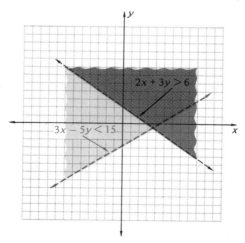

FIGURE 8.10

Example 3   Graph the solution of this system.

$3x - 4y \geq -12$

$x + 3y \geq -3$

Solution: Graph both inequalities and locate the overlapping region (see figure 8.11). This portion does include the boundary lines because both of the inequalities contain equal signs.

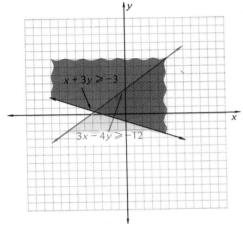

FIGURE 8.11

Graph the solution of this system.

Example 4

$x + 2y \geq 6$

$y \leq 3$

Solution: Graph both inequalities. Recall that $y = 3$ is a horizontal line crossing the y-axis at 3. The solution to the system is shown in figure 8.12 as the overlapping region. The boundary lines are included.

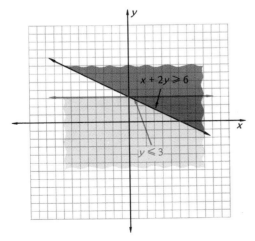

FIGURE 8.12

Exercise 8.6   Solve the following systems of linear inequalities by graphing.

1. $x + y < 5$ **2.** $x + y \leq 6$
     $x - y < 2$ $x - y \leq 1$

3. $x + y \geq 4$ **4.** $3x + 2y < 6$
     $x - y \leq 1$ $2x - y > -2$

5. $2x + 3y \geq 6$ **6.** $x + 2y < 4$
     $3x - y \geq 3$ $2x - y < 6$

7. $4x + y \geq 8$ **8.** $2x + y \leq 2$
     $x - 2y \geq -4$ $x + 3y \geq 4$

9. $3x + 2y \geq 6$ **10.** $x + y < 4$
     $x - y \geq 4$ $2x - y < 2$

11. $x - y \geq -2$ **12.** $x - 2y \geq 2$
         $x \leq y$ $2x + y \geq 1$

13. $x + y \leq 4$ **14.** $2x + y > 2$
         $y \leq 2x$ $x + y < 4$

15. $3x + 2y \geq 6$ **16.** $x + 3y > 12$
     $2x - 5y \geq 10$ $3x - 2y > -6$

17. $3x + y > 5$ **18.** $2x + 3y \geq 4$
     $2x - y > 10$ $3x - 5y \leq -5$

19. $2x + y > -2$ **20.** $3x + 3y \geq 5$
     $x + 2y < 5$ $x + 2y \leq 5$

21. $x + y < 8$ **22.** $x - y \geq 4$
     $5x - 2y < 5$ $y \geq 2$

23. $x + y \leq 3$ **24.** $3x - 5y \geq 15$
        $x \leq 2$ $y \geq -1$

25. $4x + 3y \leq 12$ **26.** $5x - 4y > 20$
        $x \geq -2$ $y > -2$

27. $x > 3$ **28.** $x \geq 0$
     $y < 2$ $y \geq 0$

29. $x - 5 < 0$ **30.** $x + 1 > 0$
     $y + 2 > 0$ $y - 3 < 0$

Summary

A system of linear equations in two variables consists of two linear equations, each containing the same two variables.

SYSTEM OF LINEAR EQUATIONS

A simultaneous solution to a system of linear equations is an ordered pair whose coordinates satisfy both equations simultaneously.

SIMULTANEOUS SOLUTION

(1) Construct the graph of both equations.
(2) Determine the point of intersection (if any).

GRAPHING METHOD OF SOLVING A SYSTEM OF LINEAR EQUATIONS

(1) Write each equation in the form $ax + by = c$. Then decide which variable to eliminate.
(2) Multiply one or both equations by appropriate numbers so that the coefficients of the variable to be eliminated are opposites.
(3) Add the two equations, obtaining an equation having only one variable.
(4) Solve the equation obtained in Step (3).
(5) Substitute this value into either of the original two equations and solve for the remaining variable. You now have the simultaneous solution, which can be checked by substituting into the other equation.

ADDITION METHOD OF SOLVING A SYSTEM OF LINEAR EQUATIONS

(1) Solve one of the equations for either x or y.
(2) Substitute the result obtained in Step (1) into the other equation.
(3) Solve the equation obtained in Step (2). This gives the value of one of the variables.
(4) Substitute this value into the equation of Step (1) and solve it for the second variable. Be sure to check the simultaneous solution.

SUBSTITUTION METHOD FOR SOLVING A SYSTEM OF LINEAR EQUATIONS

(1) A *consistent* system. The two straight lines intersect at exactly one point. Thus, the system has exactly one simultaneous solution.
(2) An *inconsistent* system. The two straight lines are parallel and do not intersect. Thus, the system has no simultaneous solutions.
(3) A *dependent* system. The two equations are equivalent and produce the same line. Thus, every point on the common line is a simultaneous solution.

THREE TYPES OF SYSTEMS OF LINEAR EQUATIONS

(1) Graph each inequality on the same coordinate system.
(2) The solution is the overlapping region.

SOLVING A SYSTEM OF LINEAR INEQUALITIES

Self-Checking Exercise

Determine whether or not the listed ordered pair is a simultaneous solution of the given system of linear equations.

1. $(3, -2)$ $2x - y = 4$
                    $4x + 3y = 6$

2. $(0, -3)$ $5x - 4y = 12$
                    $2x + y = 3$

3. $\left(\dfrac{3}{4}, \dfrac{1}{3}\right)$ $4x + 3y = 4$
                    $4x - 3y = 2$

4. $\left(\dfrac{2}{3}, 0\right)$ $6x + 7y = 2$
                    $x + 3y = 2$

Each of the following is a graph of a system of linear equations. Classify each system as consistent, inconsistent, or dependent.

5.

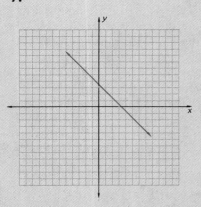

6.

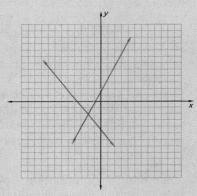

7.

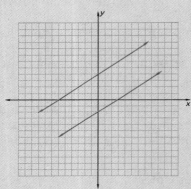

8.

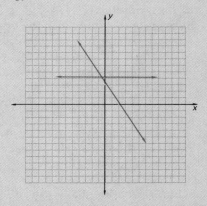

Solve the following systems of linear equations using the graphing method.

9. $3x + 2y = -4$
      $x - 3y = -5$

10. $6x - 2y = 3$
       $y = 3x + 4$

Solve the following systems of linear equations using the addition method.

11. $5x - 2y = -9$
$3x + 4y = 5$

12. $x + 3y = 1$
$2x + 6y = 2$

Solve the following systems of linear equations using the substitution method.

13. $y = 3x + 7$
$2x - 5y = 4$

14. $x - 4y = -2$
$3x + 8y = 9$

Solve the following systems of linear equations using any appropriate method.

15. $x + y = 7$
$x - y = 3$

16. $2x + 5y = 1$
$x + 2y = -4$

17. $3m - 2n = 8$
$6m - 4n = 7$

18. $3a - 6b = 9$
$a = 2b + 3$

19. $\dfrac{x}{2} - y = -4$
$x + y = 7$

20. $\dfrac{a}{4} + b = \dfrac{3b}{4} - \dfrac{1}{2}$
$\dfrac{3a}{2} + \dfrac{b}{2} = 2b - 15$

Solve the following applied problems using systems of linear equations.

21. The sum of two numbers is 17 and their difference is 5. Find both numbers.

22. A man invests $6000. Part is invested at 7% while the remainder is invested at 6%. How much is invested at each interest rate if the total annual interest is $384?

23. How much pure alcohol must be mixed with 250 ml (milliliters) of a 30% solution to produce a 50% alcohol solution?

Solve the following systems of linear inequalities by graphing.

24. $2x + y \leq 6$
$x - y \leq -3$

25. $3x + 5y > 15$
$y > 2x$

---

1. No.

2. No.

3. Yes.

4. No.

5. Inconsistent.

6. Consistent.

7. Dependent.

8. Consistent.

Solutions to Self-Checking Exercise

9.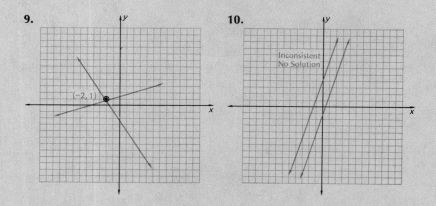

10.

Inconsistent
No Solution

11. $5x - 2y = -9$ $\xrightarrow{\text{mult. by 2}}$ $10x - 4y = -18$

$\underline{3x + 4y = 5}$ $\xrightarrow{}$ $\underline{3x + 4y = 5}$

 $13x + 0 = \overline{-13}$

$ x = -1$

Substitute -1 for x in the first equation, $5x - 2y = -9$. Then solve for y

$$5x - 2y = -9$$
$$5(-1) - 2y = -9$$
$$-5 - 2y = -9$$
$$-2y = -4$$
$$y = 2$$

The simultaneous solution is $(-1, 2)$

12. $x + 3y = 1$ $\xrightarrow{\text{mult. by } -2}$ $-2x - 6y = -2$

$\underline{2x + 6y = 2}$ $\xrightarrow{}$ $\underline{2x + 6y = 2}$

 $0 + 0 = 0$ [True]

The system is dependent. All solutions lie on the common line.

13. $y = 3x + 7$

$ 2x - 5y = 4$

Substitute $3x + 7$ for y in the second equation, $2x - 5y = 4$. Solve for x

$$2x - 5y = 4$$
$$2x - 5(3x + 7) = 4$$
$$2x - 15x - 35 = 4$$
$$-13x - 35 = 4$$
$$-13x = 39$$
$$x = -3$$

Substitute -3 for x in the first equation, $y = 3x + 7$. Then solve for y

$y = 3x + 7$
$y = 3(-3) + 7$
$y = -2$

The simultaneous solution is $(-3, -2)$

14. $x - 4y = -2$
$3x + 8y = 9$

Solve the first equation for x

$x - 4y = -2$
$\quad x = 4y - 2$

Substitute into the second equation.

$3x + 8y = 9$
$3(4y - 2) + 8y = 9$
$12y - 6 + 8y = 9$
$20y - 6 = 9$
$20y = 15$
$y = \dfrac{15}{20} \quad \text{or} \quad \dfrac{3}{4}$

Substitute $\dfrac{3}{4}$ for y in the first equation and solve for x

$x - 4y = -2$
$x - 4 \cdot \dfrac{3}{4} = -2$
$x - 3 = -2$
$x = 1$

The simultaneous solution is $\left(1, \dfrac{3}{4}\right)$

15. $x + y = 7 \longrightarrow x + y = 7$
$\underline{x - y = 3} \longrightarrow \underline{x - y = 3}$
$\qquad\qquad\qquad\qquad 2x + 0 = 10$
$\qquad\qquad\qquad\qquad\quad x = 5$

Substitute 5 for x in the first equation, $x + y = 7$. Then solve for y

$x + y = 7$
$5 + y = 7$
$y = 2$

The simultaneous solution is $(5, 2)$

16. $2x + 5y = 1 \xrightarrow{\text{mult. by } -2} 2x + 5y = 1$
$\underline{x + 2y = -4} \xrightarrow{\hspace{1.5cm}} \underline{-2x - 4y = 8}$
$\qquad\qquad\qquad\qquad\qquad 0 + y = 9$
$\qquad\qquad\qquad\qquad\qquad\quad y = 9$

Substitute 9 for y in the second equation, $x + 2y = -4$. Then solve for x

$$x + 2y = -4$$
$$x + 2(9) = -4$$
$$x + 18 = -4$$
$$x = -22$$

The simultaneous solution is $(-22, 9)$

17. $3m - 2n = 8$ $\xrightarrow{\text{mult. by } -2}$ $-6m + 4n = -16$

$\underline{6m - 4n = 7}$ $\xrightarrow{\hspace{2cm}}$ $\underline{6m - 4n = 7}$

$0 + 0 = -9$ [False]

The system is inconsistent. There are no simultaneous solutions.

18. $3a - 6b = 9$
$a = 2b + 3$

Substitute $2b + 3$ for a in the first equation, $3a - 6b = 9$. Then solve for b

$$3a - 6b = 9$$
$$3(2b + 3) - 6b = 9$$
$$6b + 9 - 6b = 9$$
$$9 = 9 \quad \text{[True]}$$

The system is dependent. All solutions lie on the common line.

19. $\dfrac{x}{2} - y = -4$ $\xrightarrow{\text{mult. by } -2}$ $-x + 2y = 8$

$\underline{x + y = 7}$ $\xrightarrow{\hspace{2cm}}$ $\underline{x + y = 7}$

$0 + 3y = 15$

$y = 5$

Substitute 5 for y in the second equation, $x + y = 7$. Then solve for x

$$x + y = 7$$
$$x + 5 = 7$$
$$x = 2$$

The simultaneous solution is $(2, 5)$

20. $\dfrac{a}{4} + b = \dfrac{3b}{4} - \dfrac{1}{2}$ $\xrightarrow{\text{mult. by } 4}$ $a + 4b = 3b - 2$

$\underline{\dfrac{3a}{2} + \dfrac{b}{2} = 2b - 15}$ $\xrightarrow{\text{mult. by } 2}$ $3a + b = 4b - 30$

Simplify the resulting equations, obtaining this system:

$a + b = -2$ $\xrightarrow{\text{mult. by } 3}$ $3a + 3b = -6$

$\underline{3a - 3b = -30}$ $\xrightarrow{\hspace{2cm}}$ $\underline{3a - 3b = -30}$

$6a + 0 = -36$

$a = -6$

Substitute -6 for a in the equation $a + b = -2$. Then solve for b

$$a + b = -2$$
$$-6 + b = -2$$
$$b = 4$$

The simultaneous solution is $(-6, 4)$

21. The system is: $x + y = 17$
$\qquad\qquad\qquad\quad x - y = \ \ 5$

The solution is (11, 6). One number is 11 and the other is 6.

22. The system is: $\quad x + \quad y = 6000$
$\qquad\qquad\qquad\ .07x + .06y = \quad 384$

The solution is (2400, 3600). $2400 is invested at 7%, while $3600 is invested at 6%.

23. The system is: $\qquad x + 250 = y$
$\qquad\qquad\qquad x + (.30)(250) = (.50)y$

The solution is (100, 350). 100 ml of pure alcohol must be used.

24.

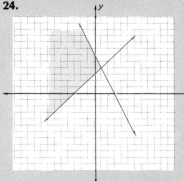

25.

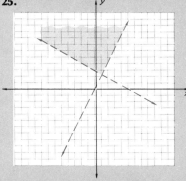

Answers to
Odd-Numbered Exercises

1. {*a, b, c, d, e, f*}
3. {16, 17, 18, 19, 20, 21}
5. {0, 1, 2, 3, 4, 5, 6, 7}
7. {6, 7, 8, 9, 10, 11}
9. Addend.
11. Sum.
13. Subtrahend.
15. Factor.
17. Product.
19. Divisor.
21. Dividend.
23. Quotient.
25. $5 + 7 = 12$
27. $0 + 5 = 5$
29. $6 + 6 + 6$
31. $1 + 1 + 1$
33. $8 \cdot 6 = 48$
35. $12 \cdot 1 = 12$
37. $0 \cdot 6 = 0$
39. 1
41. 25
43. 8
45. 128
47. True.
49. False.
51. False.
53. False.
55. False.
57. False.

1. 24
3. 1
5. 53
7. 22
9. 28
11. 8
13. 29
15. 24
17. 16
19. 47
21. 3
23. 5
25. 61
27. 45
29. 4
31. $5 \cdot (8 + 2)$
33. $(4 + 5) - (6 - 1)$
35. $3 \cdot (2 + 2 \cdot 5)$
37. $(60 - 3 + 4) \cdot 8 \div 2$ or $(60 - 3 + 4) \cdot (8 \div 2)$
39. $4 \cdot (2^3 + 1^5 \cdot 3) = 44$
      ② ①
41. $7 + 2 \cdot 9$ Sum
      ① ②
43. $3 + 9 - 4$ Difference
      ① ③ ②   ② ③ ①
45. $(4 + 3) \cdot (7 - 2)$ or $(4 + 3) \cdot (7 - 2)$ Product
      ② ①
47. $5 - (2 + 3)$ Difference
      ④ ② ③ ①
49. $2 \cdot [3 + 6 - (1 + 2)]$ Product
      ① ②
51. $8 \div 2 + 1$ Sum
      ④ ③ ① ②
53. $5 \cdot \{22 - [(13 - 2) - 4]\}$ Product

1. 8
3. 0
5. 10
7. 14

9. 13   **11.** 25
13. 6   **15.** 14
17. 18   **19.** 44
21. 40   **23.** Difference.
25. Product.   **27.** Product.
29. Product.   **31.** $x + 10$

33. $6x$ **35.** $\frac{x}{25}$

37. $x + 9$ **39.** $x - 15$
41. $7x - 15$ **43.** $2x - 10$
45. $5n$ **47.** $50(x - 4)$
49. $3(x + 5) - 2x$

Exercise 1.4

1. Positive.   **3.** Neither.
5. South.   **7.** Loss.
9. $+3$ **11.** -6
13. -2 **15.** 0
17. 7   **19.** 1
21. 1   **23.** a
25. 13   **27.** 0
29. 0   **31.** 12
33. 23   **35.** -20
37. 0   **39.** 10
41. True.   **43.** True.
45. False.   **47.** True.
49. True.   **51.** True.
53. True.

Exercise 1.5

1. False.   **3.** False.
5. True.   **7.** True.
9. False.   **11.** False.
13. True.   **15.** False.
17. 0.05   **19.** 2.2
21. 0.75   **23.** Rational.
25. Rational.   **27.** Irrational.
29. Rational.   **31.** Rational.

33.

35.

37.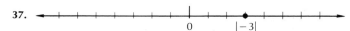

39.

41. Any number whose decimal form is nonterminating and nonrepeating.
43. Real numbers used to label points.

1. Commutative property of multiplication.
3. Associative property of addition.
5. Multiplication by zero.
7. Multiplication by zero.
9. Additive identity.
11. Distributive property of multiplication.
13. Commutative property of multiplication.
15. Associative property of addition.
17. Additive identity.
19. Multiplicative inverse.
21. Multiplicative identity.
23. Additive inverse.
25. Additive inverse.

27. $x + 3$
29. x
31. $(a + b) \cdot 2$
33. 1
35. $2(x + y)$
37. 9
39. 21
41. 0
43. 50
45. $3 \cdot 4 + 3 \cdot 2 = 18$
47. $6 \cdot 3 + 10 \cdot 3 = 48$
49. $8 \cdot 5 + 8 \cdot 0 = 40$
51. $2m + 2n$
53. $5b + 5$
55. $7m + 28$
57. $2xy + 2xz$
59. $5(x + y)$
61. $a(b + c)$
63. $2(x + 3)$
65. $6(x + 1)$
67. $5(m + 4)$
69. $3y(x + 3)$

1.

Thus, $(+2) + (+3) = +5$

3. Thus, $(-4) + (+1) = -3$

5. 11
7. 8
9. 2
11. -3
13. -8
15. -7
17. -2
19. -17
21. -52
23. -33
25. -2.5
27. -3.35
29. -8
31. 0
33. 1
35. -3
37. 12
39. -12
41. -23
43. 3.8
45. 11
47. -13
49. 16
51. 16
53. 5
55. Positive.
57. Negative.
59. \$85
61. $-25°F$
63. \$89.72
65. -36
67. -8
69. -2.557
71. -643.13

1. $x + 5$
3. $m + (-4)$
5. $-a + (-b)$
7. $2x + 1$
9. $2m^2 + 4$
11. -6
13. -4
15. 22

17. -2	**19.** -19
21. -11	**23.** 7
25. -12	**27.** -2
29. 5	**31.** -22
33. -4	**35.** -3
37. -2	**39.** -5
41. 0	**43.** 1
45. 5.5	**47.** 13
49. 16	**51.** -3
53. Positive.	**55.** $-12°F$
57. 3800 ft	**59.** $90,000
61. -23	**63.** 11
65. 9	**67.** -0.1056
69. -14.18	

Exercise 1.9

1. 10	**3.** -8
5. 0	**7.** 20
9. -100	**11.** -48
13. 2.52	**15.** -1
17. 6	**19.** 6
21. 56	**23.** 1
25. 1	**27.** -36
29. -81	**31.** 1
33. -1000	**35.** 36
37. 2	**39.** -16
41. -30	**43.** -38
45. -135	**47.** 12
49. -9	**51.** 16
53. 8	**55.** 43
57. 3	**59.** -1
61. 0	**63.** Even.
65. Positive.	**67.** 5
69. -34	**71.** 23
73. 0.2862	**75.** 21.2114

Exercise 1.10

1. 4	**3.** -4
5. Impossible.	**7.** 3
9. 0	**11.** 10
13. 1	**15.** -2
17. -2	**19.** 8
21. -5	**23.** -7
25. 5	**27.** 2
29. 9	**31.** -72
33. 0	**35.** 158
37. 16	**39.** 15
41. -3	**43.** 5
45. 2	**47.** -20
49. 3	**51.** 4
53. 1	**55.** 0
57. -1	**59.** -2
61. 9	**63.** -45
65. 5	**67.** 0
69. -1	**71.** 0
73. -2.857538	**75.** -6.4517875

1. Base is y, exponent is 6
3. Base is x, exponent is 2
5. Base is $(x + 4)$, exponent is 5
7. Base is 5, exponent is 2
9. Base is x, exponent is 4
11. y squared
13. The group $(a + 2b)$ to the sixth power
15. x^2
17. 3^5
19. a^3b^2
21. $x^2y^3z^2$
23. $4b^5$
25. $(-x)^3$
27. $\dfrac{3x^2}{4y^3}$
29. 5
31. 1
33. -1
35. 2
37. 4
39. 0
41. 13
43. 10
45. 2
47. 320
49. Zero power property.
51. Subtraction property of exponents.
53. Power to a power property.
55. Distributive property of exponents.
57. Subtraction property of exponents.
59. y^7
61. 1
63. $8a^3$
65. $(m + 3n)^9$
67. $\dfrac{4a^6}{9}$
69. $\dfrac{x^2}{9}$
71. Incorrect.
73. Incorrect.
75. Correct.
77. 18.274576
79. 0.61599998

1. $4x$ (double opposite law)
3. y^2 (double opposite law)
5. $-2a$ (opposite law)
7. $3x$ (double opposite law)
9. $-5x$ (opposite law)
11. $8x$
13. $15a$
15. $-6a^5$
17. $-6a$
19. $12x^2$
21. $10a^5$
23. $6x^3y^4$
25. $-5x^4y^3z^5$
27. $-2a^6b^3c$
29. $30a^6$
31. $12a^4b^4c^5$
33. $-40x^5y^2$
35. $-24a^3b^7c^3$
37. $-14m^3n^4$
39. $-2a$
41. $4m$
43. 3
45. 3
47. $-3x^2$
49. $-5a^2bc$
51. $-3x^2$
53. -1
55. $-5x^3y^3z^2$
57. $2x + 12$
59. $5x - 15$
61. $-6m^3 + 2m$
63. $5x^3 - 5xy$
65. $6a^2 - 2ab + 2ac$
67. $-6a + 8b - 4c + 10d$
69. $8x^3y^3 - 12x^2y^4 + 4xy^2$
71. $6x^2 + 8xy - 6x$
73. $2m^4n + 2mn^4$
75. $6a^3b + 8ab^3 - 2ab$
77. $10a^4b^3 + 15ab^3$
79. $8a^4b^2c^3 + 12a^2b^4c^4$

Exercise 2.3

1. Variables differ.
3. Exponents differ.
5. Exponents differ.
7. $10a$
9. y
11. $14n$
13. $4y$
15. $-2a$
17. $x + 3$
19. $6b$
21. $15xy$
23. r^2t
25. 0
27. $-2pq + 3p$
29. $13x^2$
31. $3x^2 - x$
33. $2ab + 5$
35. $p^3 - 2p + 7$
37. $6x - 4y$
39. $-4m^3n^2 + 7m^2n^3$
41. $-r^2 - r + 10$
43. $-a - c$
45. $5a^2 - 8a + 5$
47. $11x^2 - 5x + 3$
49. $4a^2 - 2a + 1$
51. $5x^2 + 4x - 2$
53. $-3x^2 - 3x + 9$
55. $13x^2 - 3x - 1$
57. $5n^2 - 6n - 2$
59. $-2x^2 - 3x + 3$
61. $6p^2 - 5p - 6$
63. $6x^2 + 2x - 9$
65. $6y^2 - 8y - 3$
67. $-2k^3 + 5k^2 + 2k$

Exercise 2.4

1. $3x - y$
3. $2a + 7b$
5. $2y - 3$
7. $-6a + 2b - 3c$
9. $7a - 2b - 3$
11. $6a - 2b + 7c - d$
13. $x + y + z - w$
15. $6x - 5y + 3z + 7$
17. $3a - 4b + 3c + d - 3 - 3e$
19. $6x + 3$
21. $4a^2 + b$
23. $-2x + 6$
25. $8a^2 - 2$
27. -11
29. $-a$
31. $4a - 3$
33. $-8x + 1$
35. $3a + 5$
37. $-3y + 6$
39. $9y + 7$
41. $p^2 + 5$
43. $-2x$
45. $-m + n + 5$
47. $3 - p$
49. $10t$
51. $2a - 3b$
53. $2p + 7q$
55. $7m - 8n$
57. $3a - 15$
59. $9m - 11n$
61. $-2x - 23y + 20$
63. $-9y + 3$
65. $-3x^2 - 15x + 3$

Exercise 2.5

1. 6
3. 8
5. 3
7. -10
9. -30
11. -7
13. -4
15. 0
17. 0
19. $\frac{1}{2}$
21. $\frac{-1}{5}$
23. $-\frac{2}{3}$
25. 2
27. 7
29. 2
31. -2
33. 0
35. 1
37. -8
39. -2
41. -3
43. 2
45. -2
47. $-\frac{1}{2}$
49. -2
51. $\frac{1}{2}$
53. 4.3646489
55. 22.2684

57. -6.0140759

59. $\frac{x}{3} = 15;$ $x = 45$

61. $3x = -18;$ $x = -6$

Exercise 2.6

1. 7	**3.** 4
5. 6	**7.** -2
9. -20	**11.** -4
13. 2	**15.** -3
17. 3	**19.** 8.1
21. 22.0	**23.** 1.024
25. 2	**27.** 2
29. -2	**31.** 0
33. -3	**35.** 2
37. 2	**39.** 1
41. -2	**43.** $\frac{1}{2}$
45. -4	**47.** -2
49. 6	**51.** 13
53. 1.2	**55.** 6
57. 2	**59.** -4
61. 0	**63.** 7
65. -3	**67.** 4
69. 2	**71.** 15
73. -2	**75.** $2(x + 7) = 20;$ $x = 3$
77. $5x + 3x = 56;$ $x = 7$	**79.** $10x + 3x - 5 = 47;$ $x = 4$

Exercise 2.7

1. -10	**3.** 4
5. 5	**7.** -7
9. 0	**11.** -6
13. 4	**15.** 8
17. 4	**19.** 1
21. 3	**23.** 2
25. 2	**27.** -4
29. 0	**31.** 5
33. 5	**35.** -7
37. 1	**39.** 0
41. 4	**43.** -14
45. -1	**47.** 8
49. -3	**51.** -2
53. -2	**55.** 1
57. 1.3333333	**59.** 0.22075472

Exercise 2.8

1. $x + 3$	**3.** $x + (-2)$ or $x - 2$
5. $x + 3$	**7.** $x - 10$
9. $x - 4$	**11.** $x - 12$
13. $5x$	**15.** $2x$
17. $\frac{x}{10}$	**19.** $\frac{1}{2}x$ or $\frac{x}{2}$
21. $\frac{x}{8}$	**23.** $6x - 5$
25. $4x - 10$	**27.** $5x$ cents
29. 6	**31.** 5

33. 36 and 12   **35.** 30 and 16
37. 51 and 15   **39.** 13 cm and 7 cm
41. 5, 15 and 3 amperes
43. 14,15, and 16
45. 8 cm, 16 cm, and 24 cm
47. 100, 300, and 400 liters per minute
49. 60 nickels, 10 quarters, and 9 half dollars
51. 25 nickels and 15 quarters
53. 11,100 regular seats and 5,200 box seats
55. 50 meters
57. 10 meters
59. 6 centimeters
61. 6 ft, 12 ft, and 11 ft

Exercise 3.1

1. Numerator is $3m$; denominator is 2   **3.** Numerator is $x + 7$; denominator is $x - 5$

5. $3 \div 7$; $3 \cdot \frac{1}{7}$ **7.** $2a \div (a - 5)$; $2a \cdot \frac{1}{a - 5}$

9. $\frac{2a}{3}$ **11.** $\frac{m - 3}{m + 5}$

13. $\frac{2b + 1}{7}$ **15.** a

17. $7m + 3$ **19.** 0

21. True.   **23.** False.

25. True.   **27.** Yes.

29. No.   **31.** Yes.

33. 3   **35.** -1

37. $\frac{8}{15}$ **39.** 4

41. $\frac{-23}{10}$ **43.** $\frac{23}{12}$

45. $\frac{2}{7}$ **47.** $-\frac{1}{5}$

49. 3

Exercise 3.2

1. $\frac{4}{12}$ **3.** $\frac{-6}{10}$

5. $\frac{63}{14}$ **7.** $\frac{20x}{12x^2}$

9. $\frac{6b}{10b^2}$ **11.** $\frac{-10a^3b}{25a^2b^3}$

13. $\frac{3x(x + y)}{2(x + y)}$ **15.** $\frac{3(a + 4)}{(a + 4)^2}$

17. $\frac{3}{4}$ **19.** $\frac{4}{5}$

21. $-\frac{3}{2}$ **23.** 7

25. $\frac{3}{2}$ **27.** $\frac{4}{3x}$

29. $-\frac{2a^2}{3}$ **31.** $\frac{5rt^2}{6}$

33. $-6k$ **35.** $\frac{3b^2}{5c^2}$

37. $\frac{x}{6y}$ **39.** $\frac{5}{7}$

41. $\dfrac{a + 2}{a + 3}$

43. $\dfrac{1}{2}$

45. $\dfrac{m - 5}{m}$

47. 7

49. $\dfrac{1}{4}$

51. $\dfrac{1}{2}$

53. $x - 3$

55. $\dfrac{4xy}{xy + 2}$

57. $\dfrac{-(2n + 5)}{3m}$ or $\dfrac{-2n - 5}{3m}$

Exercise 3.3

1. Yes.

3. No.

5. Yes.

7. No.

9. No.

11. Yes.

13. $\dfrac{2}{5}$

15. $\dfrac{1}{4}$

17. $\dfrac{5}{2}$

19. $\dfrac{a}{2b}$

21. $\dfrac{1}{2}$

23. $-\dfrac{7}{9}$

25. $3m$

27. $\dfrac{4}{5b}$

29. $\dfrac{1}{2}$

31. $\dfrac{m^5}{4}$

33. $\dfrac{n^5}{3}$

35. $\dfrac{s^2}{4t^2}$

37. $\dfrac{1}{2x}$

39. $\dfrac{3(m + 2)}{m}$

41. $-\dfrac{2}{3}$

43. $x + 2y$

45. $\dfrac{3}{2}$

47. $\dfrac{1}{4}$

49. $\dfrac{1}{7}$

51. $\dfrac{x - 2}{x + 2}$

53. 2

Exercise 3.4

1. $\dfrac{1}{15}$

3. $\dfrac{1}{24}$

5. $\dfrac{14}{9}$

7. $\dfrac{3}{5}$

9. $\dfrac{9}{2}$

11. $\dfrac{6x^3}{y^3}$

13. $\dfrac{5}{2ab}$

15. $\dfrac{x}{8z}$

17. $\dfrac{2}{3a^2}$

19. $\dfrac{14m}{m + n}$

21. $\dfrac{2a}{3}$

23. $\dfrac{y}{3x}$

25. $\dfrac{1}{5}$

27. $\dfrac{3m^2(m^2 + 2)}{2(m^2 + 4)}$

29. $\dfrac{8}{a + b}$

31. $-\dfrac{5}{2}$

33. $\dfrac{3}{2}$

35. $\dfrac{20x^3}{9y^2}$

37. $2xy^2$

39. $\dfrac{2s}{3r}$

41. $\dfrac{3a}{5m^2n}$

43. $\dfrac{2m^2n^2}{27}$

45. $\dfrac{m + n}{m}$

47. $2a$

49. $3a^3$

51. $\dfrac{r}{3s}$

Exercise 3.5

1. $\dfrac{3}{7}$

3. $\dfrac{11}{12}$

5. $-\dfrac{4}{7}$

7. $-\dfrac{4}{11}$

9. 0

11. $\dfrac{4m}{n}$

13. $\dfrac{2x - 1}{3}$

15. $\dfrac{2}{a}$

17. $\dfrac{4a + 7}{2b}$

19. $\dfrac{a + b}{2b}$

21. $\dfrac{2y^2 + y}{3}$

23. $\dfrac{3a^2 - a - 2}{a - 2}$

25. $\dfrac{2(m - n)}{m + n}$

27. $-\dfrac{1}{2}$

29. $\dfrac{1}{x - 1}$

31. $\dfrac{a + 9}{2}$

33. $r - s$

35. $\dfrac{2m + 4}{m - 3}$

37. $\dfrac{5a^2 - a - 4}{a - 3}$

39. 1

41. 1

43. $\dfrac{m - 18}{m - 1}$

45. $\dfrac{3x^2 - 4x - 11}{x - 2}$

47. $\dfrac{6a - 10b}{3ab}$

49. $\dfrac{10x - 9y + 1}{2x - y}$

Exercise 3.6

1. Prime.

3. Composite.

5. Composite.

7. Prime.

9. Composite.

11. 2^2

13. $3 \cdot 5$

15. $2^2 \cdot 7$

17. $2^2 \cdot 3 \cdot 5$

19. $2^2 \cdot 3^2 \cdot 5$

21. 6

23. 12

25. 36

27. 280

29. 180

31. $60x^2y$

33. x^2

35. a^2b

37. $10m^2$

39. r^2s^2

41. $2(m + n)$

43. $(a + b)(a - b)$

45. $y(y + 1)$

47. $x^2y^2(y + 1)$

49. $(r + s)(r - s)$

Exercise 3.7

1. $\frac{13}{12}$

3. $\frac{37}{30}$

5. $-\frac{8}{15}$

7. $\frac{10}{9}$

9. $\frac{16}{5}$

11. 1

13. $\frac{19}{30}$

15. 1

17. $\frac{2a+3}{a^2}$

19. $\frac{c+d}{d}$

21. $\frac{2+2m}{mn}$

23. $\frac{y+x}{y}$

25. $\frac{6a+3b}{4b^2}$

27. $\frac{9x-1}{14x^2}$

29. $\frac{15a+4}{4a^2}$

31. $\frac{5a-3}{10a^2}$

33. $\frac{-x^2-x}{(x+2)(x+3)}$

35. $\frac{-x-4}{6}$

37. $\frac{4x-2y}{(x+y)(x-y)}$

39. $\frac{5x-2}{2(x-2)}$

41. $\frac{12r+11}{24}$

43. $\frac{5}{4}$

45. $\frac{26}{3}$

47. $-\frac{4}{3}$

49. $-\frac{21}{2}$

51. $1\frac{1}{5}$

53. $3\frac{1}{3}$

55. $-2\frac{1}{2}$

57. $-4\frac{2}{3}$

Exercise 3.8

1. $\frac{1}{2}$

3. $\frac{2}{3}$

5. $\frac{1}{4}$

7. $\frac{11}{1}$

9. $\frac{3}{10}$

11. $\frac{5}{18}$

13. $\frac{1}{4}$

15. $\frac{16}{1}$

17. 50 miles per hour

19. 30 miles per gallon

21. $6.75 per ounce

23. 4 cents per ounce

25. $50 per day

27. No

29. Yes.

31. No.

33. No.

35. 5

37. 4

39. 7

41. $\frac{75}{9}$

43. 304

45. 4

47. $\frac{4}{3}$

49. -31

51. $\frac{25}{19}$

53. $\frac{7}{8}$

55. $2.70

57. 7 pounds

59. 5 inches

61. 160 sets

63. 440 miles

Exercise 4.1

1. Polynomial.
3. Polynomial.
5. Not a polynomial.
7. Polynomial.
9. Trinomial.
11. Monomial.
13. Monomial.
15. Monomial.
17. Binomial.
19. -5
21. -11
23. -89
25. 6
27. 8
29. 34
31. $3m^2 - 2m - 7$
33. $3a^2 + 3a - 5$
35. $-3m - 1$
37. $2a^2 - a + 1$
39. $6m^2 - 4m$
41. $-10a + 3$
43. $-8y^2 + 3$
45. $-4x^3 + 2x^2 + 2x - 5$
47. $-7p^2 - p + 1$
49. $3t^5 + t^4 - 4t^3 - 5t^2 + 7t + 6$

Exercise 4.2

1. $-20x$
3. $35x^2$
5. $-6m^7$
7. $-12a^3b^3$
9. $-36m^4n^5$
11. $24a^4b^3c^3$
13. $6x - 12$
15. $2m - 10$
17. $-6a^3 + 8ab^2$
19. $6x^2y - 8xy^2 + 10xy$
21. $-15r^4t + 5r^2t^3 - 10r^3t$
23. $-6x^3y^2 + 9x^4y - 12x^2y^3$
25. $x^2 + 6x + 8$
27. $6x^2 - x - 15$
29. $5a^3 + 17a^2 - 10a + 8$
31. $2r^3 - r^2 - 18r + 9$
33. $6x^4 - 13x^3 + 2x^2 + 8x - 3$
35. $6a^4 - 5a^3 - 23a^2 + 7a + 15$
37. $10x^5 - 21x^4 + 11x^3 - 13x^2 + 17x - 3$
39. $6m^6 - 5m^5 - 11m^4 + 11m^3 + 5m - 6$
41. $8x^3 - 27$
43. $x^4 - 1$
45. $y^3 + 9y^2 + 26y + 24$
47. $r^3 + 3r^2 - 4r - 12$
49. $2x^4 + x^3 - 23x^2 - 46x - 24$

Exercise 4.3

1. $x^2 + 5x + 6$
3. $a^2 + a - 20$
5. $2x^2 + 7x + 6$
7. $6m^2 + 13mn - 5n^2$
9. $6a^2 + 13a + 5$
11. $x^2 - 4$
13. $4y^2 - 25$
15. $49x^2 - y^2$
17. $4a^4 - 9$
19. $25x^2 - \dfrac{4}{9}$
21. $a^2 + 6a + 9$
23. $m^2 + 2m + 1$
25. $x^2 + 14x + 49$
27. $4m^2 - 4m + 1$
29. $9x^2 - 12xy + 4y^2$
31. $a^2 + 5a + 6$
33. $6x^2 + 5x - 6$
35. $2x^2 + 7x + 3$
37. $a^2 + 6a + 9$
39. $4x^2 - 25$
41. $20x^2 - 7x - 6$
43. $4a^6 - 25b^4$
45. $3x^2 + 13x - 10$
47. $x^2 - x - 20$
49. $6x^2 + 11x - 10$
51. $x^2 + 12x + 36$
53. $9x^2 + 24x + 16$
55. $4x^4 - y^4$
57. $a^2 - \dfrac{4}{9}$
59. $x^2 - 2xy + y^2$
61. $6 - 7x + 2x^2$
63. $a^4 - b^4$
65. $x^2 + x - 6$
67. $2a^2 - 3a - 9$
69. $36m^2 - 12mn + n^2$
71. $49 - 28z + 4z^2$
73. $\dfrac{4}{9}x^2 - \dfrac{2}{3}xy + \dfrac{1}{4}y^2$
75. $0.09y^2 - 0.04z^2$

Exercise 4.4

1. $3a + 2$
3. $2x^2 + 1$
5. $2r^2 - 3r + 1$
7. $m^3 - 3m + 4$
9. $4m^2n^4 - 2m^3n^2 + m$
11. $x + 1 + \dfrac{1}{x}$
13. $-2m - 1 + \dfrac{3}{m}$
15. $-m - 1 + \dfrac{1}{m}$
17. $-4y^2 + 3y - 1 - \dfrac{1}{2y}$
19. $ab^2 - b - \dfrac{b}{a^2}$
21. $x - 4$
23. $3a - 11 + \dfrac{37}{a + 3}$
25. $2y - 3 + \dfrac{11}{3y + 2}$
27. $a + 2$
29. $3x - 2 + \dfrac{1}{x + 2}$
31. $x + 1$
33. $3m + 2$
35. $x^2 - x + 1 - \dfrac{2}{x + 1}$
37. $2x^2 - x + 2 - \dfrac{4}{2x - 1}$
39. $2r^3 + 10r^2 + 50r + 249 + \dfrac{1251}{r - 5}$
41. $3x^2 - 2 + \dfrac{3}{x^2 + 1}$
43. $3x^2 - 5x + 1$
45. $a^3 - 2a^2 + a + 3$
47. $h^2 - 1$
49. $p^5 + p^4 + p^3 + p^2 + p + 1$

Exercise 4.5

1. $7(a + b)$
3. $2(m + n)$
5. $6(x + 1)$
7. $3a(b + c)$
9. $x(5x + 2)$
11. $a(5 + 3b)$
13. $4mn(m - 3)$
15. $6x(x - 3)$
17. $7m(3m + 2)$
19. $11(r^2 + 11)$
21. $3(2b^2 + b - 4)$
23. $5xy(3x^2 + x - 2)$
25. $3(m^2 - 2m + 1)$
27. $ab^3(a^2b + ab^2 - 1)$
29. $5y^2(5y^2 + y - 4)$
31. $2(4x^2 + 5y^2 - 2z^2)$
33. $25xy^2(x^2 - 4x + 2)$
35. $3x(2x^4 - x^3 + 4x^2 - 5x + 3)$
37. $3x^2y^2(x - 2y + 3)$
39. $3(3ab^2 - 2b + a - 4)$
41. $y^4(y^6 + y^4 - y^2 + 1)$
43. $13p^3q^4(p^3 + 2 - 3p)$
45. $20(t^2 - 5)$
47. $4a^2b^3c(6a^3 - 9a^2c + 5ab^2c^2 - 4b^3c^3)$
49. $y^2(y^4 - y^3 - y^2 - 1)$

Exercise 4.6

1. $(x + 1)(x + 2)$
3. $(x + 2)(x - 1)$
5. $(m + 4)(m + 1)$
7. $(n - 4)(n - 2)$
9. $(a + 3)^2$
11. $(p + 2)^2$
13. $(t + 5)(t + 1)$
15. $(n - 7)(n - 1)$
17. $(m + 7)(m + 5)$
19. $(k - 10)(k - 1)$
21. $(n + 6)(n + 2)$
23. $(x + 6)(x - 3)$
25. $(p + 12)(p - 4)$
27. $(h - 7)(h + 6)$
29. $(m - 9)(m + 8)$
31. Prime.
33. $(n - 12)(n + 6)$
35. $(k + 13)(k - 5)$
37. $(t - 3)(t - 9)$
39. $(a - 15)(a - 3)$
41. $(a + 2b)(a + b)$
43. $(r - 3s)(r - 4s)$
45. $(p + 2q)(p + 7q)$
47. $(x - 7a)(x + 2a)$
49. $2(a + 4)(a + 1)$
51. $4(b + 7)(b + 1)$
53. $t(t - 6)(t - 5)$
55. $m^3(m - 3)(m - 1)$
57. $3p^2(p - 5)(p - 1)$
59. $2x^3(x - 7)(x + 3)$
61. $2x^2(x + 5)(x + 2)$
63. $3y(y + 3)(y - 1)$

Exercise 4.7

1. $(2x + 1)(x + 4)$
3. $(2m - 1)(m + 2)$
5. $(2p + 1)(3p - 1)$
7. $(2y - 3)(3y - 1)$
9. $2(x + 1)(2x + 3)$
11. $3h(2h - 1)(5h + 1)$
13. $(y + 3)(y - 3)$
15. $(2m + 3)(2m - 3)$
17. $2(3n - 5)(3n + 5)$
19. $(m^2 + 1)(m + 1)(m - 1)$
21. $(x + 2)(x + 3)$
23. $(m - 4)(m - 2)$
25. $(a + 3)^2$
27. $(x + 3)(x - 2)$
29. $(x + 4)(x - 1)$
31. $(2a - 3)(a + 1)$
33. $(2r + 3)(3r - 2)$
35. $(3x + 2y)^2$
37. $(2a - b)(a + 3b)$
39. Prime.
41. $3a(2a - 7)(a + 3)$
43. $5y(3 - x^2y)$
45. $(3x + 4)(x + 1)$
47. $2(3a - 2)(2a - 3)$
49. $(4x + 5y)(4x - 5y)$
51. $(x^2 + y^3)(x^2 - y^3)$
53. $2ab(1 - 2ab - 4a^2b^2)$
55. $3rs^2(r + 1)^2$
57. Prime.
59. $\pi R(R + 5)(R - 5)$
61. $(3k + 1)(7k + 2)$
63. $2(3 - 2h)(1 - 4h)$
65. $2m^3(3m + 2)(4m - 1)$
67. $(2k + 3)(4k - 1)$
69. $3m(1 + m^6)(1 + m^3)(1 - m^3)$
71. $(a + 1)(a - 1)$
73. $(m + 3)(m - 3)$
75. $(r + 6)(r - 6)$
77. $(4a + 3)(4a - 3)$
79. $(x + y)(x - y)$
81. $4(3r + 5)(3r - 5)$
83. $5(x + 2)(x - 2)$
85. $(m^2 + n^2)(m + n)(m - n)$
87. $(n^2 + 9)(n + 3)(n - 3)$
89. $(12rt + 11)(12rt - 11)$

Exercise 4.8

1. 2
3. $\dfrac{3}{2}$
5. -6
7. 3
9. -1
11. 1
13. -5
15. -2
17. 5
19. $-\dfrac{35}{3}$
21. $6, -1$
23. 1
25. $4, -2$
27. $4, -3$
29. 4
31. $\dfrac{3}{2}, -1$
33. $3, -3$
35. $0, \dfrac{9}{2}$
37. $0, 2$
39. $1, -2$
41. $\dfrac{9}{2}, 8$
43. $3, -3$
45. $\dfrac{2}{3}$
47. $\dfrac{2}{3}, 2$
49. $\dfrac{1}{3}, 3$
51. $-\dfrac{2}{5}$
53. $-5, -2$
55. $3, -1$
57. $3, 1$
59. $5, -5$

Exercise 4.9

1. 42 ft and 58 ft
3. 2400 mi
5. 4, 6 and $-2, 0$
7. -2
9. 180 mi
11. 5 and -2
13. 2, 4 and $-2, 0$
15. 5
17. 3 ft and 5 ft
19. 8 m by 14 m
21. 5 ft by 9 ft
23. 10 cm
25. 125 mph
27. -4
29. 3 ft by 5 ft
31. 5 ft

Exercise 5.1

1. $\dfrac{9s^3rt^3}{21s^4t^4}$

3. $\dfrac{2x^2 + 2xy}{3x + 3y}$

5. $\dfrac{x^2 - 7x + 12}{x^2 - 9}$

7. $\dfrac{2x^2 + 5x + 2}{x^2 + 5x + 6}$

9. $\dfrac{2 - 3a}{1 - 2a}$

11. $\dfrac{mn}{3}$

13. $\dfrac{y^2}{2x}$

15. $\dfrac{b - c}{b + c}$

17. $\dfrac{-3}{4}$

19. $\dfrac{3}{5}$

21. $\dfrac{3a + 2}{3a - 2}$

23. $\dfrac{-1}{x + 2}$

25. $\dfrac{-5}{7}$

27. $\dfrac{x + 1}{x - 1}$

29. $\dfrac{-2 - y}{3y + 5}$

31. $\dfrac{x + 1}{x + 2}$

33. $\dfrac{-y}{2y + 1}$

35. $\dfrac{a + 3b}{3b}$

37. $\dfrac{x - 1}{x + 1}$

39. $\dfrac{-1 - a}{a - 2}$

41. $\dfrac{m}{m + 1}$

43. $\dfrac{2y - 5}{2(y - 3)}$

45. $\dfrac{2a - 1}{a - 2}$

47. $-r^2 - 1$

49. $3m^2 + 5m - 2$

Exercise 5.2

1. $\dfrac{7m}{10n}$

3. $\dfrac{8}{3a}$

5. $\dfrac{y}{45x}$

7. $2x^2y^5$

9. $\dfrac{5}{6a}$

11. $\dfrac{3x - 1}{4x}$

13. $\dfrac{-a - 2}{9}$

15. $\dfrac{m - 5}{m + 2}$

17. $\dfrac{5x(x - 1)}{6}$

19. $\dfrac{(x + 4)(x + 3)}{(x - 5)(x + 1)}$

21. $\dfrac{3a + 2}{(a - 1)^2}$

23. $\dfrac{3m + 2}{4m}$

25. $\dfrac{m - n}{9}$

27. $\dfrac{-y^2}{2}$

29. $\dfrac{x}{x + 2}$

31. $\dfrac{(x + 2)^2}{(x + 7)(x - 3)}$

33. $\dfrac{a}{a + 1}$

35. $\dfrac{(a - b)(a - 3b)}{(a - 2b)(a + b)}$

37. $\dfrac{x - 1}{x - 2}$

39. $\dfrac{m + 3}{2m^2}$

Exercise 5.3

1. $\dfrac{14b + 15a}{36a^2b^2}$

3. $\dfrac{10 + m}{6m^2n}$

5. $\dfrac{2b - 3ab + a}{a^2b^2}$

7. $\dfrac{12 - 9x - 2y}{6x^2y}$

9. $\dfrac{7}{2(m - 1)}$

11. $\dfrac{9}{4(x - 2)}$

13. $\dfrac{-x - 8}{(x + 3)(x - 3)}$

15. $\dfrac{a - 6}{a - 3}$

17. $\dfrac{-4x - 23}{(x + 4)(x - 4)}$

19. $\dfrac{12a - 46}{(a - 5)(a + 2)(a - 3)}$

21. $\dfrac{-2}{(m + 1)^2(m - 1)}$

23. $\dfrac{3x^2 + 9x + 5}{(2x - 3)(x + 2)(x + 1)}$

25. $\dfrac{a^2 - a + 1}{(3a + 2)(a + 1)(a - 2)}$

27. $\dfrac{x^2 - 12x + 7}{(3x + 4)(x + 2)(x - 3)}$

29. $\dfrac{t^2 + 8t + 3}{(t + 3)(t - 3)(t + 7)}$

31. $\dfrac{5a^2 - 17a}{(a - 5)(a + 3)(a - 1)}$

33. $\dfrac{2t^2 + 10t + 4}{(t + 2)(t + 3)(t + 1)}$

35. $\dfrac{2m^2 - 10m + 13}{(2m + 1)(m - 2)(m - 3)}$

37. $\dfrac{y^2 + 6y - 3}{(y + 3)(y - 3)(y + 7)}$

39. $\dfrac{2m^2 + 6mn + 8n^2}{(m + n)^2(m + 3n)}$

41. $\dfrac{2a^2 + 3a - 11}{(a + 1)^2(a - 1)}$

43. $\dfrac{y^2 + 4}{(y + 2)(y - 2)}$

45. $\dfrac{2m - 3n - 1}{(m + n)(m - n)}$

47. $\dfrac{6a^2 - a - 7}{(a + 3)(a - 3)(3a + 2)}$

49. $\dfrac{1}{2x - 3}$

Exercise 5.4

1. $\dfrac{21}{20}$

3. $\dfrac{35}{2}$

5. $\dfrac{5}{18}$

7. 1

9. -2

11. $\dfrac{3m^2n}{4}$

13. $\dfrac{2m + 1}{2m}$

15. $\dfrac{5}{4}$

17. $\dfrac{a^2 - 1}{a^2 + 1}$

19. $\dfrac{3rt - r^2}{3rt - t^2}$

21. $\dfrac{2}{x}$

23. $\dfrac{m}{n}$

25. $\dfrac{3ab + 2b^2}{6a + 30b}$

27. $\dfrac{2x^2 + x}{2 - 4x}$

29. $\dfrac{x - 3}{x + 3}$

31. $\dfrac{a}{4}$

33. $\dfrac{m^2}{m - 2}$

35. $\dfrac{x - 5}{x - 4}$

37. $\dfrac{x - 2}{x + 2}$

39. $-\dfrac{2}{3}$

Exercise 5.5

1. $\dfrac{25}{3}$

3. $\dfrac{3}{2}$

5. $\dfrac{10}{3}$

7. 0

9. 6

11. $\dfrac{8}{5}$

13. 4

15. $-\dfrac{1}{8}$

17. $\dfrac{4}{3}$

19. 5

21. 5

23. 2

25. -1

27. No solutions.

29. No solutions.

31. $\frac{14}{15}$

33. $\frac{2}{3}$

35. -3

37. -3

39. 1

41. 1

43. No solutions.

45. -3 and 1

Exercise 5.6

1. 21 cm²

3. 10 °C

5. 144 ft

7. 113.04 cm³

9. 3 in

11. $a = \frac{f}{m}$

13. $m = \frac{e}{c^2}$

15. $x = \frac{b}{a}$

17. $t = \frac{d}{r}$

19. $r = \frac{i}{pt}$

21. $d = \frac{C}{\pi}$

23. $l = \frac{V}{wh}$

25. $h = \frac{V}{lw}$

27. $b = \frac{2A}{h}$

29. $l = \frac{P - 2w}{2}$

31. $r = \frac{s - a}{s}$

33. $h = \frac{2A}{b + c}$

35. $a = \frac{1}{b + c}$

37. $a = \frac{bc}{b - c}$

39. $p = \frac{qf}{q - f}$

41. $m = \frac{Fqr}{v^2}$

43. $r = \frac{Sl - a}{S - l}$

45. $D = \frac{lT + 12d}{12}$

Exercise 5.7

1. [number line: open circle at 1, arrow to right; $-5\,-4\,-3\,-2\,-1\ 0\ 1\ 2\ 3\ 4\ 5$]

3. [number line: open circle at 2, arrow to left; $-5\,-4\,-3\,-2\,-1\ 0\ 1\ 2\ 3\ 4\ 5$]

5. [number line: open circle at -3, arrow to left; $-5\,-4\,-3\,-2\,-1\ 0\ 1\ 2\ 3\ 4\ 5$]

7. [number line: closed circle at -3, arrow to right; $-5\,-4\,-3\,-2\,-1\ 0\ 1\ 2\ 3\ 4\ 5$]

9. [number line: closed circle at 0, arrow to right; $-5\,-4\,-3\,-2\,-1\ 0\ 1\ 2\ 3\ 4\ 5$]

11. [number line: closed circle at 0, arrow to left; $-5\,-4\,-3\,-2\,-1\ 0\ 1\ 2\ 3\ 4\ 5$]

13. [number line: open circle at 1 to open circle at 4; $-5\,-4\,-3\,-2\,-1\ 0\ 1\ 2\ 3\ 4\ 5$]

15. [number line: closed circle at -2 to closed circle at 2; $-5\,-4\,-3\,-2\,-1\ 0\ 1\ 2\ 3\ 4\ 5$]

17. [number line: closed circle at -3 to open circle at 1; $-5\,-4\,-3\,-2\,-1\ 0\ 1\ 2\ 3\ 4\ 5$]

19. [number line: open circle at -4 to open circle at -1; $-5\,-4\,-3\,-2\,-1\ 0\ 1\ 2\ 3\ 4\ 5$]

21. [number line: closed circle at -2 and closed circle at 2; $-5\,-4\,-3\,-2\,-1\ 0\ 1\ 2\ 3\ 4\ 5$]

23. $x \le 4$

25. $-6 < x < 2$

27. $x > 10$

29. $4m > 4$

31. $x > 6$

33. $10t \ge 1$

35. $t \ge -6$

37. $3x - 3 \ge 7$

39. $x \ge 0$

41. $100 < m < 150$

45. $1000 < t \leq 1500$

49. $5.99 \leq r \leq 6.01$

43. $s \geq 45$

47. $160 < w \leq 165$

Exercise 5.8

1. $x < 3$

5. $t < 3$

9. $x > \dfrac{3}{2}$

13. $n > -5$

17. $m > -7$

21. $x \geq \dfrac{9}{2}$

25. No solutions.

29. No solutions.

33. $n \geq 0$

37. $x \leq \dfrac{9}{2}$

41. No solutions.

45. $m \geq \dfrac{12}{11}$

49. $t > 0$

3. $m \leq 2$

7. $n < \dfrac{5}{2}$

11. $y > -4$

15. $x \leq \dfrac{4}{3}$

19. $x \geq -11$

23. $x < -\dfrac{3}{2}$

27. $x \leq \dfrac{9}{4}$

31. $y \geq -\dfrac{7}{2}$

35. All real numbers.

39. $x > 0$

43. $t > 8$

47. $y \geq 6$

Exercise 5.9

1. 8 and 10

5. 10

9. $1\dfrac{3}{5}$ in. by $2\dfrac{2}{5}$ in.

13. 15ℓ

17. 16 lb

21. 15 days

25. 2 hr

29. $13\dfrac{1}{3}$ hr

33. $x \geq 87$

3. 15

7. $\dfrac{7}{12}$

11. 20 qt

15. $16\dfrac{2}{3}$ g

19. \$3300 at 6%; \$8900 at 8%

23. $3\dfrac{1}{3}$ hr

27. 40 hr; 60 hr

31. $x \geq 89$

35. $x > \$4000$

Exercise 6.1

1. $\dfrac{1}{n^2}$

5. $\dfrac{2}{m^3}$

9. $\dfrac{1}{16}$

13. $\dfrac{1}{100}$

17. $\dfrac{16}{9}$

3. $\dfrac{1}{a^4}$

7. $\dfrac{1}{8}$

11. $\dfrac{1}{32}$

15. 10

19. $\dfrac{1}{3x^2y}$

21. $\dfrac{3}{2m^4}$

23. $2ab^3$

25. $\dfrac{m^5n^2}{3}$

27. $\dfrac{c^4}{a^2b^3}$

29. $\dfrac{7}{x^3y^5}$

31. $\dfrac{2z^2}{x^3y^4w^3}$

33. $\dfrac{b-a}{ab}$

35. $\dfrac{n^2+m^2}{5m^2n^2}$

37. $\dfrac{2x^2}{1+x^2}$

39. $\dfrac{y+xy}{x+xy}$

41. $\dfrac{1}{a^3}$

43. $\dfrac{1}{x^6}$

45. $\dfrac{24}{a^2}$

47. $3mn^8$

49. $\dfrac{27b^6}{8a^9c^{12}}$

51. $\dfrac{72}{a}$

53. $\dfrac{-12c^4}{a^4b^7}$

55. $\dfrac{3x^{10}z}{y^7}$

57. 1.6556291

59. 0.43516604

61. 266,666.67

Exercise 6.2

1. 591
3. 1,100,000
5. 4.871
7. 92
9. 0.0000006
11. 0.000506
13. 3.8
15. 3.64×10^2
17. 4.02×10^{-4}
19. 2.91×10^0
21. 6.51×10^{11}
23. 1.58×10^{-1}
25. 8×10^2
27. 9.4×10^{-9}
29. 9.3×10^7
31. 1×10^{17}
33. 4×10^{-5}; 7×10^{-5}
35. 1,000,000,000
37. 24,000
39. 0.00006
41. 3×10^2
43. 8×10^1
45. 9×10^{-3}
47. 1.6

Exercise 6.3

1. A square root of a number is one of its two equal factors.
3. A cube root of a number is one of its three equal factors.
5. True.
7. False.
9. True.
11. True.
13. True.
15. True.
17. True.
19. False.
21. 9
23. 2
25. 2
27. -2
29. 1
31. 0
33. -3
35. 6
37. 4
39. Not a real number.
41. 1.414
43. 5.745
45. 2.759
47. 4.217
49. -4.380
51. 0.6
53. 0.09
55. 5.1
57. 17.464249
59. 2.3269668
61. 12.310646
63. 3.5022829

Exercise 6.4

1. $\sqrt{a}$
3. $\sqrt[5]{m}$
5. $3\sqrt{x}$
7. $\sqrt[3]{y^2}$ or $(\sqrt[3]{y})^2$
9. $\sqrt{z^3}$ or $(\sqrt{z})^3$
11. $5\sqrt[5]{m^2}$ or $5(\sqrt[5]{m})^2$
13. $b^{1/2}$
15. $z^{1/5}$
17. $(7n)^{1/2}$
19. $x^{3/2}$
21. $5a^{4/5}$
23. -6
25. $\frac{1}{6}$
27. -2
29. 0
31. $\frac{3}{4}$
33. $\frac{3}{4}$
35. 3
37. 64
39. No answer.
41. 0
43. 16
45. $\frac{1}{9}$
47. 10
49. 10
51. $\frac{1}{100}$
53. 16
55. $\frac{1}{16}$
57. $\frac{1}{27}$
59. $\frac{9}{4}$
61. 1.9004985
63. 0.94028822
65. 2.3461207
67. 528.44148

Exercise 6.5

1. $4\sqrt{3}$
3. $3\sqrt{5}$
5. $2\sqrt[4]{7}$
7. $2\sqrt{2}$
9. $-2\sqrt[3]{2}$
11. $3\sqrt{3}$
13. $4\sqrt[3]{2}$
15. $4\sqrt{3}$
17. $-4\sqrt[3]{3}$
19. $12\sqrt{3}$
21. $6\sqrt{2}$
23. $-9\sqrt[3]{2}$
25. $10\sqrt{7}$
27. 20
29. $6\sqrt[5]{2}$
31. $-12\sqrt{5}$
33. $8\sqrt[3]{10}$
35. $30\sqrt{2}$
37. $6\sqrt{5}$
39. $40\sqrt[3]{2}$
41. m
43. $x\sqrt[3]{x^2}$
45. $a\sqrt{a}$
47. x^2
49. $xy\sqrt{x}$
51. $abc\sqrt[4]{bc^2}$
53. $2mn\sqrt{m}$
55. $2x\sqrt[3]{2y^2}$
57. $6n^2$
59. $6x\sqrt[3]{2y}$
61. $10x^2\sqrt{7xy}$
63. $-10m^3n^2\sqrt[3]{10n}$

Exercise 6.6

1. $6\sqrt{5}$
3. $-4\sqrt[3]{6}$
5. $2\sqrt{5}$
7. $2\sqrt[3]{9}$
9. $4\sqrt{2} + 4\sqrt{3}$
11. $-3\sqrt[3]{5} + 3\sqrt{3}$
13. $2\sqrt{5} + 8\sqrt{7}$
15. $-\sqrt[3]{2}$
17. $25\sqrt{3}$
19. $23\sqrt[3]{2}$
21. $-5\sqrt{3}$
23. $27\sqrt{2}$
25. $-21\sqrt[3]{3}$
27. $-6\sqrt{5}$
29. $24\sqrt{6} - 6\sqrt{5}$
31. $15\sqrt{3}$
33. $3\sqrt{6} - 8\sqrt{2}$
35. $3\sqrt[3]{2} + 3\sqrt{2}$
37. $6\sqrt{x}$
39. $5y\sqrt{2}$
41. $\sqrt[3]{m}$
43. $-8\sqrt{2x} + 11\sqrt{3y}$
45. $12\sqrt{3m} + 2\sqrt{2n}$
47. $64x\sqrt{x} + 38x\sqrt{2}$
49. $13a\sqrt{3} - 13a\sqrt[3]{2a}$

Exercise 6.7

1. $\sqrt{30}$
3. $3\sqrt{5}$
5. $\sqrt{30}$
7. 7
9. $\sqrt{xy}$
11. $a\sqrt{b}$
13. $6\sqrt{30}$
15. $12\sqrt{15}$
17. $10\sqrt[3]{6a^2}$
19. $20\sqrt[3]{3}$
21. 24
23. $150\sqrt{2}$
25. $12\sqrt[3]{3}$
27. $2\sqrt{3} + \sqrt{15}$
29. $6\sqrt{6} - 8\sqrt{15}$
31. $\sqrt{6} - 4\sqrt{15} + 3\sqrt{2}$
33. 1
35. 3
37. $26 + 11\sqrt{5}$
39. $8 + 3\sqrt{2}$
41. $-24 - 5\sqrt{5}$
43. $4 + 3\sqrt{15}$
45. $17 + 4\sqrt{10}$
47. $5 + 2\sqrt{6}$
49. $\sqrt{5}$
51. $2\sqrt{2}$
53. $3\sqrt{5}$
55. $4\sqrt{5}$
57. $\dfrac{7}{8}$
59. $\dfrac{5}{3}$
61. $\dfrac{1}{4}$
63. $\dfrac{\sqrt{5}}{4}$
65. $\dfrac{\sqrt[3]{2}}{3}$
67. $\dfrac{\sqrt[4]{7}}{2}$
69. $\dfrac{\sqrt[3]{7}}{10}$
71. $5\sqrt{7}$

Exercise 6.8

1. $\dfrac{2\sqrt{5}}{5}$
3. $\dfrac{3\sqrt[3]{4}}{2}$
5. $\sqrt{3}$
7. $\dfrac{\sqrt{3}}{3}$
9. $\dfrac{\sqrt[3]{10}}{2}$
11. $\dfrac{\sqrt{6}}{2}$
13. $\dfrac{2\sqrt{21}}{21}$
15. $-2\sqrt[3]{2}$
17. $\dfrac{2\sqrt[3]{10}}{15}$
19. $\dfrac{2\sqrt{30}}{5}$
21. $\dfrac{2\sqrt[3]{6}}{5}$
23. $\dfrac{\sqrt[3]{150}}{5}$
25. $\dfrac{6 - 3\sqrt{6}}{-2}$
27. $\dfrac{4 - \sqrt{5}}{11}$
29. $\dfrac{3\sqrt{7} - 3\sqrt{5}}{2}$
31. $\sqrt{3} - \sqrt{2}$
33. $\dfrac{3\sqrt{7} + 3\sqrt{3}}{4}$
35. $\dfrac{2\sqrt{3} - 3\sqrt{2}}{-6}$
37. $\dfrac{6\sqrt{14} + 70}{-157}$
39. $\dfrac{3 + \sqrt{15}}{-6}$
41. 0.577
43. 0.630
45. 1.260
47. 1.225
49. -6.674
51. 0.639

Exercise 6.9

1. 9
3. 1
5. 12
7. 20
9. 4
11. $\dfrac{9}{4}$
13. 1
15. 7
17. -5
19. 3
21. 7
23. 7

25. $\frac{4}{7}$ **27.** 6

29. 5   **31.** 7

33. 12   **35.** −1 and 0

37. 9   **39.** 7

41. 2   **43.** 3

45. 4   **47.** 7

49. 1

Exercise 6.10

1. 36   **3.** 9

5. 100   **7.** 49

9. $\frac{1}{4}$ **11.** $\frac{9}{4}$

13. $\frac{81}{4}$ **15.** $\frac{9}{64}$

17. $\frac{k^2}{4}$ **19.** ± 3

21. ± 6 **23.** $\pm\sqrt{7}$

25. No solutions.   **27.** 8 and 2

29. 2 and 1   **31.** −1 and −3

33. 3 and −1   **35.** $2 \pm \sqrt{5}$

37. $-3 \pm \sqrt{13}$ **39.** −3 and −2

41. $\frac{3}{2}$ and −1   **43.** 2 and $-\frac{5}{2}$

45. 3 and $-\frac{7}{2}$ **47.** No solutions.

49. 1 and $-\frac{3}{5}$ **51.** $\frac{-1 \pm \sqrt{5}}{2}$

53. $\frac{-7 \pm \sqrt{17}}{4}$ **55.** $3 \pm 2\sqrt{2}$

57. $\frac{9 \pm \sqrt{21}}{6}$ **59.** 2

Exercise 6.11

1. $2x^2 + 5x + 1 = 0;\ a = 2, b = 5, c = 1$

3. $4x^2 - 3x - 5 = 0;\ a = 4, b = -3, c = -5$

5. $x^2 - 3x - 7 = 0;\ a = 1, b = -3, c = -7$

7. $x^2 - 5x + 2 = 0;\ a = 1, b = -5, c = 2$

9. $4x^2 + 0x + 5 = 0;\ a = 4, b = 0, c = 5$

11. −3 and −2   **13.** 3 and 2

15. ± 3 **17.** 3 and $-\frac{1}{3}$

19. $\frac{3 \pm \sqrt{5}}{2};$ 2.618 and 0.382   **21.** No solutions.

23. 1 and $-\frac{5}{3}$ **25.** $\frac{5 \pm \sqrt{145}}{6};$ 2.840 and −1.174

27. $1 \pm \sqrt{11};$ 4.317 and −2.317   **29.** $\frac{1}{2}$ and $-\frac{2}{3}$

31. $\frac{1 \pm \sqrt{7}}{2};$ 1.823 and −0.823   **33.** $\pm 2\sqrt{5};\ \pm 4.472$

35. No solutions.   **37.** $\frac{12 \pm \sqrt{142}}{3}$

39. No solutions.   **41.** 2 and −3

43. 10 and $-\frac{1}{2}$ **45.** $\frac{5}{2}$ and −1

47. Length is 9 in.; width is 5 in.   **49.** 9 and 6

51. 8 ft

53. No real solutions.

55. Two real solutions.

Exercise 7.1

1.

x	y	
−2	−10	(−2, −10)
0	0	(0, 0)
1	5	(1, 5)
3	15	(3, 15)

3.

x	y	
−4	−13	(−4, −13)
−2	−7	(−2, −7)
0	−1	(0, −1)
1	2	(1, 2)

5.

x	y	
−3	9	(−3, 9)
−1	1	(−1, 1)
0	0	(0, 0)
2	4	(2, 4)

7.

x	y	
−4	48	(−4, 48)
−1	15	(−1, 15)
0	8	(0, 8)
3	−1	(3, −1)

9.

x	y	
−3	0	(−3, 0)
−1	$2\sqrt{2}$	(−1, $2\sqrt{2}$)
0	3	(0, 3)
2	$\sqrt{5}$	(2, $\sqrt{5}$)

11.

x	y
−5	−5
−3	−3
−1	−1
0	0
2	2
4	4
6	6

13.

x	y
−6	−12
−4	−8
$-\dfrac{3}{2}$	−3
0	0
$\dfrac{1}{2}$	1
2	4
4	8

15.

x	y
−4	−11
−2	−5
−1	−2
0	1
$\dfrac{1}{3}$	2
$\dfrac{4}{3}$	5
2	7

17.

x	y
−3	2
−2	1
$-\dfrac{1}{2}$	$-\dfrac{1}{2}$
0	−1
2	−3
$\dfrac{5}{2}$	$-\dfrac{7}{2}$

19.

x	y
−4	0
−2	1
−1	$\dfrac{3}{2}$
0	2
2	3
3	$\dfrac{7}{2}$
4	4

21.

x	y
−4	−12
−3	−10
−1	−6
0	−4
2	0
4	4

23.

x	y
−4	17
−3	10
−2	5
−1	2
0	1
1	2
2	5
3	10

25.

x	y
−3	28
−2	13
−1	4
0	1
1	4
2	13
3	28

27.

x	y
−2	12
−1	3
$-\frac{1}{2}$	0
0	−2
1	−3
2	0
3	7

29.

x	y
−18	−3
−8	−2
−2	−1
0	0
−2	1
−8	2
−18	3

31.

x	y
20	−2
6	−1
0	$-\frac{1}{3}$
−2	0
−4	1
0	2
10	3

33.

x	y
−100	$-\frac{1}{10}$
−10	−1
−5	−2
−2	−5
$-\frac{1}{2}$	−20
1	10
2	5
10	1

35.

x	y
−10	10
−5	5
−2	2
$-\frac{3}{2}$	$\frac{3}{2}$
−1	1
0	0
3	3

37.

P	V
$\frac{1}{10}$	1000
$\frac{1}{5}$	500
$\frac{1}{2}$	200
1	100
2	50
5	20
10	10

39. Yes.

41. No.

43. Yes.

45. Yes.

47. No.

49. Yes.

51. $y = .15x + 10$ where x represents miles and y represents the daily rate in dollars.

53. $y = 15x + 75$ where x represents the number of checks and y represents the monthly charge in cents.

Exercise 7.2

1. (4, 5)

3. (−6, 0)

5. (8, 0)

7. (0, −2)

9. (9, −4)

11. (−4, 4)

13. (−2, −1)

15. (0, 0)

17. (7, 4)

19. (0, −5)

21. (−6, 3)

23. (−1, 4)

25. (0, 0)

27. Abscissa is −5; ordinate is 1

29. Abscissa is 3; ordinate is 0

31. Abscissa is 2; ordinate is −1

33. Abscissa is $\frac{7}{3}$; ordinate is $-\frac{1}{6}$

35. I

37. II

39. None.

41. I

43. III

45. IV

47. (0, 0); origin

49. 0

Exercise 7.3

1.

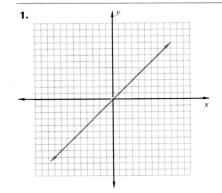

3.

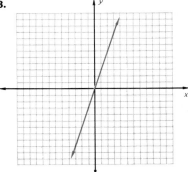

5.

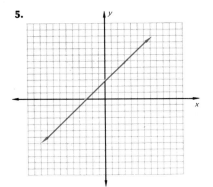

7.

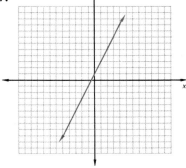

9.

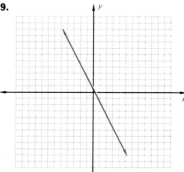

11.

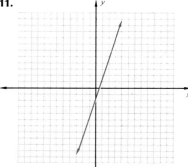

13.

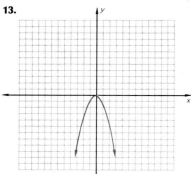

15.

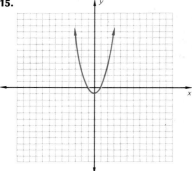

17.

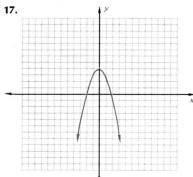

19.

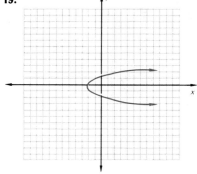

21.

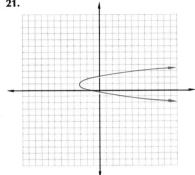

23.

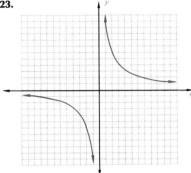

25.

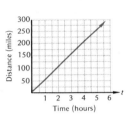

27.

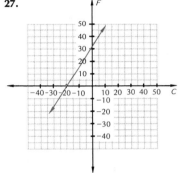

29. 10 in³.

31. 10 lbs/in².

33. Decreases.

35. 3.5 A.

37. 1 A.

39. Increases.

41. About 250 ft.

43. Increases radically.

Exercise 7.4

1. $y = -x + 6$; $m = -1$ and $b = 6$

3. $y = -2x + 7$; $m = -2$ and $b = 7$

5. $y = 5x + 1$; $m = 5$ and $b = 1$

7. $y = -2x + 0$; $m = -2$ and $b = 0$

9. $y = \frac{1}{3}x - 3$; $m = \frac{1}{3}$ and $b = -3$

11. $y = \frac{3}{4}x - 3$; $m = \frac{3}{4}$ and $b = -3$

13. $y = 0x + 5$; $m = 0$ and $b = 5$

15.

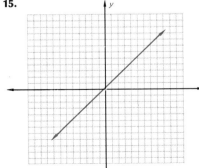

17.

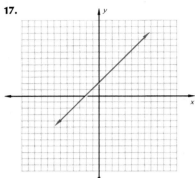

19.

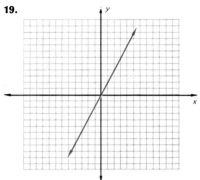

21.

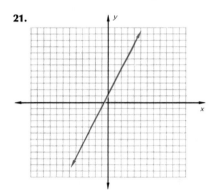

23.

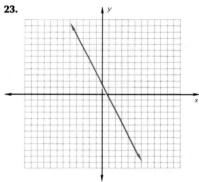

25.

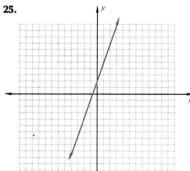

27.

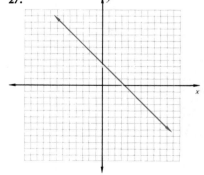

29.

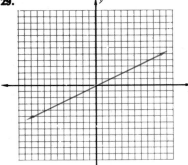

31.

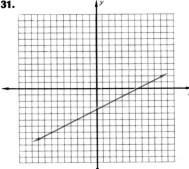

33.

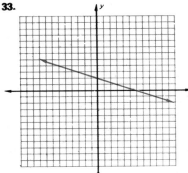

35.

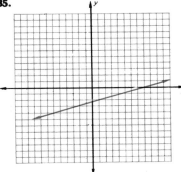

37.

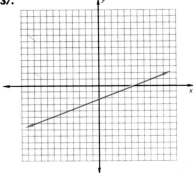

39.

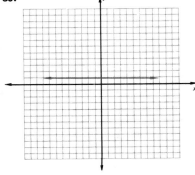

41.

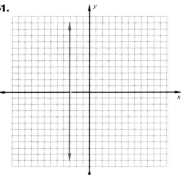

43.

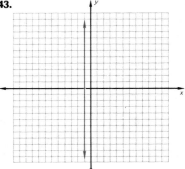

45.

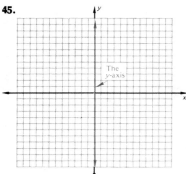

47.

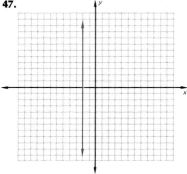

49.

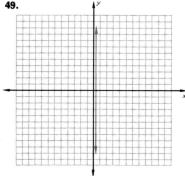

51. $y = 3x$

53. $2x = 3y + 6$

55. $x = \frac{1}{2}y + 3$

57. $y = 3$

Exercise 7.5

1. Rise = 6; run = 3
3. Rise = 4; run = 2
5. Rise = −1; run = 3
7. Rise = 6; run = 3
9. Rise = 9; run = 2
11. Rise = 5; run = 8
13. Rise = 8; run = 0
15. 1
17. −2
19. −1
21. −3
23. No slope.
25. 1
27. $\frac{1}{3}$
29. $-\frac{5}{2}$
31. $y = x - 2$; $m = 1$
33. $y = 2x - 3$; $m = 2$
35. $y = \frac{1}{2}x - 2$; $m = \frac{1}{2}$
37. $y = \frac{1}{2}x - 3$; $m = \frac{1}{2}$
39. $y = -5x$; $m = -5$
41. False.
43. True.
45. False.

Exercise 7.6

1. $x - \text{int} = 1$; $y - \text{int} = 1$
3. $x - \text{int} = 5$; $y - \text{int} = 5$
5. $x - \text{int} = 3$; $y - \text{int} = -3$
7. $x - \text{int} = 4$; $y - \text{int} = 2$
9. $x - \text{int} = 6$; $y - \text{int} = 2$
11. $x - \text{int} = 2$; $y - \text{int} = 4$
13. $x - \text{int} = 2$; $y - \text{int} = 10$
15. $x - \text{int} = 5$; $y - \text{int} = -\frac{5}{3}$
17. $x - \text{int} = 3$; $y - \text{int} = -2$
19. $x - \text{int} = 2$; $y - \text{int} = -5$
21. $x - \text{int} = 7$; $y - \text{int} = -2$
23. $x - \text{int} = \frac{8}{3}$; $y - \text{int} = -4$
25. $x - \text{int} = \frac{1}{2}$; $y - \text{int} = -2$
27. $x - \text{int} = 2$; $y - \text{int} = -4$
29. $x - \text{int} = -\frac{3}{4}$; $y - \text{int} = 3$
31. $x - \text{int} = -4$; $y - \text{int} = 4$
33. $x - \text{int} = 1$; $y - \text{int} = \frac{1}{3}$
35. $y = 2x + 3$; $m = 2$; $b = 3$
37. $y = \frac{1}{2}x - 2$; $m = \frac{1}{2}$; $b = -2$
39. $y = -x + 3$; $m = -1$; $b = 3$
41. $y = -\frac{1}{2}x + \frac{1}{4}$; $m = -\frac{1}{2}$; $b = \frac{1}{4}$
43. $y = \frac{1}{5}x - 2$; $m = \frac{1}{5}$; $b = -2$
45. $y = 4x - 2$; $m = 4$; $b = -2$
47. True.
49. True.
51. False.
53. True.
55. True.

Exercise 7.7

1.

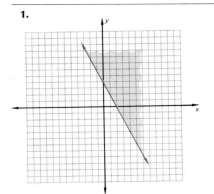

3.

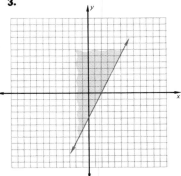

5.
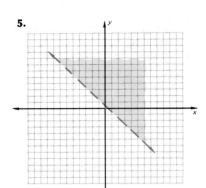

7.
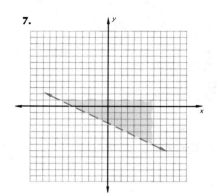

9.
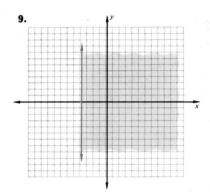

11.
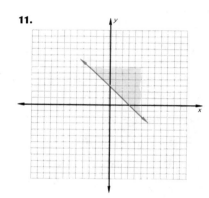

13.
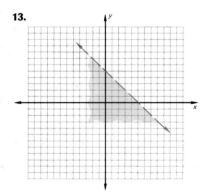

15.
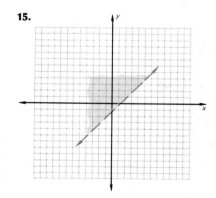

17.

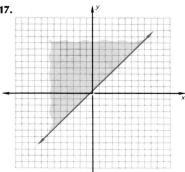

19.

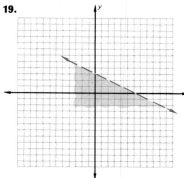

21.

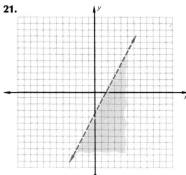

23.

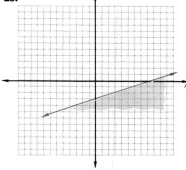

25.

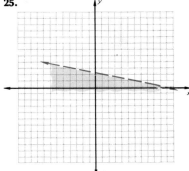

27.

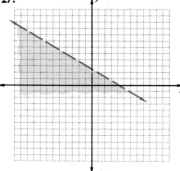

29.

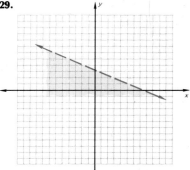

31.

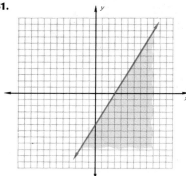

33.

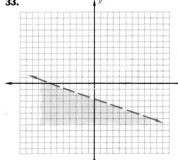

35.

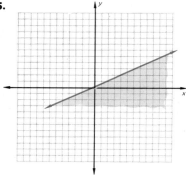

37.

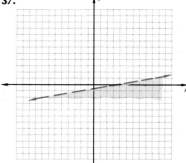

39.

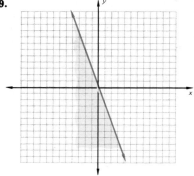

41.

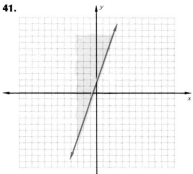

43.

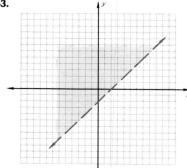

45.

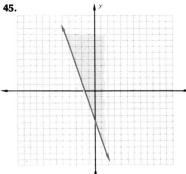

47.

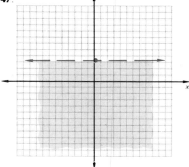

49.

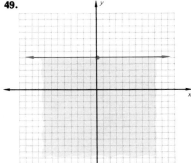

51.

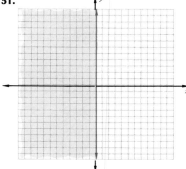

Exercise 7.8

1. $a = 1$; upward.
5. $a = 1$; upward.
9. $a = -3$; downward.

3. $a = -3$; downward.
7. $a = 1$; upward.
11. $a = -3$; downward.

13.

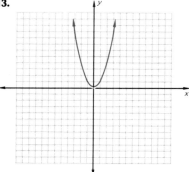

15.

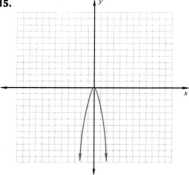

17.

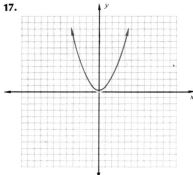

19.

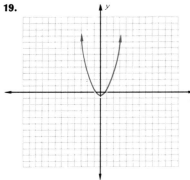

21.

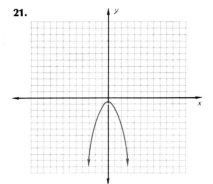

23.

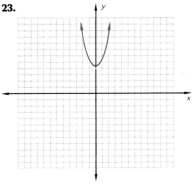

25.

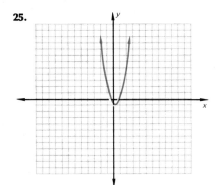

27.

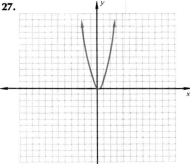

29.

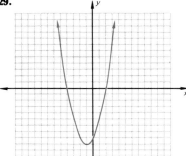

31.

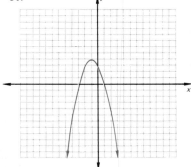

33.

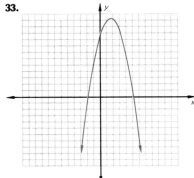

35.

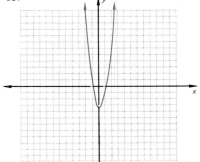

1. $\{(x, y) \mid y = 3x - 4\}$

3. $\{(x, y) \mid y = x^2 - 1\}$

Exercise 7.9

5. $\{(x, y) \mid y = \sqrt{x}\}$

7. $\left\{(x, y) \mid y = \dfrac{1}{\sqrt{x}}\right\}$

9. $\{(x, y) \mid 3y^2 - x^2 = 0\}$

11. $D = \{x \mid x$ is any real number$\}$; $R = \{y \mid y$ is any real number $\}$

13. $D = \{x \mid x$ is any real number$\}$; $R = \{y \mid y$ is any real number$\}$

15. $D = \{x \mid x \geq 0\}$; $R = \{y \mid y$ is any real number$\}$

17. $D = \{x \mid x$ is any real number$\}$; $R = \{y \mid y \leq 0\}$

19. $D = \{x \mid x \geq 0\}$; $R = \{y \mid y \geq 0\}$ **21.** $D = \{x \mid x \geq 1\}$; $R = \{y \mid y \geq 0\}$

23. $D = \{x \mid x \neq 0\}$; $R = \{y \mid y > 0\}$ **25.** $D = \{x \mid x \geq 0\}$; $R = \{y \mid y \geq 2\}$
27. $D = \{x \mid x$ is any real number$\}$; $R = \{y \mid y$ is any real number$\}$
29. $D = \{x \mid x = 3\}$; $R = \{y \mid y$ is any real number$\}$
31. $D = \{x \mid x$ is any real number$\}$; $R = \{y \mid y \geq 1\}$
33. $D = \{x \mid x$ is any real number$\}$; $R = \{y \mid y \leq 0\}$
35. $D = \{x \mid x$ is any real number$\}$; $R = \{y \mid y \geq -9\}$
37. $D = \{x \mid -6 \leq x \leq 6\}$; $R = \{y \mid -6 \leq y \leq 6\}$
39. $D = \{x \mid -3 \leq x \leq 3\}$; $R = \{y \mid -3 \leq y \leq 0\}$
41. $D = \{x \mid -8 \leq x \leq 8\}$; $R = \{y \mid -4 \leq y \leq 4\}$

43.

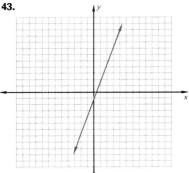

$D = \{x \mid x$ is any real number $\}$
$R = \{y \mid y$ is any real number $\}$

45.

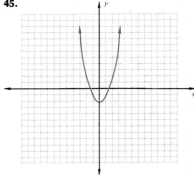

$D = \{x \mid x$ is any real number$\}$
$R = \{y \mid y \geq -2\}$

47.

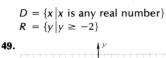

$D = \{x \mid x \geq 0\}$
$R = \{y \mid y \geq 0\}$

49.

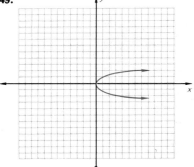

$D = \{x \mid x \geq 0\}$
$R = \{y \mid y$ is any real number$\}$

Exercise 7.10

1. Yes.
5. No.
9. Yes.
13. No.
17. No.
21. Yes.
25. Yes.
29. No.
33. -1

3. Yes.
7. Yes.
11. Yes.
15. Yes.
19. No.
23. Yes.
27. No.
31. No.
35. 2

37. 15
41. 3
45. 8
49. −5

39. −16
43. −6
47. $\sqrt{3}$

Exercise 8.1

1. Yes.
5. No.
9. Yes.
13. Yes.
17. Yes.
21. No.
25. Yes.
29. Yes.
33. Dependent.
37. Parallel.

3. Yes.
7. Yes.
11. Yes.
15. No.
19. No.
23. Yes.
27. No.
31. Inconsistent.
35. Consistent.
39. Consistent.

Exercise 8.2

1. (4, 2); consistent.
5. Same line; dependent.
9. No solutions; inconsistent.
13. (0, 3); consistent.
17. (2, 6); consistent.
21. No solutions; inconsistent.
25. (3, −3); consistent.
29. No solutions; inconsistent.
33. (1, 4); consistent.
37. $\left(6\frac{1}{2}, -1\frac{1}{2}\right)$

3. (−6, 1); consistent.
7. (3, −2); consistent.
11. Same line; dependent.
15. (−3, 4); consistent.
19. (3, 0); consistent.
23. (−3, −5); consistent.
27. No solutions; inconsistent.
31. (0, 5); consistent.
35. (−4, −1); consistent.
39. $\left(1\frac{3}{4}, -3\frac{5}{8}\right)$

Exercise 8.3

1. (3, 1); consistent.
5. (4, 7); consistent.
9. (4, −1); consistent.
13. (−1, 5); consistent.
17. (1, −1); consistent.
21. No solutions; inconsistent.
25. Same line; dependent.
29. (3, 1); consistent.
33. No solutions; inconsistent.
37. $\left(\frac{13}{2}, -\frac{3}{2}\right)$; consistent.
41. (0, 3); consistent.
45. (1, −2); consistent.
49. $\left(-\frac{1}{2}, 2\right)$; consistent.

3. (6, −1); consistent.
7. $\left(\frac{11}{3}, \frac{4}{3}\right)$ consistent.
11. (2, 0); consistent.
15. (7, −6); consistent.
19. $\left(\frac{10}{3}, \frac{4}{3}\right)$ consistent.
23. (1, 1); consistent.
27. (−1, 4); consistent.
31. (11, −7); consistent.
35. (−4, −3); consistent.
39. (−5, 2); consistent.
43. $\left(-\frac{1}{3}, \frac{5}{4}\right)$; consistent.
47. (3, −2); consistent.

Exercise 8.4

1. (2, 6); consistent.
5. (2, −2); consistent.
9. (1, 7); consistent.

3. (−2, 6); consistent.
7. $\left(\frac{1}{2}, -\frac{5}{2}\right)$; consistent.
11. (7, 4); consistent.

13. (−1, 8); consistent.

15. No solutions; inconsistent.

17. $\left(\dfrac{10}{3}, \dfrac{4}{3}\right)$; consistent.

19. No solutions; inconsistent.

21. (−2, 5); consistent.

23. No solutions; inconsistent.

25. (−2, 1); consistent.

27. (−2, 5); consistent.

29. (3, −1); consistent.

31. (−3, 4); consistent.

33. (1, 2); consistent.

35. $\left(-\dfrac{3}{4}, \dfrac{13}{12}\right)$; consistent.

37. (5, 2); consistent.

39. Same line; dependent.

41. (3, 1); consistent.

43. $\left(\dfrac{20}{3}, -\dfrac{13}{3}\right)$; consistent.

45. (1, −1); consistent.

47. (−5, −1); consistent.

49. $\left(1, \dfrac{1}{2}\right)$; consistent.

51. Same line; dependent.

53. (2, 0); consistent.

55. (−2, 4); consistent.

57. (4, −6); consistent.

Exercise 8.5

1. 44; 18

3. 57; 19

5. 80 V; 40 V

7. 28 m by 60 m

9. 36 in. by 54 in.

11. 8 ft by 14 ft

13. $6000 at 5%; $4000 at 6%

15. $2500 at 6%; $1500 at 8%

17. 12 ℓ

19. 10 oz

21. 16 lb

23. 17 ft; 37 ft

25. 42 in.; 54 in.

27. 42 mph; 56 mph

29. 575 mph

31. 20 nickels; 25 dimes

33. 63 $5 bills; 22 $10 bills

35. 39 yr; 13 yr

37. 32 yr; 4 yr

Exercise 8.6

1.

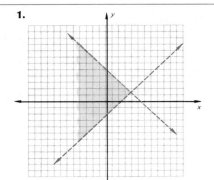

3.

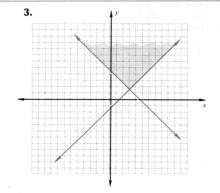

5.

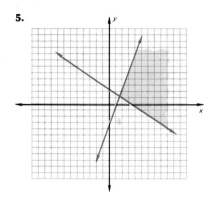

7.

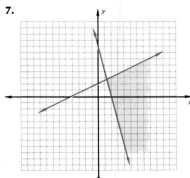

9.

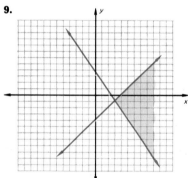

11.

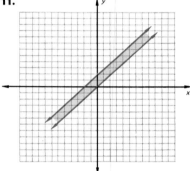

13.

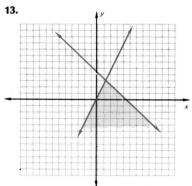

15.

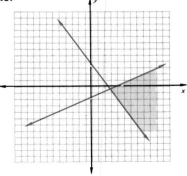

17.

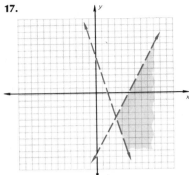

19.

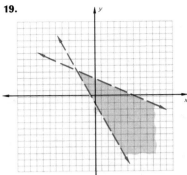

21.

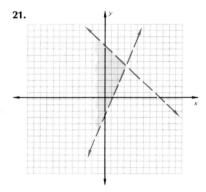

23.

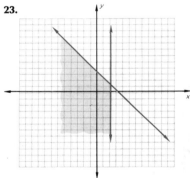

25.

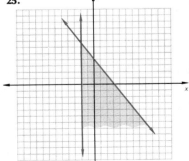

27.

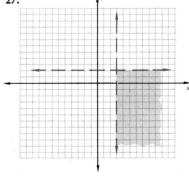

29.

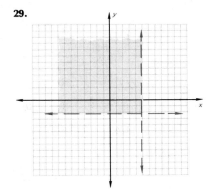

Powers and Roots

n	n^2	n^3	$\sqrt{n}$	$\sqrt[3]{n}$	n	n^2	n^3	$\sqrt{n}$	$\sqrt[3]{n}$
0	0	0	0.000	0.000	50	2 500	125 000	7.071	3.684
1	1	1	1.000	1.000	51	2 601	132 651	7.141	3.708
2	4	8	1.414	1.260	52	2 704	140 608	7.211	3.733
3	9	27	1.732	1.442	53	2 809	148 877	7.280	3.756
4	16	64	2.000	1.587	54	2 916	157 464	7.348	3.780
5	25	125	2.236	1.710	55	3 025	166 375	7.416	3.803
6	36	216	2.449	1.817	56	3 136	175 616	7.483	3.826
7	49	343	2.646	1.913	57	3 249	185 193	7.550	3.849
8	64	512	2.828	2.000	58	3 364	195 112	7.616	3.871
9	81	729	3.000	2.080	59	3 481	205 379	7.681	3.893
10	100	1 000	3.162	2.154	60	3 600	216 000	7.746	3.915
11	121	1 331	3.317	2.224	61	3 721	226 981	7.810	3.936
12	144	1 728	3.464	2.289	62	3 844	238 328	7.874	3.958
13	169	2 197	3.606	2.351	63	3 969	250 047	7.937	3.979
14	196	2 744	3.742	2.410	64	4 096	262 144	8.000	4.000
15	225	3 375	3.873	2.466	65	4 225	274 625	8.062	4.021
16	256	4 096	4.000	2.520	66	4 356	287 496	8.124	4.041
17	289	4 913	4.123	2.571	67	4 489	300 763	8.185	4 062
18	324	5 832	4.243	2.621	68	4 624	314 432	8.246	4.082
19	361	6 859	4.359	2.668	69	4 761	328 509	8.307	4.102
20	400	8 000	4.472	2.714	70	4 900	343 000	8.367	4.121
21	441	9 261	4.583	2.759	71	5 041	357 911	8.426	4.141
22	484	10 648	4.690	2.802	72	5 184	373 248	8.485	4.160
23	529	12 167	4.796	2.844	73	5 329	389 017	8.544	4.179
24	576	13 824	4.899	2.884	74	5 476	405 224	8.602	4.198
25	625	15 625	5.000	2.924	75	5 625	421 875	8.660	4.217
26	676	17 576	5.099	2.962	76	5 776	438 976	8.718	4.236
27	729	19 683	5.196	3.000	77	5 929	456 533	8.775	4.254
28	784	21 952	5.292	3.037	78	6 084	474 552	8.832	4.273
29	841	24 389	5.385	3.072	79	6 241	493 039	8.888	4.291
30	900	27 000	5.477	3.107	80	6 400	512 000	8.944	4.309
31	961	29 791	5.568	3.141	81	6 561	531 441	9.000	4.327
32	1 024	32 768	5.657	3.175	82	6 724	551 368	9.055	4.344
33	1 089	35 937	5.745	3.208	83	6 889	571 787	9.110	4.362
34	1 156	39 304	5.831	3.240	84	7 056	592 704	9.165	4.380
35	1 225	42 875	5.916	3.271	85	7 225	614 125	9.220	4.397
36	1 296	46 656	6.000	3.302	86	7 396	636 056	9.274	4.414
37	1 369	50 653	6.083	3.332	87	7 569	658 503	9.327	4.431
38	1 444	54 872	6.164	3.362	88	7 744	681 472	9.381	4.448
39	1 521	59 319	6.245	3.391	89	7 921	704 969	9.434	4.465
40	1 600	64 000	6.325	3.420	90	8 100	729 000	9.487	4.481
41	1 681	68 921	6.403	3.448	91	8 281	753 571	9.539	4.498
42	1 764	74 088	6.481	3.476	92	8 464	778 688	9.592	4.514
43	1 849	79 507	6.557	3.503	93	8 649	804 357	9.644	4.531
44	1 936	85 184	6.633	3.530	94	8 836	830 584	9.695	4.547
45	2 025	91 125	6.708	3.557	95	9 025	857 375	9.747	4.563
46	2 116	97 336	6.782	3.583	96	9 216	884 736	9.798	4.579
47	2 209	103 823	6.856	3.609	97	9 409	912 673	9.849	4.595
48	2 304	110 592	6.928	3.634	98	9 604	941 192	9.899	4.610
49	2 401	117 649	7.000	3.659	99	9 801	970 299	9.950	4.626
					100	10 000	1 000 000	10.000	4.642

INDEX

8/95   26
4/98   36   10/97